Roberto Bermejo

Handbook for a Sustainable Economy

Springer

Roberto Bermejo
Faculty of Economics & Business Studies
University of the Basque Country
Bilbao, Spain

Kind permission was given for use of parts of chapters published in 2011 by Los Libros de la Catarata, Madrid from "Manual para una economía sostenible", by Roberto Bermejo, in chapters 1–3 in this volume. Chapters 4, 7–12, 14–18, and 20 are updated versions of the Spanish book.

Translated by Daryl Hanlon

ISBN 978-94-017-8980-6 ISBN 978-94-017-8981-3 (eBook)
DOI 10.1007/978-94-017-8981-3
Springer Dordrecht Heidelberg New York London

Library of Congress Control Number: 2014938354

Printed on acid-free paper

Springer is part of Springer Science+Business Media (www.springer.com)

Preface

The *World Scientists' Warning to Humanity,* signed by some 1,700 of the world's leading scientists (with 102 Nobel laureates, including the majority of Nobel laureates in the sciences), declares: "Human beings and the natural world are on a collision course", (UCS 1992). The information we now have regarding the process of that collision and its seriousness is overwhelming. This handbook ventures to explain the main elements of a Sustainable Economy, the development of which is an unavoidable endeavour if we are to retain our civilisation.

This handbook is an amended, updated and adapted translation of my original Spanish version (Manual para una economía sostenible). The adaptation consists of reductions or elimination of the original focus on aspects of Spain's specific situation. On the other hand, all of the references mentioned, which appear emphasized (in bold face), are taken from the original text.

The book is divided into 2 large blocks containing 20 chapters divided into 4 thematic sections. The first block is dedicated to a critical analysis of theoretical assumptions and appropriate tools that proponents of the conventional economy propose as the solution to the "environmental problem". The second block explains the main elements (concepts and tools) of an alternative approach through creation of a Sustainable Economy.

Part I contains five chapters that present a critical discussion of the main elements of orthodox economical thinking and tools: review of essential premises of the dominant paradigm (inspired by orthodox economic theory) (Chap. 1); analyses of the efficiency of a free market, the historical process through which the capitalist market emerged, the commodification of nature, its repercussions on the environment and the prevailing theories on the supply of natural resources (Chap. 2); theory and tools of Environmental Economics (a branch of orthodox economics) in order to deal with the task of commodification of nature (Chap. 3); analysis of the theoretical and practical contradictions which appear when attempts are made to combine environmental protection and free trade (Chap. 4); analysis of the concept of sustainable development premises which appears in the Brundtland Report, and its two main distortions (the theory of triple sustainability and the theory of dematerialization) (Chap. 5).

Part II contains four chapters that deal with principles and tools needed to build a sustainable economy. Chapter 6 begins with a brief description of the Complex Adaptive Systems (CAS), and it explains the main elements of nature's behaviour. Taking into account that human economy is a subsystem of the general economy of nature, we describe its functional principles in order to deduce the principles of a sustainable human economy. Chapter 7 has three sections. The first one explains the adaptive cycle of ecosystems; the second one analyses the validity of the concepts useful for studying SESs, and the third one explains the concept of transformability and the factors that determine it. Chapter 8 studies two issues: it broadly evaluates the capacity for transformation of National Sustainable Development Strategies (NSDS), and it analyzes the essential requirements of an ecological tax reform. Chapter 9 offers a critical vision of the dominant paradigm of science and technology, and defines the bases of the paradigm of their sustainability.

Part III deals with sustainable production and consumption; it has ten chapters, five related to energy, one dedicated to transport, three describe a circular economy of materials, and the last is dedicated to sustainable consumption. Chapter 10 analyses the factors that determine the limits of fossil fuels, the current and future development of their respective offers, focusing in particular on oil and to a lesser extent on natural gas. Chapter 11 analyses the geostrategic conflicts caused by the distribution of fossil fuels. Then it reviews the causes of the current crisis and analyses the economical repercussions of peak oil. Finally, it studies the peak oil structural effect and its sectoral impacts. Chapter 12 contains the following issues: an analysis of the current model of transport; a study of the economic impact of the construction of new infrastructures; an assessment of the impact of rising oil prices on the transport system; a critical review of the EU strategy on transport; and finally some basic elements of a transport strategy aimed at achieving sustainability. Chapter 13 defines master lines for a sustainable electric system, assessing the current development and foreseeable evolution of the three main technological systems, besides other general elements: efficiency, grids and storage. Chapter 14 deals with alternative fuels: biofuels, electricity and hydrogen. After discarding the initial two, only hydrogen associated to fuel cells is left as the sole broad alternative to oil. Chapter 15 studies the following issues: the origin, development and characteristics of societies in energy emergency (SEE); a comparative analysis of the two organisations that form the movement; a study of the process phases in the design and implementation of transformative strategies; and an evaluation of the SEE movement. Chapter 16 analyses the concepts and principles of a circular economy, critical metals, and non-metallic materials, the EU's policy on materials, and ultimately the basis for a circular economy. Chapter 17 is dedicated to analysing the concept of Industrial Ecology (IE), and its two main fields: Material Flow Accounting and Industrial Symbiosis (IS). Chapter 18 analyses the so-called Integrated Product Policy (IPP) and its limits, and proposes the basis of an Integrated Product Strategy (IPS). Chapter 19 studies, on the one hand, the structural causes of the current high-level consumption model through the lens of motivation, provision and access systems. On the other hand, it describes the policies that must be adopted as part of a sustainable consumption strategy.

Part IV contains the final chapter, which undertakes the task of analysing the capability of our societies to transform themselves to reach sustainability. To do so we broadly evaluate each factor, as a prior step to carrying out an overall evaluation. However, we have to emphasise that we accomplish this task based only on the information contained in this book, as a first approach. In order to carry out a broad and in-depth analysis, a multidisciplinary group is necessary.

Bilbao, Spain Roberto Bermejo

Acknowledgements

I am grateful to the many persons who made elaboration and publication of this book possible. I warmly thank Iñaki Arto for his contribution to Chap. 17 and to Charo Martinez and Juan Pablo Benavides, both documentalists who were always willing to seek information for which I asked. On the other hand, I am deeply grateful to David Hoyos for assuming management of the translation process, including raising of its necessary funds. In this regard, I would like to extend my thanks for the financial support of Hegoa (Institute of International Development Studies) and Ekopol (research group on Ecological Economics and Political Ecology), both at the University of the Basque Country. I would also like to thank Daryl Hanlon for his translation.

Contents

Part I
Critical Review of the Orthodox Economy

Chapter 1
Paradigms

Keywords Conventional paradigms • View of nature • Homo economicus • Linear and mechanistic thinking • Paradigm of sustainability

1 The Collision Process with Nature Is Ignored

The *World Scientists' Warning to Humanity,* signed by some 1,700 of the world's leading scientists (with 102 Nobel laureates, including the majority of Nobel laureates in the sciences), declares: "Human beings and the natural world are on a collision course", (UCS 1992). The information we now have regarding the process of such a collision and its seriousness is overwhelming. Consequently, the warnings about the urgency of change multiply. The United Nations has been warning us for more than two decades. The *Brundtland Report* (WCED 1987) affirms that human survival is at risk. The *Millennium Declaration* (United Nations 2000) declares: "We must spare no effort to free all of humanity, and above all our children and grandchildren, from the threat of living on a planet irredeemably spoilt by human activities, and whose resources would no longer be sufficient for their needs". We already have a detailed report from the UN on the state of the biosphere: *The Millennium Ecosystem Evaluation* (MEA), conducted by more than 1,300 experts in the first half of this decade. Among its findings, two are to be emphasized: "Human activities have taken the planet to the edge of a massive wave of species extinctions, further threatening our own well-being" (MEA-SB 2006: 3), and "Land use changes are perhaps the most critical aspect of anthropogenic global change in influencing the future of ecosystems and their services" (MEA-V2 2006: 450).

Yet we are not witnessing a threat that will only affect future generations. This generation is already starting to suffer, and hence calls to act swiftly are increasingly recurrent. The BR declares in *A Call for Action*: "We are unanimous in our conviction that the security, well-being, and very survival of the planet depend on such changes, now" (WCED 1987: 17, 18). This warning further declares: "No more than one or a few decades remain before the chance to avert the threats we now

R. Bermejo, *Handbook for a Sustainable Economy*,
DOI 10.1007/978-94-017-8981-3_1,

confront will be lost and prospects for humanity immeasurably mutilated" (UCS 1992). However, the cited reports do not consider the natural limit that is drawing nearest: the end of the age of fossil fuels.

However, the vast information that we dispose of with regards to the seriousness of the collision process and of the first tangible signs of its impact has not produced any significant policy changes. Numerous reasons can be put forward for this: society's inertia, the relative shortness of politicians' terms in office, a partial knowledge of the issues at hand, etc. Yet the ultimate cause, the dominant paradigm, is the existing web of beliefs and worldviews.

2 Presentation and Critique of the Dominant Paradigm

The dominant paradigm (inspired by orthodox economic theory) is founded on several essential premises that are interrelated. They form a body of ideas and views that determine a concrete view of the meaning of life and the relation between humankind and all other species. Eighteenth and nineteenth century thinkers (Hobbes, Descartes, Locke, Newton, etc.) contributed decisively to the formation of this paradigm. However, the system of power chose the elements that it required for its formation from statements that were frequently in contradiction with each other.

For the first time in the history of humankind, the human species claims to be the chosen one and to be separate from the rest. Humankind asserts ownership over nature, which it regards as hostile and chaotic, a perception which is embedded in the idea of the *law of the jungle*. It follows, that progress can only be achieved through the domination and humanization of nature (Schutz 1999). This premise rests on a mechanistic assumption since it assumes that the behaviour of the entire social and natural system can be inferred through the study of its constituent parts.

When we view the human being as *Homo economicus,* we assume a totally rational being that seeks to maximise his wellbeing, where wellbeing is understood as the possession of goods and services in increasing quantities. These persons therefore appear as one-dimensional beings that, endowed with infinite necessities, take action to satisfy those needs. This is how Mumford (1971) describes this change in values from the old world to industrial civilization: "Happiness was the ultimate objective of mankind. It was based on achieving the greater good for the greater amount of people. Ultimately, it was thought that the perfection of human institutions could be measured by the amount of commodities a given society was able to produce: expanded necessities, expansion of the market, expansion of businesses [. . .] happiness and expanded production were equivalents".

As a result, the economy needs to grow endlessly (something that requires a planet with infinite resources). This premise is made credible by arguing that technical- scientific developments will allow for the elimination of any form of scarcity, the finding of new resources to substitute the depleted ones and, finally, the de-materialization of the economy. Therefore such a premise is fundamental in the dominant paradigm. Another such premise is the belief in the virtues of the free market, since it is considered to be the only mechanism capable of maximizing growth and boosting technical-scientific developments.

2.1 View of Nature and of the Position of Mankind in It

Western thinking constructs a hierarchy that sets mankind above all other species. Biology establishes that primates are at the top of the hierarchy ("primate" comes from the Latin word *primus*, first), and the *Homo sapiens sapiens* occupies its apex. This is nothing more than a desacralized heritage from Greek and Judaeo-Christian traditions. For the former, the hierarchy is in a descendant order: gods, man, woman, slaves, animals and vegetables. For the latter it is: God, angels, humans, animals, and vegetables (Schutz 1999).

The relationship between humanity and nature is presented as a struggle to overcome the supposed initial yoke of the latter, so that humankind moves from the role of dominated to that of dominant. His command of nature will allow humankind to prevail over the limitations imposed by nature on the desire of mankind to satisfy their unlimited needs. Since nature is seen as chaotic and dangerous, humankind grants to itself the duty of putting its chaotic essence in order and of eliminating its dangers so as to obtain from it the resources it needs. As Schütz (1999: 24) argues, humankind is "changing the environment according to one's desire". This is made possible due to the development of science and technology. This leads us to another premise: the maximisation of production through the manipulation of natural systems can supposedly increase limitlessly without endangering their natural stability (Walker 2005: 79). Genetic engineering would allow us to modify genetic codes to produce animals and plants with the properties that we choose, control sicknesses, etc. So, "Biotechnology is extended across the biosphere, making us, over time, in the words of environmental historian J. R. McNeill (2000), 'lords of the biosphere'" (Allenby 2009: 169).

This view currently held by the dominant paradigm has no scientific basis. Anthropology affirms that, in general, primitive societies maintained a harmonious relationship with nature and that they were societies characterised by abundance. This relationship was made possible because nature was considered to be a legacy that humans were obliged to transmit to the following generations; nature had a spiritual value, it was the source of their wellbeing, etc. For Goldsmith this harmony is based on two principles: "The first is that the living world or ecosphere is the basic source of all benefits and hence of all wealth, but we only dispense these benefits to ourselves if we preserve its critical order. From this fundamental first principle follows the second, which is that the overriding goal of this behaviour pattern of an ecological society must be to preserve the natural order of the natural world or of the cosmos" (1996: XV).

After Darwin, "We cannot suppose that man is anything but a precocious primate, a denizen of the Earth and a member of its community of life" (Callicott 1999:

335). The hierarchical pyramid with humankind at the top has no scientific basis. Lynn Margulis (1998: 3, 120) claims that, "These ideas are rejected as obsolete nonsense by the scientific worldview. All beings alive today are equally evolved. All have survived over three thousand million years of evolution from common bacterial ancestors". So, "The planet is not human, nor does it belong to humans [. . .] Humans are not the center of life, nor is any other single species. Humans are not even central to life. We are a recent, rapidly growing part of an enormous ancient whole". The MEA declares that, "Humans are an integral part of ecosystems" (MEA-CF 2006: 27). It is shocking that humankind, which has never played a key role in nature, bestows upon itself the right to dominate her (Rammel and Staudinger 2004).

Only a being that has lost its vital connection with nature can identify her with violent chaos. Science is increasingly showing how nature is extremely complex, accurate, harmonious and coherent, an amazing order embedded in an extremely complex reality. For example, instantaneous and coherent behaviour takes place among tens of thousands of genes, and hundreds of thousands of proteins and other macromolecules that constitute a cell. The same behaviour takes place among cells that constitute an organ, tissue or a living organism. This order is extremely accurate, to such an extent that if the vital signs were to change by an infinitesimal quantity, this entire order would collapse. The same situation is witnessed in the cosmos and in its atoms: They would both collapse if there was to be a minimal change in any of the more than 30 parameters that they depend on (velocity of expansion of the universe, relation of the mass between proton and neutrons, relation between the electrical charge of electrons and protons, etc.) (Laszlo 2007: 60, 79). In the Vedic texts, the idea of subjugating or exploiting the Earth is incomprehensible: "Destruction of forests is taken as destruction of the state, and reforestation an act of rebuilding the state and advancing its welfare. Protection of animals is considered a sacred duty" (Weeramantry 2007).

Our needs are basically the same as for the rest of the species. We need air, water, clean food, materials to construct shelter, energy, etc. We depend on the cyclical processes of nature: "Our health depends upon the purity of the air we breathe and the water we drink, and it depends on the health of the soil from which our food is produced", a service that depends on the sound health of our ecosystems. Hence the collision between mankind and nature is, in reality, a collision of nature with nature, endangering the first. So, "In the coming decades the survival of humanity will depend on our ecological literacy: our ability to understand the basic principles of ecology and to live accordingly" (Capra 2002: 230, 231).

2.2 *Linear and Mechanistic Thinking*

Conventional thinking assumes that the behaviour of all systems (social, economic etc.) can be determined by studying its constituent parts: "The common belief has been that if we know everything about the parts, we will understand the whole", so the "human conventional thinking model is based on a mechanical image of the world and a linear causality to explain the phenomena" (Hjorth and Bagheri 2005–6: 75, 77). Linearity means order, predictability, and comprehensive universal laws so that the linear interaction of several factors will produce a predictable result (Rihani 2002: 3). For example, if we want to produce an event D, we must manipulate the known factors of A and B which we have studied separately. It is unacceptable that between A, B and C there may exist interconnections that create a dynamic that is not entirely predictable (Hjorth and Bagheri 2005–6).

Such mechanistic and linear thinking is applied in all aspects of life. Right after World War II, Rostow considered that all nations will develop in accordance with five stages already completed by developed countries. Fukuyama announced the end of history when the Soviet bloc collapsed. Structural adjustment programs carried out by the IMF assume that the same economic program is valid for any country (Rihani 2002: 4).

Linear and mechanistic thinking belongs to the nineteenth century; it was during this century that the neoclassical revolution in economic theory was forged. And it is a theory that continues to determine current orthodox economic thinking. Its objective is to convert economic science into something similar to Newtonian mechanics or to the astrology of the nineteenth century. Surprisingly enough, it takes place at a moment when both Newtonian mechanics and astronomy are being profoundly revised. Later, neurosciences, psychiatry and psychology will also question the *Homo economicus*. Also throughout the world, complex systems (cells, organisms, ecosystems and societies etc. are) proliferate, making them incomprehensible to linear thinking. A system or functional whole possesses properties that cannot be inferred from its parts. When such systems have the ability to evolve, adapting themselves to the changes in their surroundings, then they are defined as complex adaptive systems (CAS). Through nonlinear interactions among its components the CAS organise themselves hierarchically in structures that determine, and are reinforced by, the flux of people, commodities, energy and information (Matutinovic 2002; Nielsen 2007).

2.3 The View of Human Beings: 'Homo Economicus'

2.3.1 Rational Person

Orthodox economic theory is based on the assumption that human beings behave in a way that aims to maximise individual satisfaction, and that therefore we can order our different desires or needs according to a scale of preferences. Also, it is assumed that such a state is permanent, so that decisions are not affected by the social context or by the frame of reference: "the policy recommendations of most economists are based, explicitly or implicitly, on the *rational actor* model of human behaviour. Behaviour is assumed to be self-regarding, preferences are assumed to be stable, and decisions are assumed to be unaffected by social context or frame of reference" (Gowdy 2007). The free market is the framework in which the individual can best manifest his rationality. Such a view cannot be sustained in light of the studies about human behaviour that have been conducted in the fields of neurosciences, psychology, sociology and, even, in game theory and behavioural economics (Gowdy 2007).

Hodgson (1992: 44 onwards) considers that cognitive processes are complex and the result of multiple interrelations. These processes are carried out with varying degrees of consciousness, from actions with little or no conscious control to other actions with a high level of conscious deliberations. From Freud, we know that our actions are not fully determined by rational calculations or conscious deliberations, because the unconsciousness (the information of our life experiences which we are not conscious of) conditions our behaviour. We know that concepts and theories are constructed in social contexts and that, in accordance with our ideology, we filter the information received, discarding whatever does not fit. Apart from the tendencies towards accumulation and greed, human beings possess others: aversion to loss, cooperation, pure altruism, altruist punishment, habituation, etc. Altruistic punishment is a common behaviour in society and it shows that there is great disposition to punish those that do not follow norms of cooperation, even in cases where the punisher bears substantial costs for such action. Such behaviour dissuades selfish behaviours and stabilizes cooperation. Rational behaviour is common both to animals with limited cognitive ability and to the simplest of human choices (Gowdy 2007). Hence, "The person is viewed as a judging self which examines its urges and evaluates them by various criteria, the most important of which are the moral/social values [. . .] and that these wants cannot be neatly ordered or regulated by prices" (Etzioni 1992: 49).

2.3.2 Selfish Person

Orthodox economic theory falls into a contradiction with regard to values. On the one hand, it argues that economics is a science alien to moral values because, just as in the case of physics and astronomy, it deals with persons as they relate to things. However, at its basis, Adam Smith's economic theory asserts that greed is innate to economic activity. Today, it is argued that competition is central to economic activity. In *The Wealth of Nations* it is recognized that we need the help of our fellow men, yet Smith adds, "It is in vain for him to expect it from their benevolence only", it must be found in the individual search for his own benefit. "It is not from the benevolence of the butcher, the brewer, or the baker, that we expect our dinner, but from their regard to their own interest" (Smith 2005: 40; 206).

And this selfishness, that assumes that each individual aims to maximise his own interests, does not produce chaos but harmony. Mandeville defended in *The Fable of the Bees* that, in economics, private vices become collective virtues. And the fact that a Middlesex jury ruled in 1723 that this book was detrimental to the public interest clearly shows the shift in values that took place under capitalism. Adam Smith agreed with this theory but, since he could not find an objective explanation to such a paradox, he opted for a rather unscientific explanation that ascribed the aforementioned harmony to the workings of a supposed invisible hand. Each person is "led by an invisible hand to promote an end which was no part of his intention" (Smith 2005: 46). But according to L. Dickey, "Towards the end of the 1780s, Smith was becoming increasingly alarmed by what he referred to as 'the depleting of moral legacy' of commercial society" (Goldsmith 2003: 298).

The neoclassical revolution does not question this view of human nature, although neither does it present it with the harshness described by Adam Smith. Today it is presented as a defence of individualism, of consumer sovereignty and of competition, and as an essential requirement for economic efficiency. In the previous section we have shown that this vision of mankind is unscientific, so now we will cite some of the most outstanding economists who have disagreed with the orthodox position. The classical economist J. S. Mills held that mid nineteenth century England had enough economic potential so as to satisfy the basic needs of the English population without having to experience fierce competition. Keynes believed in the prevalence of greed and usury, but he did not consider them to be permanent and immutable features of human nature. He thought that in a hundred years society will reach a state of abundance, in which unsupportive behaviours in the economy will cease to be necessary. Marx aimed to create a state of abundance and freedom from exploitation that he defined as communism (Schumacher 1978).

2.3.3 Competitive Person

An individualistic and selfish being can only be competitive. It is assumed that competition not only is one of the distinctive characteristics of all living beings, but that it is an indispensable element of achieving economic efficiency. In many other fields of science there have been outstanding supporters of competition since Spencer formulated his principle of "the struggle for survival". "If Adam Smith showed that the competitive principle applied to economics, and Herbert Spencer did the same for sociology, Malthus for demography and Darwin for evolutionary ecology, academic ecologists made sure that they were not left out". Yet, "it is sociobiologists who have taken up the most extreme position. For them, it is individual self-interest that prevails in every sphere", (Goldsmith 1992: 204). Despite this, competition has not attained broad dominant status in any scientific field apart from the economic one.

Orthodox economic theory affirms that the entrepreneurial struggle for survival is the directing force of economic activity. J. Bhagwati (World Trade Organization) considers this to be the behaviour of nature and believes that the economy tends to follow such path as the latter is deregulated: "The phenomenon of spider web (global integration) increasingly exemplifies that one tends to be in everybody else's backyard, producing import competition in one's markets, and export competition in each other's markets and in that of third parties, each time in a more fierce way, in an almost identical fashion as the struggle for the sun in a thick tropical wet forest" (Anderson et al. 1995: 53). J. D. Rockefeller goes further, and presents the struggle for survival even as a divine law: "The growth of large business is merely a survival of the fittest (...). It is merely the working out of the law of nature and the law of God" (Goldsmith 1992: 202).

M. J. Wheatley and G. Creneam (Wheatley and Creneam 2004), experts on management, state that, "Aggression is the most common behaviour used by many organizations, a nearly invisible medium that influences all decisions and actions (...) Aggression is evident in the consistent use of war and sports metaphors. There is constant use of these images as we 'bring in the big guns', 'dominate the field', plan a 'sneak attack', or 'rally the troops'."

No other species shows such aggressive behaviour towards its own kind. The twentieth century has witnessed two world wars and numerous local ones. Keynes had no doubt that states that did not know how to solve internal conflicts tend to shift them abroad. Even in the conclusions of his *General Theory* he claims that the economic causes of war are "the pressure of population and the competitive struggle for markets. It is the second factor, which probably played a predominant part in the nineteenth century, and might again" (Keynes 1964: 381–2).

Based on the studies cited above, the statements that attempted to back up the theory of the struggle for survival appear groundless. The "divine" law of competition mentioned by Rockefeller cannot be maintained in light of the texts of major religions and traditional cultures. For example, Buddha taught that, "Those who wish to follow his Path should practice loving kindness, not to harm the life of any beings – not only to protect mankind, but also to protect animal and vegetation (...) He saw that all beings in the universe were equal in nature" (Quang 1996). It does not seem plausible either to present the "struggle for survival" in the context of a tropical wet forest as an example of economic competition. It is ignored, as we will see in the second part, that nature has the virtue of transforming competition among individuals into basic collaboration among different species, that translates into the formation of ecosystems perfectly adapted to their surroundings and able to maintain their identities.

2.3.4 Consumist Person

The sacralization of work: The dominant ideology portrays history as a long march from scarcity to abundance, in which human beings gradually provide themselves with technologies that allow them to subjugate nature and, therefore, move away from extinction, while accessing an increasing number of commodities and services. It is difficult to find in our societies an assumption with more support. Ayres sustains that the technological revolution happily condemns primitive, prescientific and preindustrial cultures to extinction: "Since the technological revolution is itself irresistible, the arbitrary authority and irrational values of pre-scientific, pre-industrial culture are doomed" (Sbert 1993: 194).

However, anthropology and history define a very different landscape to that of progress viewed as a long march from primitive scarcity to relative current abundance. Numerous anthropologists and economic historians describe primitive societies relatively removed from extreme scarcity. Shalins (1986: 14) considered hunter-gatherer societies to be societies of abundance, referring to the latter in the sense of the satisfaction of all essential needs with minimum effort. Tainter (2005) shows how, historically, peasants have steadily reduced their working hours because of a consciousness that their productivity decreases with every increase in the working time. Mumford (1971) and Sahlins explain that in precapitalist societies working time decreased as gains in productivity were realised. This process took place in Europe until the sixteenth century, in which more than half of the days were bank holidays. From that moment onwards a regression in salaries and working conditions has taken place and, although this tendency will be reversed in the nineteenth century, even today annual working hours in the developed world are greater than in the ancient world (Durning 1994: 35). The fact that they cannot satisfy their most basic needs is also a step backward for millions of people.

In primitive societies human beings do not work to maximize the accumulation of material goods but to secure their position, their rights and concrete social achievements. Neither production nor distribution processes are linked to economic interests. Each society articulates itself in relation to a series of social interests that change as society evolves. Production and distribution are regulated by the principles of reciprocity and redistribution. The principle of reciprocity constitutes a social contract whereby donations of labour or commodities oblige the receptors to return them. The redistribution principle means that societies organise the redistribution of goods in such a way that nobody is left out and condemned to poverty. "Broadly, the proposition holds that all economic systems known to us up to the end of feudalism in Western Europe were organized either on the principles of reciprocity or redistribution, or householding, or some combination of the three" (Polanyi 1989: 34).

The incipient bourgeoisie was obliged to fight this culture since it was incompatible with the necessity of the endless expansion of the capitalist system. Work becomes something praised and it is endowed with inherent qualities that are independent from its objective of satisfying basic needs: It is transformed into a virtue. As a result, the bourgeoisie will lash laziness in workers for centuries. The following texts, cited by Marglin (1974), show this behaviour. The first one is of J. Smith: "It is a fact well known (...) that scarcity, to a certain degree, promoted industry, and that the manufacturer (worker) who can subsist on three days of work will be idle and drunken for the remainder of the week". The second is from Ure, a Scottish academic, who was amply cited by Marx: "It is found nearly impossible to convert persons past the age of puberty, whether drawn from rural or from handicraft occupations, into useful factory hands. After struggling for a while to conquer their listless or restive habits, they either renounce the employment spontaneously or are dismissed by the overlookers on account of inattention".

Although work is transformed into a virtue, it presupposes suffering (tiredness, danger in some activities). Hence, consumption becomes the reward for the virtue of work. The reward must be on a par with the suffering: happiness produced by access to a greater number of goods and services. Yet this reasoning is pure ideology and hides the need of capitalism to grow endlessly to continue to increase entrepreneurial profit margins. But ideological it is, as overwhelming empirical evidence in sociology and psychology has also shown that beyond the satisfaction of basic physical needs, increases in disposable income do not increase happiness. In some countries happiness may actually be reduced (Mulder et al. 2006).

Major religions and philosophies share the view that material wealth does not make us happier, and that unhappiness is the result of dysfunctional behaviours far removed from the true human essence.

In ancient Greek, "sin" means "missing the point, to act against human essence". Therefore, the key must be in elucidating the true human essence or the inherent destiny of mankind. So as to speed up the rise in the number of persons that search for their essence by their own means, although with the guidance of the masters that they choose (Tolle 2005: 19).

2.4 *View of the Economy*

Georgescu-Roegen (1971: 1) declares: "No science has been criticized as openly and constantly as economics". Although there were plenty of reasons for this, "the most important pertains to the fiction of *Homo economicus*", since "that fiction strips man's behaviour of every cultural propensity, which is tantamount to saying that in his economic life man acts mechanically". In this section, we will analyse three essential features of classical economics: economic science, the role of the market and the view on technological development.

2.4.1 Economic Science

The so-called neoclassical revolution that took place at the end of the nineteenth century (especially personified in Jevons, Walras and Menger) attempted to overcome any ideological and ethical stand with the contention that economics is a science with a degree of accuracy equivalent to that of Newtonian mechanics or astronomy. Therefore, it was above ethics. In order to reinforce this supposedly scientific basis they provided economics with a powerful mathematical tool. Jevons says: "To me it seems that *our science must be mathematical, simply because it deals with quantities*" (Roll 1966: 378). Jevons aimed at unveiling the mechanics of self-interest and utility, while Walras opted to unveil the mechanics of satisfaction. For the latter the economy was morally neutral because it refers to the relation between persons and things. Menger compares economic science with astrology, suggesting that through the study of visible movement of prices (planets), we could deduce the nature of the invisible force that governs the maximization of utility (energy) (Cole et al. 1983).

In this way, politics is pulled out of the economy in such a way that, as a result of the neoclassical revolution, political economy (the term used until then) is transformed into pure economy or simply, economics (Soderbaun 1992). Friedman,

Nobel Prize laureate and leader of the liberal Chicago school, states the same as Jevons, Walras and Menger with a modern terminology: "Positive economics is, in principle, independent of any normative judgment or ethical position. Its objective is to provide generalizations that can be used to produce accurate predictions about the consequences of any changes in the economic environment. Its performance must be judged by the accuracy, scope and its conformity with the reality of the predictions produced. In short, positive economics is, or can be, an objective science just as any of the sciences that study the physical world" (Naredo 1987: 383). Many outstanding economists have criticised the overuse of mathematical models. Among these prestigious economists there are some mathematicians, such as Hanh and Morishima. Hanh finds it "shocking" that there are so many economists dedicated to refining "the analysis of economic situations that there is no reason to believe that they have existed or will ever exist at some point in the future". Moroshima claims that "mathematical economic theory has recently turned into something that is increasingly abstract, watered down and sterile" (Gómez Uranga 1997). The attempt by orthodox economists to copy the methodologies of hard sciences and of mathematics is doomed to fail. The economy is a social science and it is therefore conditioned by ideologies. Also, sciences that represent the purest scientific reference, such as physics or astronomy, have developed to such an extent that they have entered into fields that challenge the current world view (Laszlo 2007).

2.4.2 Technology Central Role

The technical-scientific development has a central role in orthodox economics because in a finite world the need to grow endlessly can only be sustained if it is invested with great dematerializing qualities, and with the ability to establish order in the chaos of the natural world. The World Business Council for Sustainable Development (WBCSD 2001) declares: "Companies have used technology as a motor of progress since, at least, the industrial revolution, which have given us notorious models to marshal the physical world for human benefit".

It is certainly true that in the past two centuries enormous advancements in science and technology have taken place. Many technologies have produced outstanding improvements in the welfare of many societies, such as electricity, telecommunications, transportation, some productivity enhancing technologies, etc. But many technologies consume resources in a non-sustainable manner because they deplete non-renewable resources, generate negative environmental impacts and reduce our quality of life. This is due to the fact that traditional technological systems are at the disposal of the prevailing economic system. Despite all this we are told of great technological breakthroughs that will improve our standards of living.

However, these promises are being questioned by well-known organizations of scientists and technologists, by reality and by the protests carried out by sectors of the population that are severely affected by the direct or collateral damages produced by technical-scientific progress: "The scientific system, thus, faces a crisis of confidence, of legitimacy, and ultimately of power, as there is a growing feeling from many quarters that science is not responding adequately to the challenges of our times, and particularly, those posed by the quest for sustainable development" (Hjorth and Bagheri 2005–6: 75). In opposition to the myth of the correct solution based on rigorous proofs, since the social and natural systems are quite complex and, therefore, show high degrees of randomness, multiple factors of uncertainty are seen. There are huge gaps in our knowledge, particularly as to how nature works and, consequently, the degree of uncertainty with regard to the effects of our actions over the natural system, over our health, over future generations, over the social cohesion, over the future of the economy, etc., is high. Besides, numerous human groups (aboriginal peoples, fishermen, farmer communities, etc.) have developed knowledge and frameworks for action (in medicine, sustainable production, biotic materials, ability to foresee changes in their local environment, etc.) that together form a very valuable scientific heritage. That is why science cannot hold a monopoly over definitions of technical-scientific policies. Instead, all social agents are experts in different ways and in relation to the multiple sides of each issue, and they are legitimized to participate in the definition of the solution as well as in its practical application: "Citizens (as well as scientists) become both critics and creators, providers and recipients in the knowledge production process" (Spangenberg and O'Connor 2003). This scientific vision is known as "post normal science".

The role of the Market as an evolutionary process and as a determinant of history: The construction of the free market is the key element that explains and justifies the neoclassical revolution. The former will be the instrument through which individuals will be increasingly better off. In this regard, the limited primitive market constitutes a stage in the natural evolution towards the capitalist market and the type of social behaviour that best fits rationality: "According to the orthodox history, right from the days of the early marketplace to the present era, and the price-setting principle, there has been or should have been a progressive evolution, from strictly limited market to a limitless one" (Berthoud 1993: 82).

If it is accepted that the free market is the historical culmination of the market, then societies with a free market economy must constitute the highest stage of a supposed natural evolution. Hence we must find ourselves witnessing the end of history. In fact, for the neoliberal academic Alain Minc, "Capitalism cannot collapse, it is society's natural state" (Ramonet 1997: 59). For this reason Abalofia and Biggart (1992: 317) state that, "The neoclassical model of a competitive market is transhistoric and a-cultural". On the other hand, it is argued that mercantilist rationality represents an inspirational model that can be found in the rest of the features of human behaviour. Altvater believes that orthodox economists' interpretations of the history of the economy determine some universal "ordering principles" without putting "into question whether market rationality is a real possibility on Earth, since it is 'placed' outside the coordinates of time and space" (1993: 74). Therefore, "In this new era, the market is not considered merely as a technical device for the allocation of goods and services, but rather as the only possible way to regulate society" (Berthoud 1993: 70).

To apply mercantilist rationality to the organization of social and political structures is a totalitarian approach, since to support positions that are not in line with the free market is to put oneself in the margins of – or in opposition to – what is rational and natural. It implies also placing orthodox economics and its economists in an inquisitorial position. For this reason, they accuse their critics of keeping positions that are pathological, conservative, anti-democratic etc. Martin Wolf, associate editor and chief economics commentator at the Financial Times, claims that anti-liberals "have three motivations: hatred of markets; fear of foreigners; and concerns about wages, jobs and economic activity. The first two attitudes are pathological. The last is at least rational." Peter Martin, editor of the international edition of the same publication, argues that his critics "base their arguments on a visceral desire to preserve the status quo, to retain the hegemony of their profoundly conservative ideology", and this undermines a fundamental "democratic rights, including that most precious right, the right to be left alone" (*Le Monde Diplomatique*, June 1997). In view of the totalitarian dangers of monolithic thinking, some orthodox economists, like Besso, accept that "perhaps the old error has been made of interpreting the historical phase in which we are living as the end of history" (Ravaioli and Ekins 1995: 124).

3 Sustainability Paradigm

The dominant paradigm is unscientific, arrogant and suicidal (because we have no future if we continue the collision course); it is typical of less developed beings and in direct conflict with the traditional wisdom of pre-capitalist societies and most major religions and philosophies in regards to human nature and its relation with the rest of nature. The confluence of dysfunctional beings and the ownership of highly destructive technologies produce a very dangerous situation. Consequently, a paradigm shift from the current one to one aimed at servicing sustainability is urgently needed. Its fundamental features may be inferred from the critique that we have carried out of the dominant paradigm. We will briefly express them on four main axes: our position in relation to the rest of nature, the meaning of life, sustainable economy, and the science and technology of it.

Humankind is part of nature. The rest of the species generate growth and maintain life on the planet, so they are not there to serve us. That is why we must respect nature, and behave like its managers or administrators. It is not chaos filled with violence, but a wonderful and extremely complex order built over thousands of millions of years. We need to live in harmony with nature to preserve the meaning and quality of life. To live in a non-degraded natural environment is crucial for the maintenance of our standards of living. It has been amply demonstrated that we work better, we feel more in equilibrium, and get healthier faster in natural environments. Children with grave relational problems improve considerably in monitored contact with animals. Some authors called biophilia the innate bond between human beings and nature (Orr 2002: 25).

Table 1.1 Modernist paradigm versus sustainability paradigm

Modernist paradigm	Sustainability paradigm
Reductionist	Interconnected
Simplicity	Complexity
Determinacy	Indeterminacy
Atomistic	Holistic
Mechanistic	Organic
Anthropocentric	Biocentric
Individualistic	Communitariam
Quantitative	Qualitative
Disenchantment	Enchantment
Competition	Cooperation
Geo-political boundaries	Natural boundaries
Linear, predictable	Nonlinear, unpredictable
Equilibrium	Steady-state

The meaning of life is a much more difficult concept to comprehend. It has multiple dimensions (self-awareness, awareness of the roles we must play in relation to the natural environment, in society, etc.) that go beyond this text. Therefore, we will only deal with the second dimension so that, "Instead of using nature as a mere tool for human purposes, we can strive to become tools of nature who serve its agenda too" (McDonough and Braungart 2002: 156), regenerating, when possible, destroyed or degraded natural environments and integrating nature in the cities, creating "buildings that, like trees, produce more energy than they consume and purify their own waste water" (McDonough and Braungart 2002: 90).

If, as claimed by ancient wisdom, we are dysfunctional beings, unhappiness is not outside (in the unlimited accumulation of commodities, in our conflicts with others) but in ourselves (in our view of who we are, in our feelings and, ultimately, in our idea of happiness). Hence it is of paramount importance to gain self-awareness of the root causes of our dysfunctionality, because it is the only way to solve them. Many civilizations have captured this view of human nature and its relation to the cosmos in remarkably similar concepts that have been named differently: "R'ta" in the India of the Vedas; the similar concept of "Dharma", also in India; "Tao" in China; "Maat" in ancient Egypt; "Nomos" in ancient Greece, etc. Many other civilizations, like the Maori, define it as "the way" (Goldsmith 1992: 300).

Human economy is just a subsystem of the general economy of the resources and energy of nature. General System Theory shows us that a subsystem cannot survive if it behaves differently to the system to which it belongs. Therefore, we must infer from natural behaviour the guiding principles of sustainability. We will see in Chap. 5 that these principles are divided into abiotic and biotic. The abiotic principles are the closure of the flow of materials, solar energy and the maintenance of the physicochemical balance of the inert world. The biotic principles emerge from the behaviour of the ecosystems: evolution, diversity, hierarchy of services, decentralization, self-sufficiency, and predominance of cooperation versus competition. These principles are essential to achieve the abiotic ones. The market should service these principles, because it is “a good servant but a bad master and a worse religion” (Hawken et al. 1999: 261).

Biomimicry also needs to be applied to the fields of science and technology. Since our technologies are vastly inferior to those of nature, science should constantly expand its knowledge of the basic behaviour of the biosphere and its ecosystems. This will contribute to the development of biomimicry technologies, that is, technologies with a high level of efficiency and sustainability. Yet it is not only about imitating nature’s technology but also its organizational structures. Hence, for example, we must create industrial ecosystems. The following table summarizes the two paradigms (Ehrenfeld 2007: 77) (Table 1.1).

Chapter 2
The Commodification of Nature and Its Consequences

Keywords Free market premises • Free market efficiency • Commodification of nature • Resources scarcity • Environmental impacts

The defence of the free market is based on the premise that it is the most efficient economic instrument and, therefore, the one that delivers the most economic growth. In addition, it is said that it carries out a fair distribution of the wealth produced, as it does so depending on each person's contribution, and that nature is an unlimited resource that the market uses with efficiency. But geology shows us the process of resource depletion. In this chapter we will analyse the efficiency of the free market, the historical process through which the capitalist market emerged, the commodification of nature, its repercussions on the natural environment and the theories on the supply of natural resources.

1 The Efficiency of a Free Market

Adam Smith arrived at the paradoxical conclusion whereby, when each individual pursues the maximisation of their usefulness, the maximum social satisfaction possible is achieved. So a private vice (selfishness) becomes (through "an invisible hand") a collective virtue (social well-being). Today the orthodox economy explains this paradox according to a series of virtues that it attributes to the free market:

- It determines, due to the free action of supply and demand, the real costs (prices), which are essential information for efficient economic action.
- It distributes resources efficiently.
- It satisfies people's desires, as businesses are always ready to satisfy demand, which expresses these desires.
- It avoids the need for complex planning, as it works automatically.
- It adapts to changes quickly and flexibly.

R. Bermejo, *Handbook for a Sustainable Economy*,
DOI 10.1007/978-94-017-8981-3_2,

However, in order to express these virtues it has to meet the following premises:

- The existence of institutions capable of establishing property rights and guaranteeing the usufruct of the same.
- The existence of a clear and precise structure of property rights.
- Freedom and equality for all the members that intervene in the market. This means, among other things, that both parties have the same information and that there is freedom to enter and exit the market.
- This is achieved through the action of a high number of buyers and sellers.
- The existence of a balance between supply and demand.
- Divisibility of the production factors and of the products.
- Absence of public assets.
- Absence of environmental impacts.
- Absence of absolute lacks of resources.

The requirement that the economic agents be free and equal is systematically unmet, because the system is characterised by the private appropriation of the means of production and the tendency to concentrate them in a few hands. This creates a relationship of domination, of the owners over the dispossessed. The consumer's right to choose freely presupposes that their choice does not affect other members of society and of the world. As long as goods are reproducible and do not affect other goods and other consumers, the proposal makes sense, provided the consumer is perfectly informed. But when we factor in the social and environmental repercussions of the actions of economic agents, we realise that the supposed sovereignty of the economic agents often hides very strong impacts on other people, on societies and, above all, on future generations. For example, the ownership of a car, according to the sovereignty mentioned, contributes to the depletion of resources (particularly oil) and produces environmental impacts that generate a loss of well-being in cities and climate change.

The existence of oligopolies, their symbiotic relationships with political power, agreements to limit competition, pressure groups, multiple obstacles to enter a growing number of sectors (due, mainly, to the strong investments that are necessary to enter very monopolised sectors), show how far the market's behaviour is from that determined by the requirements indicated above. For E. Altvater (1993: 67), "markets are necessarily sources of inequality (...) if the market actors are unequally endowed with physical and economic or political power (...) then the procedures will not be neutral but will tend to reinforce the power disparity".

If these requirements were met, there is no doubt that the market would be a much more efficient instrument than what it is today, but even then it would still be inadequate to sustainably satisfy the vital needs of all the population. The market only services solvent demand. It does not distinguish between vital and non-vital needs. Thus, the OECD market usually offers countless manufactured products at low prices, while many people cannot afford decent housing or a quality health system. They are also denied the right to clean air and water, to be in contact with a non-degraded natural environment, etc. And these deficiencies become more evident the more the market is deregulated. Thus, in the United States (USA), where public regulatory action is minimum, they have the worst health and education indicators of the OECD.

Last of all, to propose the market (a micro-economic instrument) as the adequate option to solve general problems (economic, social and environmental) means being oblivious to the hierarchical order of complex systems. In the biosphere, the biggest ecosystems are at the summit of the hierarchical system. As a result, instead of solving problems, it magnifies them: degrading nature, polarising the distribution of incomes and encouraging social disintegration. So "markets are only tools. They make a good servant but a bad master and a worse religion" (Hawken et al. 1999: 261).

2 The Emergence of a Capitalist Market and the Consequences

We have seen in the previous chapter that the defence of the free market, that is, one entirely guided by its internal forces, without the mediation of States, leads to the affirmation that the economic liberalisation process is a natural process; so those who oppose it should be treated as dangerous people, people who conspire to subvert the natural order of things. So such a premise culminates in totalitarianism. But economic history does not back the theory of the natural evolution of the market. Many authors [such as Polanyi (1989), Mumford (1971), Thompson (1995), Altvater (1993), etc.] firmly maintain that there is no such natural evolution, but that there was a qualitative rupture in the capitalist market and that it was the result of the strong interventionism of the bourgeois governments that emerged after the success of the revolutions against the Ancien Régime.

The pre-capitalist market had two features of its own. First of all, it was limited in volume and in the type of goods, in space and in time. It was limited in volume because only a fraction of the goods produced were sold, as an immense majority of the population practiced an agriculture that was basically of subsistence. Polanyi considers that until the second half of the eighteenth century "so-called nations were merely political units, and very loose ones at that, consisting economically of innumerable smaller and bigger self-sufficing households and insignificant local markets in the villages. Trade was limited to organized townships which carried it on either locally as neighbourhood trade or as long-distance trade – the two were strictly separated, and neither was allowed to infiltrate the countryside indiscriminately" (1989: 39). Even in the Europe of the late eighteenth century there was hardly any trade between regions within the same State (Sachs 1992: 7). The types of goods being bought and sold were also limited. Land and labour were not considered goods, meaning they could not be exchanged on the market. In some societies not even food was sold, as it was considered of such importance that it was not considered a commercial asset. The primitive market only took place in a location with well-defined physical and temporal boundaries, "and clearly differentiated from ordinary life, very often the market lies at a distance from the inhabited area, and functions as a neutral meeting place" (Berthoud 1993: 75). The temporal limitation means that it exclusively operated on certain days of the year. Secondly, the pre-capitalist market was heavily regulated. The authorities not only established order in the markets, but they also controlled the various aspects involved in the transaction, in particular the price and weight, as shown by historians Braudel and Thompson. In Great Britain, emergency measures were prepared for the periods of food scarcity between 1580 and 1630 that were codified in the Book of Orders. This "granted magistrates the power (with the aid of local courts) to inspect the stocks of cereals in chambers and granaries; to order the dispatch of certain amounts to the market; and to impose with severity all the regulatory rules on licences and hoarding. Cereals could not be sold outside the public market" (Thompson 1995: 256).

Medieval guilds, as is known, regulated the quality and price of the products, the number and the way workers were promoted, and many other aspects of life within the guilds, some of which were not strictly economic, such as in the case of certain services that are currently included under the term of social security. At least until the late eighteenth century intermediaries were considered as suspicious in the eyes of the law and their activities were very limited. On the other hand, millers and bakers were seen as servants of the community who did not work to make a profit but to earn a reasonable income.

Therefore, there was no "free price formation". The authorities sought price stability as a means of guaranteeing social peace. For precisely this reason, the pre-capitalist economy was focused on satisfying needs. R. H. Tawney considered that "the economy of the medieval borough, was one in which consumption held somewhat the same primacy in the public mind, as the undisputed arbiter of economic effort, as the nineteenth century attached to profits" (Thompson 1995: 286). Despite this, pre-capitalist economies grew and became inter-related with the growth of the economy. But these factors "did not, however, transform markets into a market economic system, since the commodity form was not yet the universal ordering principle of social regulation, and private property in the means of production had not yet been firmly established" (Altvater 1993: 58).

3 The Commodification of Nature

The existence of nature is taken for granted and its appropriation is the result of a legal act, which historically has been frequently preceded by a violent conquest, because this appropriation means a loss for the rest of society or for other communities. In primitive societies land was not normally considered a good, because it was their territory, a shared asset that provided the food and materials necessary for their survival and which was the burial ground of their ancestors. Its commodification means an assertion of human control over nature and the de facto negation of the systemic character of nature (it ceases to be considered an ecosystemic entity and becomes a succession of privatised plots of land), which opens the path to its destruction, because the land becomes a private means of production and therefore, subject to the mercantile logic of profit maximisation.

Economics handbooks state that a good is an asset produced for being sold, something that does not happen with natural resources; therefore, nature is not a good. Each authentic good is an individualised unit, perfectly separate from the others, meaning that the destruction of one does not affect the rest; for example, the destruction of one car does not affect the rest of the cars in circulation. The Earth formed 4 billion years ago; life appeared 500 million years later. And the human species is the latest product of the evolution of life on the planet. Evidently, we did not produce the Earth, but rather it produced us, so therefore we have no right to take over and commodify it. Its consideration as a good requires a valuation (mercantile) and a use that is incompatible with the vastly complex network of life. Natural systems have a holistic, indivisible character that rebels against mercantilist reductionism. So, for example, as Leipert says, "a tree or a species is part of a local ecosystem, which is part of a regional ecosystem, and this is in turn is part of a network of ecosystems in the whole country, continent, the world. And all of these ecosystems interact with each other in such a complex way that they constitute together a single, indivisible, collective, public good, which belongs to the human race" (Ravaioli and Ekins 1995: 38). A couple of examples will be enough to illustrate the consequences of ignoring the systemic character of nature. The Chinese authorities banned cutting down trees anywhere in the Yangtze river basin, in addition to implementing a broad reforestation plan, due to the fact that the catastrophic floods in the summer of 1998 were worsened by the deforestation carried out during previous decades. From a more general perspective, the serious global ecological problems we face (climate change, ozone layer depletion, ocean degradation, etc.) are a consequence of the accumulation of many alterations produced at a local level. For this reason, Aldo Leopold (a naturalist considered to be the father of conservationism) states that the conservation of nature is an act of harmony between man and the land and that the commodification of nature destroys it: "We abuse the land because we regard it as a commodity belonging to us. When we see land as a community to which we belong, we may begin to use it with love and respect" (Meeker-Lowry 1995: 158).

Despite what has been said until now, nature provides us with goods and services that cannot be appropriated or mercantilised but are meant to be public or free assets. They are assets whose enjoyment by one person does not prevent others from benefiting from them. In addition they cannot, in general, be privatised. Many natural assets are public assets (rivers, oceans, the atmosphere, wildlife, etc.), and even those that have been privatised (land, aquifers, etc.) are public in nature, because their use affects the rest of the biosphere and, therefore, humanity. But nature represents much more than a public economic asset. It is a totality of many aspects: it is our habitat, which means that respect for it is a guarantee for survival; and it provides us with many vital services, apart from being a source of resources necessary for sustaining life. De Groot (1992: 263) considers that nature provides functions for regulation, support, production and information, which include 36 sub-functions.

Capitalist society tends to value the things that the market values and dismisses and does not respect whatever has no value for the market. This is what happens with public environmental goods and services, even if they are vital not only for the economy, but also for the survival of humanity. But as the deregulated market is the organising and valorising principle, it tends to dismiss everything that does not have a market value, which translates into a growing process of environmental deterioration and destruction: "Valorisation'– or, in other words, extension of the market's formal rationality and scarcity principle to previously free resources – always entails a largely hidden definition of the 'non-valorisable' or 'valueless' objects whose destruction is permitted" (Altvater 1993: 69).

The process of collision with nature is produced because the market is not the appropriate instrument to achieve harmony with it. Nature tends to shorten the life cycle of materials, minimising the transport and energy consumption they entail, meaning that it recycles most of them at a local level. As an ecosystem evolves towards maturity, it gradually reduces its need for external contributions of material. On the contrary, the market economy tends towards globalisation and, therefore, towards distancing the transportation of goods. This causes very important environmental damage and a considerable consumption of resources. Nature generates growing biodiversity, as it is a guarantee of stability and survival. The market tends towards the specialisation of countries and to the standardisation of products and production techniques. This is the case of industrial agriculture, where intensive specialisation is greatly reducing the genetic base and, therefore, the capacity to develop new varieties that can adapt to a changing reality. But this depletion (as Swanson declares) "is spoiling 'a uniquely formulated insurance policy against shocks to the life system itself (. . .) because existing life forms encapsulated a history of successful adaptation within a changing physical environment'" (Gustafsson 1998: 266).

4 Free Market and Natural Resources

4.1 The Impact on Natural Resources by Establishment of a Capitalist Market

For a political system to transform itself into a market economy system, land, work and money need to become goods and, therefore, separate property. The market goes from being a mere instrument for exchanging objects, as it had been in pre-capitalist societies, to becoming the universal regulating medium of society, determining social relations and classes. Money goes from being an instrument that facilitates exchanges to becoming the measure of all things. And the increase of its amount, based on the appropriation of the surplus value generated by labour, becomes the aim that gives meaning to economic activity. As Hume says, "it is an infallible consequence of all industrious professions, to (...) make the love of gain prevail over the love of pleasure" (Altvater 1993: 60). So, if the world had been governed by the laws of the free market, it would have prevailed through the force of events, in a process of natural evolution.

These ideas became widespread from the 1820s and were defended by the bourgeoisie with religious fervour: "Born as a mere penchant for non-bureaucratic methods, it evolved into a veritable faith in man's secular salvation through a self-regulating market" (Polanyi 1989: 78). But the discovery "of the laws that governed the world of men" did not lead to an acceleration of the "natural evolution", because society presented fierce resistance: far from dismantling the strongly interventionist State of the absolute monarchies to give way to a market that regulates itself, while also regulating economic life and determining social organisation, the public administration was strengthened, to establish the market "by fire and sword, resorting to the full force of the State apparatus" (Naredo 1990).

Apart from the systematic use of force, destruction of the old order required suppression of its legal framework. In western Europe, between 1830 and 1850 many laws were passed that abolished the regulations of primitive society. In addition, it was necessary to create a legal and institutional framework that outlawed any defensive reaction. Thus, the appropriation by the bourgeoisie of common land was carried out according to new laws. Railways were developed under protectionist laws that, for example, eliminated the right to demand payment for damages arising from fires caused by steam engines. After the civil war, the American government donated to the railway companies a land surface greater than the states of California and Florida (Gorelick 1998: 15). A French decree in 1810 established the automatic authorisation of hazardous, unhealthy or unpleasant installations, with the aim of keeping them legally immune to the claims of those affected, who were receiving favourable rulings from magistrates. According to the French Institute, this "arbitrariness of the magistrates [...] disrupts the environment in which the manufacturer can operate freely and safely" (Naredo 1987: 272–273).

But the deregulation of early capitalism and the reduction of work and natural resources to mere goods soon showed their life-threatening nature, endangering even the survival of the system itself: "The idea of a self-adjusting market implied a stark utopia. Such an institution could not exist for any length of time without annihilating the human and natural substance of society; it would have physically destroyed man and transformed his surroundings into a wilderness" (Polanyi 1989: 7).

4.1.1 Repercussions on Labour and on Nature

The market economy brought private appropriation of the means of production, turning most of the active population into wage-earners and their labour into a good. In the orthodox economy, labour power appeared as just another good whose price (wage) was determined by the laws of supply and demand, and that, once incorporated into the production process, behaved as such a good. However, labour is but one of the manifestations of human life, meaning that it cannot be a good. Mothers have children and do not look after them with the intention of supplying productive resources to society. Unlike goods, the sale of labour does not entail a change of property; the business owner can only appropriate the value produced by the labour, which is one of human beings' expressions, inseparable from the rest of its expressions. The commodification of labour, therefore, degrades people.

Once labour became a good, it had to find its price in the market. Any labour price that had not been established in this way was considered anti-economic. But converting labour into a good means wiping out the organic relationships that existed in the pre-capitalist world, based on kinship, proximity, trade, etc., that linked labour with the other dimensions of life. In addition, it involved limiting the freedom of individuals to organise themselves. Chatelier's law in Napoleon's France banned worker's associations because they threatened the freedom of the employers. As people did not want to abandon the countryside and its community relations, they had to be forced, and the method usually consisted of undermining their livelihood base. In Europe this was usually achieved by taking land away from the peasants. In the colonies many methods were used to break the resistance of the natives: the best land was taken away from them, they were forbidden to manufacture goods of particular importance, they were heavily taxed, etc. And when these practices failed, they were turned into slaves. The French Minister of Commerce stated in 1901: "The black does not like work and is totally unaccustomed to the idea of saving; he does not realise that idleness keeps him in an state of absolute economic inferiority. It is therefore necessary to use (. . .) slavery to improve his circumstances and afterwards lead him into an apprenticeship of freedom". (The Ecologist, July–August 1992)

Treating labour as a good means turning it into something that is perfectly exchangeable, like parts of a clock. It is therefore necessary to divide it and trivialise it so that any worker can take the place of another. The commodification of labour entails, therefore, the structural need to eliminate any knowledge that could give workers the capacity to work autonomously. Thus, the implantation of capitalism brought about a process of elimination of knowledge among workers. American engineer Taylor was the one who systematised the techniques to achieve it and Ford completed the process with development of the assembly line. Adam Smith, despite defending the division of labour, acknowledged its degrading nature, although he only lived at the start of the British industrial revolution: "The man whose whole life is spent in performing a few simple operations (. . .) has no occasion to exert his understanding (. . .) He naturally loses, therefore, the habit of such exertion and generally becomes as stupid and ignorant as it is possible for a human creature to become" (Schumacher 1980: 60–61). Apart from the loss of control over the production process, the commodification of labour caused other known effects: a reduction of wages, longer work shifts, worsened occupational health, child labour, etc. Equally known are its side-effects of moral degradation: alcoholism, prostitution and delinquency.

On the other hand, the result of the commodification of nature is the start of a process of degradation that is still ongoing, because it is intrinsic to the system and only slows down when the market is heavily regulated, that is, when its supposed mercantile nature is questioned. At the beginning, capitalist deregulation did not cause ecological problems comparable to today's (population, economic activity and technological development levels were much lower than today's), although its effects are far from insignificant. One of the most serious is widespread deforestation. In England many communal forests disappeared through Parliament laws and the same happened in Italy and Spain with the expropriation processes. In the USA, colonisation ended in the late nineteenth century and, once the forests in the east were destroyed, the country strived to cut down what was left in the rest of the territory (Ramos Gorostiza 2009).

Another anti-ecological process was the gradual substitution of sustainable primitive agriculture with another that was industrial in nature. These effects were soon made evident. Liebig is known because of his development of chemical fertilisers, but he also criticised productivist agriculture because it breaks the cycle of matter of traditional agriculture, as the food waste from cities does not return to the countryside. Marx stated that capitalistic agriculture exhausted the land: "Moreover, all progress in capitalistic agriculture is a progress in the art, not only of robbing the labourer, but robbing the soil; all progress in increasing the fertility of the soil for a given time, is a progress towards ruining the lasting resources of that fertility (. . .) Capitalistic production, therefore, develops technology, and the combining together of various processes into a social whole, only by sapping the original sources of all wealth – the soil and the labourer" (Marx 1990: 463).

Meanwhile there was a two-sided process that had serious repercussions on the countryside, and which did not cause an ecological disaster because it was interrupted. On the one hand, introduction of the land market, which was established in Europe between 1830 and 1860, brought the elimination of traditional renting rights, causing a widespread exodus of landless peasants to cities, which became overcrowded, similar to what is currently happening in non-OECD countries. On the other hand, there was a sharp drop of agricultural prices as a result of the massive introduction into Europe of American cereals at very low prices, brought about by development of the railway and the steam boat. Both processes threatened the survival of most peasant farms and the abandonment of fields was an ecological disaster.

Last of all, the establishment of capitalism caused a demographic explosion in all societies, in particular due to the disappearance of the birth controls that primitive societies had gradually designed. Malthus developed his population theory by observing the exponential growth of the British population during his time. Europe soon became an overpopulated continent and looked for a safety valve in emigration. Between the early nineteenth and twentieth centuries 50 million Europeans emigrated (Crosby 1988: 16).

4.1.2 The Regulating Reaction

The story of the free market is, at the same time, the story of the State's aggression on the community bases of pre-capitalist societies and the spontaneous reaction of the latter. This reaction was aimed during a first phase against the inclusion of land and labour in the market, and during a subsequent phase, against the most degrading effects this inclusion had. Wage flexibility and the mobility of labour had to be reduced, minimum wages guaranteed, regulation of the urban environment to prevent it from becoming totally unhealthy, regulation of natural resource management and limits on the activities that were most destructive for the environment. But regulation went beyond the strict framework of the problems dealt with here. The USA, industrialised European countries, and in particular Japan, protected their emerging industries from international competition by establishing strong customs barriers. It is curious that, in the name of economic liberalism, this right and need is denied to non-OECD countries.

While the destruction of the pre-capitalist economic and social structures was the result of a conscious and systematic action by the States, the reaction against the terrible side effects of this intervention was spontaneous in nature. This is proven by the fact that regulation came about through the adoption at the same time of very similar formulas in countries with very different political regimes, with not only central governments but also local governments working in the same direction. The process was simultaneous in Victorian England, in the Germany of Bismarck, in the French Third Republic and in the Habsburg empire. In addition, its promoters were people from across the ideological spectrum: "In Protestant England, Conservative and Liberal cabinets laboured intermittently at the completion of factory legislation. In Germany, Roman Catholics and Social Democrats took part in its achievement; In Austria, the Church and its most militant supporters; in France, enemies of the Church and ardent anti-clericals were responsible for the enactment of almost identical laws" (Polanyi 1989: 85).

Laws on work accidents were approved in 1880 and 1897 in England, in 1879 in Germany, in 1887 in Austria and in 1899 in France. Factory inspections were established in England in 1883, in Prussia in 1853, in Austria in 1883 and in France in 1874 and 1883. Laws were approved to limit the age for children to start working and to regulate hygienic conditions in factories; workday hours were limited, social security systems were established, etc. Meanwhile, and under pressure due to epidemics and unhealthy living conditions, cities established sewage, waste collection and food inspection systems, measures were taken to improve housing for workers and public parks were built. In the USA, in the late nineteenth century, Congress successively approved the Pure Food and Drug Act, the Federal Meat Inspection Act and the Historic Sites Act. By around 1880 most cities had built wastewater systems. In Europe, a powerful farmer's movement opposed to agricultural deregulation managed to push for protectionist measures from the 1870s, which allowed a stabilisation of the peasant population. Import duties were established for imported cereals, the transferability of land and direct and indirect economic aid for farmers was limited (Fabe and O'Connors 1990).

In addition, it was necessary to protect nature from the rapid expansion of capitalism and important conservationist movements emerged, whose action has led to the creation of many protected areas. In the United Kingdom a powerful movement against forest clearing saved many of them. In the USA, after the civil war, a strong conservationist movement emerged, a result of concerns about the rapid process of destruction of nature. The movement was very heterogeneous and was led by very different people. This lack of clear and common goals and the anti-environmentalist policy of president Taft (1909–1913) led to their decline in the early twentieth century, but their legacy is evident in many fields: the creation of protected areas, territorial planning, public water management, the nationalisation of large extensions of forests. The Forest Service was founded to manage them sustainably (Fabe and O'Connors 1990; Ramos Gorostiza 2009). This environmentalist movement saw a powerful resurgence during the recession of the 1930s with Roosevelt's New Deal policy and meant that millions of young people were hired for the conservation of protected areas. In Spain, 8 years after the second expropriation law was published (Madoz Law of 1855), which like the first law was disastrous for its forests, published the Forest Act of 1863, "which responds to a desire to restrict the expropriation policy undertaken years earlier" (Aunos 1991). From 1917 natural parks began to be created.

Thus, the regulations for the protection of nature during capitalism's first phase referred mainly to the agricultural use of land (and was very determined by social problems), and in some cases to the protection of high-value ecosystems, with the promulgation of laws on protected spaces. But there was not much progress in what would later be known as environmental policies: the limitation of emissions and discharges. It would be necessary to suffer the strong degradation caused by the shockwave of the postwar period for these policies to begin their development.

4.2 The Scarcity of Resources in Classical and Neoclassical Thought

Despite the destructive process of natural resources during the historical period analysed here, there was no general concern about their possible exhaustion. In Europe, the defence of forests was due to the fact that they were a key resource for the economy of small peasants. In the USA there was a conservationist motivation in very active minorities, without there being (at least directly) an interest in preserving a resource of high economic value. This concern is logical, because in the nineteenth century most of the Earth's natural resources were still available, as the industrialised economic activity and the population were much smaller than today's. The world's population reached one billion in the mid-nineteenth century.

An exception to the general panorama described is the problem of Great Britain's natural resources. The fact that the industrial revolution started in the late eighteenth century led to serious shortages of resources as the nineteenth century progressed.

The systematic clearing of forests decided by the British parliament led to a severe shortage of wood. Coal was another reason for concern. Although Jevons was one of the fathers of the neoclassical revolution, he was concerned about the exhaustion of coal in Great Britain (Bradley 2007). It is also very possible that the Spanish Forest Act was motivated by the scarcity of wood after most forests were cut down, due to the demand caused by the development of the railway and mining. But, except for the cases described, it is only in the late twentieth century when the scarcity of natural resources became manifestly clear.

The classical authors, (mainly economic thinkers from the first half of the nineteenth century) were influenced by the process of destruction of natural resources that took place during their time and they had, in general, a clear vision of the existence of natural limits to unlimited growth. They were therefore concerned about population explosion. Malthus believed that human beings tended to reproduce explosively, and therefore, to exhaust resources. In reality, what was happening was a phenomenon inherent to the establishment of the first capitalism: a population explosion motivated, among other factors, by the disappearance of the birth control culture of previous cultures. These used individual abortive mechanisms (abortifacient products, use of natural preservatives, etc.) and social mechanisms (delaying the age of marriage, sending many children to live a celibate religious life, such as in the case of Buddhism and Catholicism). With the industrial revolution in Great Britain, the age of marriage among young women dropped from 28 to 22. Based on this vision of population dynamics, some economists considered the problem that could arise regarding food. Malthus argued that food scarcity would lead to a competition that only the fittest would survive. The same concern led David Ricardo to develop a theory that proposed an innate tendency of deregulated capitalism towards stagnation. Population pressure would to lead to an increasingly marginal use of land, which would cause the price of food to rise, and with them salaries, but the latter would see their purchasing power reduced to covering the bare survival of the worker and their family, the moment at which the population would stabilise because families would only have the children that they could afford to feed. On the other hand, rising salaries would reduce company profits, investment would fall and the economic system would stagnate. And to avoid this structural tendency, States should systematically intervene to encourage economic growth. This theory was refined by Keynes in the nineteenth century and is known as the neo-Ricardian or Keynesian school (Roll 1966). Keynes was concerned about the population problem. In his "General Theory" he states that one of the causes of war is "the population pressure" (Keynes 1964: 382).

For J. S. Mill and K. Marx the problems of the capitalist system's unsustainability lead to the need to replace it. J. S. Mill understood during his mature stage the impossibility of limitless growth and argued in favour of a stationary state economy with a more equitable distribution of resources. In his Principles of Political Economy he wrote: "No man made the land. It is the original inheritance of the whole species. Its appropriation is wholly a question of general expediency. When private property in land is not expedient, it is unjust. But, it is some hardship to be born into the world and to find all nature's gifts previously engrossed"

(Mill 1996: 233). K. Marx was convinced that there is a contradiction between capitalism and the conservation of nature, that the search for profit in the shortest term possible brings the destruction of farmland, due to overexploitation, and a lack of nutrient recycling. But the solution to this contradiction, as well as to others, should wait for the proletariat to overcome the main contradiction (between work and capital) by means of the proletarian revolution (Marx and Engels 1975).

On the contrary, with the so-called neoclassical revolution, any concerns about physical limits disappeared. They focused on the micro-economy (company economy) and lost any general perspective. And the fact that at a company level there are no problems of resource exhaustion or environmental impacts implicitly means that the Earth is an inexhaustible source of resources and a sinkhole for waste with limitless capacity. The orthodox economy maintained this vision for 100 years: "Between 1870 and 1970, mainstream economists (with some notable exceptions) appeared to believe that economic growth remained feasible (a growing economy need not run out of natural resources)" (Pearce and Turner 1990: 13). And up until now this has been the widely dominant opinion regarding the economy, to the extent that no economy manual takes this problem into account. However, the enormous environmental impacts that were produced during the period after World War II, due to the intensive growth of the economy and, in particular, the chemical industry, led to the birth and development of environmental policies from the 1970s onwards. But they continued to ignore that there could be an exhaustion of resources. The historical trend of falling prices of raw materials seemed to support this approach, but in reality it was mistaken. We will see later that the market only sends out signals (in the shape of rapid price rise processes) during the final phase of resource exhaustion. This is the scenario that emerges in the late twentieth century and during the following decade there were steep rises in fossil fuel prices and in most strategic metals, in addition to monopolisation phenomena.

Despite the fact that the dominant thought in the orthodox economy does not acknowledge the scarcity of natural resources, there has been an academic discussion about the "Economics of Natural Resources" where there are opposing points of view. And it is convenient to analyse the academic discussion because there are beginning to be swift changes of opinion faced with the evident growing lack of resources. There are, at least, four types of position: denial of the existence of scarcity; the scarcity of a resource raises prices and leads to a more efficient use of it, postponing the problem when not solving it; acknowledgement that resources, though finite, are abundant due to technological development or human ingenuity, which could mean that, although some resources may be exhausted, they will be replaced by others; and, there is a scarcity of resources. So the first three defend the capacity of the markets to face situational scarcities.

Among the pure deniers we have Adelman, Seaborg, Brooks and Andrews. Adelman states that the idea of "finite limited resources . . . is an empty slogan" [. . .] "but inventories of 'proved reserves', constantly renewed by investment in finding and development" (Bradley 2007). For Seaborg the abundance of energy allows us to "recycle any waste [. . .] to extract, transport and return to nature, whenever necessary, all materials in an acceptable way". Brooks and Andrews state that the idea of "running out of minerals is ridiculous because the entire planet is composed of minerals" (Carpintero 2006: 147).

The positions of Samuelson, Friedman or Frances Cairncross (managing editor of The Economist) fit in with the second group. Samuelson states that "when resources start to be scarce [. . .] prices rise" and their use becomes more efficient (Carpintero 2006: 181). For Cairncross (1996: 7) "the environmental resources least in danger of exhaustion are those that are privately owned and traded. As they start to become scarce, their price will rise. This is likely to encourage their owners to conserve the supply". For Friedman there are no limited resources, because "when resources are really limited prices go up, but the prices have gone down and down. Suppose oil became scarce: the price would go up, and people would start using other energy sources. In a proper price system the market can take care of the problem" (Ravaioli and Ekins 1995: 33). DeGregori affirms that "the central role of knowledge as the ultimate resource and creator of minerals is -or should be- a fundamental principle of economics" (Bradley 2007).

The third group is the most widespread. Schumpeter declares that "there is not a law of diminishing returns in relation to technological progress". Zimmermann affirms that "the problem of resource adequacy for the ages to come will involve wisdom more than limits set by nature". Also adding: "Human wisdom is the principal resource". Simon rejects Malthusian ideas of scarcity and states that "a theory of endogenous invention is more persuasive in my view". Hotelling became the focus of the academic debate about natural resources since publication, in 1931, of the article "The economics of exhaustible resources". It is based on the assumption that natural resources are exhaustible, as the title indicates, but also replaceable, and reflects on what exploitation rate can maximise profit, which depends on the following premises: perfect knowledge of reserves, extraction technologies, prices, interest rates, alternative resources, etc. (Bradley 2007).

The fourth type states that resources are exhaustible, meaning they should be adequately managed. Authors who defend this approach coincide with many others who have spoken out about the exhaustion of resources and who are today associated with ecological or sustainable economy. Paradoxically, Jevons (one of the three people responsible for the neoclassical revolution, which determined the development of the economy of nature) expressed his concern about the exhaustion of coal in Great Britain in his book The Coal Question, of 1865. He was worried that England extracted 50 % of the world's coal, both for domestic consumption and for exports, while the country had 0.04 % of the world's surface area and 2.5 % of the population. Based on the definition of the principle of thermodynamics: the law of entropy (the energy used cannot be recycled because it degrades over time, the temperature is reduced so much that the energy cannot be used: entropy increases), many scientists have insisted on the unsustainable nature of the exponential growth in non-renewable energy resource consumption, such as Carnot, Clausius, Cournot, Podolinsky, etc. At an economic level, for Georgescu-Roegen the law of entropy is irrefutable proof that non-renewable energy runs out. He also defined what he called the fourth law of thermodynamics, with which he sought to express the impossibility of totally recycling materials, because the dissipation of many of them makes it impossible to recycle them from an economic point of view. However, the fourth law has been rejected, even by his followers. And the science of thermodynamics never accepted it (Bradley 2007).

Positions such as those held by Adelman tend to disappear when faced with the growing evidence of the scarcity of many resources. But the approaches that combine market and technological development still prevail among orthodox economists, with authors differing greatly in their emphasis on one factor or another. So orthodox economists live, once more, with their back turned to other sciences and, in this case, to the Earth sciences. But we will later see that geologists are showing us clear processes of exhaustion.

Chapter 3
Foundations and Instruments of Environmental Economics

Keywords Environmental economics • Environmental valuation • Polluter pays principle • Internalization of externalities • Market tools for internalising externalities

In the 1960s the environmental problems caused by the strong economic growth of the post-war period and the explosive development of the chemical industry started to become evident. As a result, the governments of OECD countries, under pressure from public opinion, started to define regulations to limit environmental impacts. Faced with this situation, the traditional economy adopted the position of considering that such expenses were a restriction on growth and the creation of jobs, because it was said that they increased inflation, curbed innovation and were an obstacle for the development of trade. Therefore, there was an admission, albeit implicit, of the incompatibility between economy (identified with growth, as usual) and ecology. An International Conference on Environment and Economics was held at OECD Headquarters in 1984 and in the report on its results many of the attendees (Delegations appointed by Member Governments) stated that the idea that environment protection regulations negatively affected economic growth was one of the past: "Fifteen years ago there was great concern that environmental action could impose a heavy, if not intolerable burden on economies, slowing growth, aggravating unemployment, adding to inflation, inhibiting innovation and distorting trade". But other attendees argued that the idea was still dominant in some sectors ("in particular industry representatives") (OECD 1985: 66).

As the environmental problem worsened and reached planetary dimensions, the traditional economy was unable to keep maintaining the theory of incompatibility, because to do so meant the need to replace the capitalist economy with some other one, as it was suicidal to fail to adopt measures to reduce the level of unsustainability. As this approach was unacceptable for the supporters of the system, the only way out left was to defend the compatibility between unlimited growth and free trade, on the one hand, and the protection of the environment, on the

R. Bermejo, *Handbook for a Sustainable Economy*,
DOI 10.1007/978-94-017-8981-3_3,

other: "Continued environmental improvement and sustained economic growth are essential, compatible and interrelated policy objectives for OECD Member countries. This, the major conclusion of the Conference, means that the environment and the economy, if properly managed, are mutually reinforcing" (OECD 1985: 10). The theory that attempts to give coherence to this discourse is the valuation of *free* or public assets (the atmosphere, rivers, oceans, etc.); that is, those that cannot be privatised and commodified. In order to deal with this task, it segregated a branch: Environmental Economics (EE). This allows the rest of the traditional economists to maintain their discourse, because *there is already someone taking care of the environmental problem.*

1 Polluter Pays Principle

The theoretical meaning of the polluter pays principle (PPP) is that the costs of nature restoration have to be borne by the polluter. It is a direct consequence of the theory of the valuation of free assets, as the amount to be paid by society is based on this valuation. However, we will see that the payment required does not correspond to the total value assigned, but to the optimal value where the total of public and private costs is minimised. This means that there is a private right to contaminate up to the optimal level: "In practice, it is more feasible to think and design policy in terms of 'acceptable' levels of pollution. Accordingly, the OECD PPP is formulated in broader terms of making the polluter bear the costs of standard-setting". And this means that "the polluter is obliged (...) to pay the necessary costs of getting to that standard" (Pearce et al. 1994: 158). The Amsterdam Council introduced the principle in the European Treaty. And the EU Revised Strategy of Sustainable Development established the policy of making sure that the "polluter pays", and that means to: "Ensure that prices reflect the real costs to society of consumption and production activities and that the polluters pay for the damage they cause to human health and the environment" (European Council 2006: 5).

This principle should be the object of at least three theoretical criticisms and a further three related to its application. From a theoretical point of view, the polluter pays principle is based on the premise that the system basically works well and that the cases of environmental impact generation can be corrected to sustainable levels, obliging the agent to comply with emissions regulations. There is therefore no need to change the economic model. Secondly, this principle commodifies dimensions of life that from an ethical point of view are unacceptable. Lastly, it legitimises pollution for those who can pay for it. Resolution A3-0317/92 of the European Parliament, on the Community Programme of Policy and Action in relation to Environment and Sustainable Development, "calls on the Commission to review the 'polluter pays' principle in the light of the precautionary and preventive action principles laid down in the Treaty of European Union, since pollution can in no case be legitimised by counter-payment" (European Parliament 1993: 12).

The three criticisms of its practical application are with reference to who pays, how much is paid and the extent of compliance. Regarding the first point, polluters can be producers and consumers. Consumers are forced to pay once economic mechanisms are established (normally taxes) and the extent of the contamination is largely determined by the design of the product. The corporation, in addition to being responsible for the design of its products, has many ways of evading their responsibility: "If polluters can pass on an increase in the cost of production to *consumers*, they will do so". And it is evident that monopolies have the power to pass on the environmental costs to consumers, via prices. Which is at odds with the PPP principle: "Making the consumer of the polluting product pay some of the clean-up cost may seem at odds with the PPP principle but in fact it is exactly what should happen" (Pearce et al. 1994: 158). Secondly, corporations have ex ante and ex post mechanisms to avoid or reduce pollution costs. The US and the EU systematically subsidise corporations so that they adopt anti-pollution measures. The Fourth Environmental Action Programme of the EU (1987–1992) proposed that, in the case of regular polluters, they would pay for the control and clean-up costs. In addition, the lobbying power that business associations have in Brussels is well known.

Finally, the polluter pays principle is not complied with at all at an international level. There are many states that suffer environmental aggressions from neighbouring countries, as is the case of acid rain and river and coastal pollution. There is no international regulation that forces the aggressor to pay. However, the Rio Declaration supports the application of the PPP at an international level. Principle 13 states: "States shall develop national law regarding liability and compensation for the victims of pollution and other environmental damage. States shall also cooperate in an expeditious and more determined manner to develop further international law regarding liability and compensation for adverse effects of environmental damage caused by activities within their jurisdiction or control to areas beyond their jurisdiction" (UNCED 1992).

2 The Theory of Environmental Valuation

To explain the theory of the valuation of free assets we will base our observations on two books that widely dominate the EE panorama: *Blueprint for a Green Economy*, by Pearce, Markandya and Barbier (whose first edition in 1989 was hailed by the British press as an intellectual landmark), and *Economics of Natural Resources and the Environment* by Pearce and Turner, which is the most widely used manual on environmental economy. Both books have been a great publishing success and make Pearce, a co-author in both, the most influential environmental economist there is.

If we start from the premise that the free market is efficient, the only reason behind the environmental problem is that the environment contains *free assets*, that is, assets that cannot be privately owned, so the market acts without taking them into account. And economic activity will produce effects (externalities), normally negative, without having to pay for them. Therefore, the solution would be to disruptions, particularly those that involve non-linear processes and irreversibilities, have been hardly explored. Ecologists have little more than a premonition of the repercussions and interactions inherent to human activities, such as the pollution of oceans and coastal wetlands and the production of fisheries" (Daily et al. 1996).

As a result, we have an instrument conceived for solving exceptional situations, when in reality environmental impacts are a normal manifestation of economic activity, and it is based on premises that are not fulfilled. This is what Pearce thought before becoming a dominant figure in environmental economics: "A policy for the correction of externalities cannot in any way guarantee the preconditions for the survival of species [. . .] If this is correct, the economists who have analysed environmental problems within the context of the theory of externalities may not only be prescribing incorrect policies, but also using mistaken conceptual foundations for their analysis" (Aguilera Klink 1992).

4 Internalisation of Externalities

4.1 Environmental Valuation Methods

Environmental valuation methods for public goods proposed by the EE is the same as that used by the traditional economy to define the value of commercial goods and services. As in the case we are dealing with there is no market, the idea is to create an artificial market by estimating people's preferences (declared preferences), valuated by their readiness to pay for environmental improvement or by their readiness to receive compensation for its loss: "At its simplest, what we seek is some expression of how much people are willing to pay to preserve or improve the environment. Such measures automatically express not just the fact of preference for the environment, but also the intensity of that preference. Instead of 'one man a vote', then, monetization quite explicitly reflects the depth of feeling contained in each vote" (Pearce et al. 1994: 55). There are two basic methods for valuation:

- Looking for a *surrogate market*. Looking for a commodity the price of which is supposed to be influenced by an environmental variable (Hedonic property prices and travel costs models).
- Using *experimental* techniques. Creating a virtual market where the value of the environmental asset is determined through questionnaires that ask people, using more or less elaborate techniques, how much they are prepared to pay for a certain good or what compensation they would accept for losing it. (Contingent valuation) (Pearce et al. 1994: 64, 69).

The practical reality of environmental valuation is more complex and several techniques are usually combined, such as considering the expenses brought about by illnesses or premature deaths caused by environmental impacts or assessments by the research team of the costs of complex environmental impacts, such as climate change.

4.1.1 Hedonic Pricing Method

This technique tries to identify how much of a property value differential (frequently of houses) is due to a particular environment quality, and to infer how much people are willing to pay for the better quality. Three factors are normally taken into account: air pollution, noise nuisance and water quality deterioration. Although there is overwhelming evidence of a positive correlation, "the accuracy with which we can quantify such effects is, however, much more debatable (. . .) this matter of fuzziness is (. . .) inherent to the problem being considered" (Pearce and Markandya 1989: 26).

4.1.2 Travel–Cost Approaches

This method is used to define the demand of natural spaces, valuated depending on an assessment of the different costs involved in a visit to a natural space (travel costs, time invested travelling and the stay), assessed by means of the salary value of the time invested. For this reason D. Pearce et al. affirm that "special attention is paid to the value time. That time is valuable is self-evident" (1989: 71). It is considered that the main specific inconsistency of this technique is that it is based on the valuation of the time used, when it is evident that visits are carried out during people's leisure time and that for many, travelling is a pleasure.

4.1.3 Contingent Valuation

The contingent valuation method (CVM) "uses a direct approach – it basically asks people what they are willing to pay for a benefit, and/or what they are willing to relieve by way of compensation to tolerate a cost" (Pearce et al. 1989: 69). The valuation can be carried out through answers to a questionnaire on a specific case, previously known by the people interviewed, or through the information facilitated by the interviewer. Although different surveying techniques are used, they all have elements in common. In the case of willingness to pay, the interviewer makes a payment proposal, which will supposedly be accepted because it is small. From here on, and as the interviewee accepts the proposals, the interviewer raises them until one is refused. The above is the top price they are willing to pay for an environmental asset. In the case of willingness to accept compensation, the first proposal is high, and to the extent that it is positive, it is lowered until it is rejected. The last one accepted is the minimum predisposition to accept compensation for tolerating environmental damage. This method "has been applied with increasing frequency" and "one major attraction of CVM is that it should, technically, be applicable to all circumstances", because "it will frequently be the *only* technique of benefit estimation". But this technique has many "biases": lack of accuracy, "vehicle bias" (instrument of payment), "information bias", "hypothetical bias" (difference between real markets and hypothetical markets), and "operational bias" (operating conditions) (Pearce and Markandya 1989: 35–37).

4.2 Analysis

We will focus our analysis on contingent valuation as it is the most widely used technique and the one that best fits the premises of EE; and we will do so under the light of four parameters: ideological premises, scientific measurement requirements, environmental law and discriminatory effects.

4.2.1 Ideological Premises

In the techniques that we refer to there is a mercantilist vision, according to which everything has a price and people are only moved by economic motives. There is also another central idea of the traditional economy: people behave rationally in the market: they maximise its utility. The EEA extends this axiom to the valuation of environmental assets that have no market value.

4.2.2 Premises of Scientific Measurement

It is not the same to say how much one is prepared to pay as to actually pay. In the hypothetical context in which the question is made, people tend to show more willingness to pay than what they would in a real situation. Here the customer may have access to important information about the good on sale; they know its price and know that their decision will not influence that of other possible buyers. This does not happen with collective assets, and knowledge about the predisposition to pay of others influences the decision of those surveyed. Last of all, there are many environmental problems and it does not seem feasible to ask people about each and every one of them.

One of the requirements of a scientific measurement is the consistency of the measurements; that is, the level of coincidence of the measurements made with the same method or instrument. If the results differ greatly, the measurement cannot be considered scientific. This is what happens with contingent valuation. It has been found that if slight variations are introduced in the method used for the surveys, such as for example varying the magnitude of the first valuation offer, the final results differ greatly. On the other hand, when valuation studies are carried out on one aspect, such as the external costs of transport, with a difference of several years, it turns out that the valuations of some factors change radically, because the impacts are better known. This has occurred with the valuation studies of the external costs of road transport made for the EU (Hueting 1991).

It is normally considered that the predisposition to pay is equal to the predisposition to receive compensation because it is the only way for the cost functions to be continuous. However, psychology shows that the second is much higher than the first. Shorgen and Nowell (1992) point out that in fifteen studies carried out it has been found that the predisposition to accept compensation is normally around five times greater than the predisposition to pay. Ward and Duffield report similar results in other experiences. In one of the studies the ratio found was in the region of 75 to one (Bromley 1995: 133).

4.2.3 Environmental Law

Since at least the 1980s developed countries have introduced changes in environmental Law that question the legality of contingent valuation. During the first period of environmental policies there were no laws that acknowledged the environmental rights of people. Despite this, many people thought that they were entitled to a healthy and safe environment, and did not want to respond to the question of how much they were prepared to pay. For example, in the late 1970s a group of economists from the University of Wyoming attempted to valuate, by asking people, the aesthetic cost of the visibility lost because of the fumes produced by an electrical power plant. Around half of those interviewed refused to collaborate (*The Ecologist*, July–August 1992). But later, laws were approved that acknowledged the right to a healthy environment. In the US, laws such as the CERCLA (*Comprehensive Environmental Response, Compensation, and Liability Act*, 1980) and the OPA (*Oil Pollution Act*, 1990) acknowledged these rights. The same happened in the EU with the Single Act and the Maastricht Treaty. Principle 1 of the Rio Summit Declaration states: "Human beings are at the centre of concerns for sustainable development: They are entitled to a healthy and productive life in harmony with nature" (UNCED 1992: 277). In this situation, to ask people how much they are prepared to pay to avoid certain environmental ills is like asking them how much they are prepared to pay to avoid being mugged on the street (Bromley 1995: 129 and subsq.).

4.2.4 Environmental Justice

Last of all, predisposition to pay is directly related to income levels, so according to this criterion, environmental impacts in poor areas will have less value and entrepreneurs and governments will tend to install industries and infrastructures with a higher impact in these areas. This fact is evident in the US, which has led to the environmental justice movement.

Pearce and Turner accept, at least, some of the criticism exposed: "A good many types of damage did not prove capable of 'monetisation', so that, if the monetised figures are accepted, actual damage exceeds the estimates shown" (123). But they insist on the subject: "Money units remain the best indicator we have" (1990: 121).

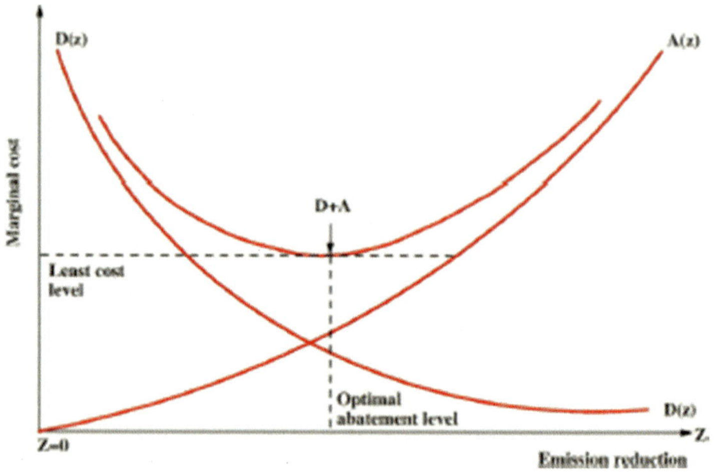

Fig. 3.2 Private and social costs due to pollution (Source: Ayres (2008: 283) Ecological Economics 67)

4.3 *It Is Not Possible to Measure the 'Environmental Optimal'*

The EE, ignoring the right of people to live in a non-degraded environment, considers that not only should the costs of environmental degradation for the population be taken into account, but also the business costs of impact reduction. For this reason it seeks the minimum costs of both, for which purpose it uses the classic cost and benefit curves shown in Fig. 3.2. It shows that business costs of pollution control (represented by the D curve) grow exponentially as emissions are reduced. The social costs (defined by the A curve) of pollution also grow exponentially with it. Therefore, both curves intersect at one point, which is what approximately determines the minimum total cost, so the *economic and environmental optimal* are the same.

As a result, the optimal solution is not to eliminate pollution, but rather to guarantee the optimal amount of pollution, which will maximise the value of production. This entails taking into account the costs of all the agents involved and that means "to have to take into account the costs involved in operating the various social arrangements" (Coase 1960: 23). From a theoretical point of view, it can be proven that this has no validity. Let us consider some possible scenarios. If the pollutants are not bioaccumulative, it can lead to two situations. One is that the pollution corresponding to point X is less than the load capacity of the ecosystem concerned; in this case it is not necessary to act, at least from the point of view of the traditional economy. The second situation is produced when the discharge level of point X is greater than that corresponding to the load capacity; then the designed taxes will not be capable of stopping the degradation of the ecosystem. Last of all, if the pollutants are bioaccumulative, they will concentrate in the ecosystem until it is degraded, whatever the optimal pollution point is.

On the other hand, when trying to establish the environmental offer and demand curves, we meet practically insurmountable barriers. The business costs can be found, but only approximate figures, due to the reluctance of companies to provide reliable information. On the other hand, we have seen that by means of contingent valuation, it is impossible to build the social costs curve, due to the differing valuations obtained regarding predisposition to pay and receive compensation. The curve is discontinuous when the aim is to accept a reduction in quality. And if it cannot be built, the optimal point cannot be calculated either.

5 Market Tools for Internalising Externalities

The OECD (1994: 15) defines economic instruments as those which "affect estimates of costs and benefits of alternative actions open to economic agents". There are four types of instruments: taxes/charges, subsidies, tradable pollution permits, and returnable container systems.

This organisation has been insistently recommending the use of economic instruments in environmental policy since 1984. The OECD "Council Recommendation on the Use of Economic Instruments in Environmental Policy" recommends that Member countries "make a greater and more consistent use of economic instruments as a complement or a substitute to other policy instruments such as regulations" (OECD 1997b: 7). And for this reason it has published many reports on the subject.

The defence of economic instruments is based on two motives: the deficient application of administrative instruments and, above all, the better efficiency attributed to the former. Some of the —reasonable— criticisms of the administrative instruments due to their deficient application are: multiplication of regulations with little coherence among them, which makes their application difficult; a lack of coordination of the regulations, as in general they are aimed at one environment without taking into account the effect on others; an excessively punctual and locally-oriented approach that does not take into account the effects at a wider scale; a more reactive than preventive nature; dissuasive instruments that are too small or not applied often enough, which leads to widespread non-compliance; the negative effect of other sectoral policies that do not take into account the ecological dimension, etc. The arguments in favour of economic instruments are:

- "They could reduce the economic cost of achieving a given level of environmental protection, by allowing polluters greater flexibility". The reason is that the market instruments allow the economic agents to choose the most efficient method to respond to the price signal.
- "They may stimulate more rapid innovation in pollution abatement technologies", because businesses face costs whenever they emit pollutants, which does not happen when they are only required to comply with an administrative regulation.

- “Also, in some cases (. . .) economic instruments may raise revenues which may allow other taxes to be reduced” or they can be used to finance environmental policies.
- The moral problems that accompany the administrative instruments disappear. When a company has to pay a fine it means they have committed an offence.
- The political and economic problems (lengthy litigation processes) that accompany the assignment of scarce environmental resources disappear; the market is left to search for the most efficient solution (OECD 1997d: 9, 15).

5.1 Environmental Taxes

There are many terms in taxation literature: tax, charge, fee, levy, fine, penalty, etc. The most widely used terms are tax and charge. A tax is defined as “any compulsory, unrequited payment to general government levied on tax bases deemed to be of particular relevance”. The term of charge is “defined as compulsory and requited payments to general government or to bodies outside general government, such as environmental funds or water management boards” (EEA 2005: 41). However, “in the area of environmental taxation different meanings are often given to similar terms in different Member States, and no precise definitions are offered by EU legislation” (EEA 2005: 40). But the widely used definition of environmental taxes (by international organizations as the European Union and the OECD) is: “A tax whose tax is a physical unit (or a proxy of it) of something that has proven, specific negative impact on the environment” (EEA 2005: 41). The EEA (1996: 8) classifies three types of environmental taxes:

- “*cost-covering charges* – e.g. designed to cover the costs of environmental services and abatement measures, such as water treatment”.
- “*incentives taxes*- designed to change the behaviour of producers and/or consumers”.
- “*fiscal environmental taxes* – designed primarily to raise revenues”.

Although the last type and the previous definition of environmental taxes (“physical units” and designed “to raise revenues”) could be understood to represent a broader vision that could include taxes on resources used (e.g. taxes on petrol), revenue statistics show that they refer exclusively to taxation for environmental reasons. This is seen clearly in the case of the energy tax that some countries use, where on top of the traditional taxes on energy use there are others added that tax the CO2 emissions associated with energy consumption.

Below is a brief summary of the reasons provided by those that support environmental taxes. The most important argument is that they internalise externalities. In addition, the following arguments are used: efficiency, dynamic nature, revenue and jobs creation. The dynamic nature is manifested in that emissions are paid for by units, even if the total volume emitted is within regulations, which means a permanent incentive to improve environmental quality and technological innovation.

The creation of jobs is defended because revenue is used to lower income tax: "Some payroll tax reductions could be financed by new or increased eco-taxes, thus providing a *double dividend* in terms of both improved environmental protection and lower unemployment" (EEA 2005: 45, 46; OECD 1997b).

However, environmental taxes are a market regulation mechanism, because they require a regulatory framework and administrative control, apart from the fact that they modify the established market prices. The following is the opinion of the European Economic and Social Committee: "Economic instruments of environmental policy, particularly in the form of charges, levies and taxes, require a regulatory framework as well as supervision by the authorities. By their very nature, such measures entail intervention in the free market economy mechanism" (CES 1992: 160).

Despite the unconditional support of bodies and institutions such as the OECD, the EU, the UNEP and ecologist organisations, environmental taxation has not been consolidated; it has actually retreated from the positions several EU countries reached in 1999. The EU has been incapable of promoting environmental taxation, because fiscal changes need the unanimity of all its members. This has led to a situation where in these countries environmental taxation has been aimed at private consumption, because companies rejected the measures claiming that they undermined their competitiveness.

As for the capacity to substantially change behaviours, studies indicate that the outreach of the existing environmental taxation is manifestly insufficient to produce any substantial transformations in today's reality. The reasons: its low degree of application and the difficulty of defining its effectiveness. The second argument is reflected, for example, in the study by the EEA *Environmental Taxes: Recent Developments in Tools for Integration*: "Assessing the effectiveness of environmental taxes is by no means a simple task. Firstly, it is not always clear how effectiveness should be defined and measured (. . .) Secondly, environmental taxes are almost always one element in a package of policy measures (. . .) And thirdly, lack of relevant data is frequently a bottleneck" (2000: 49). The logical conclusion is that the results are by no means clear. The EEA (2008: 47) acknowledges that the evaluation carried out in various countries shows that "the effects of the tax in relation to the national objective provided mixed results". But there are other drawbacks:

- The tax system may become very complex and costly. In particular, if it is to be adapted to the different local and regional situations: "If we are to set taxes with at least a broad relationship to damage done, it will be necessary to vary the taxes by source since different receptor points will have different assimilative capacities for pollution" (Pearce and Turner 1990: 115).
- The effects of taxes are uncertain because the response of the economic agents is *ex ante* unknown.
- The effectiveness of taxation is diminished by inflation.
- Environmental taxes are indirect and therefore regressive: they affect lower income more than higher income brackets.

- There is a contradiction between the goals of improving environmental quality and revenue collection: when one factor rises, the other falls.
- Environmental taxes are unfair, because they constitute a right to pollute and to use scarce resources for people with high incomes.

5.2 *Marketable Pollution/Emission Permit Systems*

In this part of the chapter we will analyse two systems: tradable pollution permits and tradable emission certificates. But before doing so it is necessary to state that the general approach of these systems is based, like the regulations, on setting physical standards that the agents must comply with. The difference is in the way of achieving it.

5.2.1 Marketable Pollution Permits

D. Pearce and E. Turner explain this approach: "The authority allows only a certain level of pollutant emissions, and issues permits (. . .) for this amount. However, whereas standard-setting ends there, the pollution permits are tradable, they can be bought and sold on a permit market" (1990: 110). Although their implementation is only extensive in the US (where they have seen significant development for air), here we will analyse these systems quite extensively, given the importance they have for the traditional economy. They have the following advantages attributed to them:

- Their efficiency is considerable in comparison to other methods. The OECD states: "A major characteristic of this instrument is its cost-saving potential" (1997b: 17).
- They avoid uncertainties about the standards reached, because the number of permits issued determines the environmental quality.
- Inflation does not erode the incentivising capacity of this instrument, because the market takes it into account when setting the prices of the permits (Tietenberg 1998).
- Better environmental quality can be achieved if aware citizens decide to purchase permits. But in this case, D. Pearce and R. Turner think that it alters "the level of pollution it had decided was optimal or acceptable" (1990: 114).
- "The emission permits system (EPS) is much simpler. It simply issues permits on the basis of source emissions and ignores what effects those emissions have on the receptor points". But, "one area is likely to experience some concentration of pollution in specific small areas (so-called *hot spots*) where actual concentrations exceed the standard" (Pearce and Turner 1990: 116).
- They are compatible with direct regulation, such as technical requirements for new or modified equipment (OECD 1997d: 17).

- The creation of new companies or the expansion of existing ones does not cause more environmental deterioration, because the number of permits is fixed (Tietenberg 1998).
- They adapt to the falling costs of less polluting technologies by lowering their prices (Tietenberg 1998).

The US Environmental Protection Agency (EPA) has established a programme for implementation of the system which affects the main sources of five atmospheric pollutants: hydrocarbons, nitrogen oxides, particles, sulphur oxides and carbon monoxide. The system is the usual: companies receive emission permits and can buy and sell them. For its application a complex regulatory framework for this market has been designed. It establishes various categories of emissions sources: sources of pollution existing in the mid 1970s; new sources, which correspond to new plants; and modified sources, which correspond to plants that existed in the mid-1970s and which have expanded their installations, leading to an increase in emissions. It also distinguishes between areas that exceed air quality regulations and areas that have adequate air quality. These factors limit the marketing options. The permitted exchanges are:

- *Offsets*. A new plant can enter an area that does not comply with the quality regulations if the company buys the corresponding permits for that same area. Some states force them to buy more permits than necessary to improve air quality.
- *Netting*. It allows a modified plant to increase its emissions if another plant of the same company reduces its emissions in such a way that the total emissions do not increase. This programme is controlled at a state level.
- *Banking*. It allows a company that reduces its emissions to save the corresponding permits for possible future needs.
- *Bubbles*. In this case a plant with multiple sources is made to meet its emissions commitments globally, in such a way that excess emissions at some sources may be compensated with reductions at others (Tietenberg 1998: 243).

Estimates of the results of applying the system vary greatly, but it is admitted that an immense majority of savings made are from permit exchanges between plants of the same company or between sources at the same plant, from the *netting* and *bubbles* programmes. The OECD considers that "the savings in costs is almost entirely due to internal trading", and adds that "it is virtually impossible to determine how much of the air quality improvement, if there has been any, can be directly attributed to the tradable permits programme" (1997c: 59–60). But what is called the internal market is not a market as such, that is, the sale of something from one economic agent to another. Tietenberg states that "*netting* is more properly considered a regulatory relief than regulatory reform" (1998: 243), and the same applies to the *bubbles* system. The procedure to have a transaction approved is lengthy and expensive, and only 20 % of the requests are admitted under the terms proposed. 40 % are rejected without any further considerations. W. E. Oates and P. R. Portney (1996: 367) state that "unfortunate restrictions on trading" have been produced, which "have clouded definitions of property rights and raised serious uncertainties about the ability to obtain these rights in the marketplace when needed".

The cost of the procedure often exceeds that of the pollution permits and is caused by the need for registration, verification, certification, buying and selling of certificates, follow-ups, etc. But if the companies had to pay for the pollution permits initially handed out, the total costs would be far greater. For this reason free distribution is dominant (OECD 1998: 248). As in all cases of the application of this instrument, in the US a high concentration of permits has occurred.

On the other hand, if the aim were – as would seem logical – to achieve environmental quality at all points, it would be "judged to be a largely unworkable system because of its complexity" (Pearce and Turner 1990: 117). Tietenberg acknowledges that "when emission location matters, the dominance of economic instruments over the traditional command–and–control is less clear cut in practice than it might appear in theory" (1998: 249).

This system rewards the historically most polluting companies. It discriminates against new companies as opposed to those that are already established, as they have to buy the permits from them to be able to establish themselves, which becomes an initial barrier and restrains the technological innovation that new companies in particular provide. The permits concede a property right because they are indefinite and are not linked to the lifetime of the plants, whose existence has given rise to the right. This is backed by court rulings. Last of all, the efficiency of the system is yet to be proven. In the US the SO_2 emissions system reduced pollution by 50 %. But in Germany, a regulatory policy reduced SO_2 and NOx emissions by 90 % (Maro 2007).

Because of these results, the OECD has to acknowledge that there is "very limited evidence on the efficiency or effectiveness of economic instruments". However, "much of the explanation for this (. . .) would appear to be the extremely restrictive terms on which trades were permitted" (1997d: 53). The EPA reaches the conclusion that "the price of *tradable emissions permits* under cap-and-trade systems will almost never meet the requirements for using cost as a proxy for value" (http://yosemite.epa.gov/Sabproduct.nsf/Web/Tradable%20emission%20permits).

5.2.2 The EU Greenhouse Emission Trading Scheme

Among the mechanisms contemplated in the Kyoto agreement to reduce greenhouse gas emissions (GHG), the US introduced a mechanism of tradeable emission rights based on the same philosophy as the above systems. It did not, however, ratify it.

The Kyoto Protocol is a 1997 international treaty which came into force in 2005, binding most developed nations to a cap and trade system for the six major greenhouse gases (GHG). They must reduce their overall emissions by 5.2 % with respect to their 1990 levels by the end of 2012. But the EU decided on an 8 % reduction of GHG emissions by means of a *bubble system*. Member States have quotas that range from reductions far higher than the general goal to permits allowing them to substantially increase their emissions, but if each country meets its goals the general goal is achieved. Apart from those countries (Annex I countries), many other non-OECD countries adhered to the protocol, but without reduction commitments (Greño 2004).

The Kyoto Protocol

Fulfilment of the commitments was facilitated by establishing "flexible mechanisms", which means that countries can fulfil their quotas by means of the following: in the "joint implementation (JI) scheme nations that emit less than their quota can sell emissions credits to nations that exceed their quota (especially Russia, because it was assigned emissions rights due to the strong reduction caused by the collapse of the USSR)"; and in the "clean development mechanism (CDM)" Annex I countries can invest in Non-OECD countries in projects which reduces GHG emissions. The production of emission reductions generated by the CDM and JI can be used by Annex I countries in meeting their emission limitation commitments (Greño 2004).

The protocol has positive aspects: the broad international agreement reached whereby quick action must be taken to solve the problem, for which reduction goals were defined; it has led to a great leap in the treatment given to it by the media; the creation of a body that, in addition to the agreement reached, continues to work on further agreements. On the other hand, the agreement is clearly insufficient: the reduction goals are small and multiple mechanisms are contemplated for signatory countries to avoid full compliance with them; it does not take into account that most of the CO2 emitted historically is the responsibility of industrialised countries and that, therefore, there is an historical debt with all other countries; the clean development mechanism relies on the planting of fast-growing trees and there are doubts as to their capacity for CO2 sequestration, due to the speed with which they are cut down; and, bearing in mind that the necessary reduction of emissions involves transforming energy and transport models in particular, it requires multiform and integrated policies, instead of punctual and sometimes ephemeral actions.

The EU Trading Scheme

The EU Directive (2003/87/EC) introduces a scheme for GHG allowance trading within the Community. It has two phases: the 2005–2007 and the 2008–2020 phases. In the first phase only the most energy-intensive industrial sectors have been considered (electricity generation, oil refineries, steel and metal, cement, etc.). The second system was approved within the "energy package" on which an agreement was reached in December 2008 and which included goals to be reached in the contribution of renewable energy sources (20 % by 2020) and in efficiency improvement (20 % by 2020). It also establishes 20 % of CO2 equivalent emissions (the greenhouse effect of all other gases is reduced to CO2 emissions) (European Commission 2008a). Later, the reduction goal of 30 % was approved subject to a global reduction agreement.

The first period was a failure, because so many emission permits were handed out that the price of the CO2-equivalent tonne fell to almost zero in the final phase. The new policy established in 2008 improves on the previous one in various aspects: a 21 % reduction in emission permits is established; the industrial sectors involved are extended to others (petrochemical, ammonia and aluminium); air and sea transport are included (although no measures are established, pending an international agreement; and if it does not happen, they shall be defined in 2012); more gases are included (nitrous oxide and an industrial gas); a policy whereby most permits are auctioned is established. In this case the aim is to move from the situation in 2008, where 90 % of the permits were distributed without any cost (*grandfathering system*), to one where 88 % are auctioned by 2013, 10 % will be distributed depending on criteria of solidarity and 2 % will be assigned to a group of Eastern countries. There is permission for 3 % of the 2005 emissions to be achieved through the mechanisms established with countries outside the EU, provided that the total of emissions certificates achieved does not exceed 50 % of the reductions that each country must reach. For the rest of the sectors that are not part of the system (particularly road transport) there will be taxes on emissions and obligatory efficiency improvements (European Commission 2008a).

This system undoubtedly improves on the first one. But it leaves out road transport and it is not probable that fiscal measures will be approved, which require the unanimous support of countries. The system is very complex (apart from the fact that there are 27 countries involved), because it contemplates many exceptions, it conditions the measures in some sectors to the achievement of international agreements, etc. The verification of the application of such a complex system requires an enormous administrative apparatus, which will not be created. We have seen that in the US system, which is much simpler, the administrative costs borne by the companies were far higher than the price of the permits. Last of all, in the future we will see that the extraordinary impact of the escalating price of oil that will take place has such a capacity to reduce CO2 emissions that it will ridicule the enormous and costly set-up that the EU has created.

Chapter 4
Free Market and Sustainability

Keywords Free market versus sustainability • Sources of contradiction • GATT-WTO system • Multilateral environmental agreements • Government Proposals to Change WTO

This chapter starts with an analysis of the theoretical contradictions that are fallen into to combine environmental protection and trade liberalisation. It then analyses the contribution of the GATT/WTO system to the unsustainability of the economic model, the role of Multilateral Environmental Agreements (MEA), their conflicting nature, and the proposals for the strengthening of the international nature protection system. It concludes with a reflection on the factors that invalidate the globalising process and the GATT/WTO system itself.

1 Contradictions Between Nature Monetisation and Economic Liberalisation

The traditional economy defends the valuation of free assets, which is obtained basically from three factors: scale of the problem, income level and level of ecological awareness. The resulting valuation has to be introduced into the market, and the usual proposal is that it is done by means of applying equivalent taxes. But the OECD admits that situations (and therefore valuations) vary widely. They depend on differing environmental circumstances, on incomes, etc.: "The 'right' level of environmental protection will vary by country; different local circumstances will lead to different values being placed on environmental protection. Most countries will prefer a higher level of environmental quality, the higher their income levels" (OECD 1997c: 24).

If the factors that determine the valuation are very different from one country to another, taxes will also vary and, in a context of trade liberalisation such as

R. Bermejo, *Handbook for a Sustainable Economy*,
DOI 10.1007/978-94-017-8981-3_4,

the one we have today, companies in countries with high taxes will see a fall in their competitiveness. This situation is frequently acknowledged in the texts of the OECD and the EU, but they cannot accept its consequences, so they are always looking for a false way out: the coordination of environmental policies. The European Commission acknowledges that "differences in environmental standards can be due to (. . .) differences in capacity of absorption of ecosystems" but rules out "differences in environmental policies". The problem "can often be addressed by international harmonisation or co-ordination of environmental policies" (1996: 6).

But this solution lacks rigour, because if the costs are different, no coordination or harmonisation will prevent them from continuing to be so. In no case is this proposal gone into in depth; it is left as a mere declaration, without practical consequences. The OECD and the MEA admit that the loss of competitiveness has forced some north European countries to not apply their high energy taxes on their most energy-intensive companies (OECD 1993: 96; EEA 1996). This is the case of Sweden, Denmark, Norway and the Netherlands. And this occurs within the EU, where there is a common environmental policy and institutions to coordinate sectoral policies.

If environmental taxes cannot be applied to the companies that consume the most resources and pollute the most to avoid a loss of competitiveness, the theory of the internalisation of externalities ceases to be, in general, operative, as it comes into open confrontation with the free market. The market, therefore, will not receive the *correct signals* and behaviour will continue to be anti-ecological, according to the logic analysed here. There is, as a result, a contradiction between asking them to internalise environmental costs and forbidding customs barriers for goods that do not bear them. This is evident for Daly and Goodland: "There is a inconsistency between a national policy of cost internalisation and an international policy of deregulated trade with non cost-internalising nations" (Daly and Goodland 1994: 78).

2 Contradiction Between Liberalisation and Sustainability

There are abundant declarations by economic institutions and governments according to which, in principle, there is no contradiction between free trade and sustainability. And the same arguments are being repeated endlessly. The European Commission (1996: 11) declares that "trade rules allow countries to take any measures necessary for protecting the environment within their own territory if they are not discriminatory, arbitrary and do not result in a disguised restriction on international trade". The *World Development Report 1992* by the World Bank states that "the primary cause of environmental problems is not liberalised trade but the failure of markets and governments to price the environment appropriately" (WB 1993: box 3.1). Although here part of the blame is attributed to "trade", in reality the only responsibility that is in essence being suggested is that of governments. The European Commission (1996:5) states that "**the impact of trade on the environment depends mainly on the environmental policies and sustainable**

development strategies implemented at a national and international level". And it goes further, assigning to the market the role of **"magnifier of policy failures".**

However, it is evident that liberalisation reduces the power of governments to rule over societies: "The globalisation of economic activity means that individual governments are less 'in control' to resolve environmental problems than they used to be" (OECD 2008b: 9). Although this liberalising trend questions the theory of governments' responsibility, it is necessary to maintain it, because if not, the whole theoretical framework collapses. The argument used is that there is still a broad margin for autonomy. The European Commission (1996: 7) declares: "It is widely recognised that the margin of manoeuvre available to countries is already large". And this margin is due to the fact that "WTO are allowed to take measures to protect the environment within their own territory". But, as usual, the measures have to be "not discriminatory or arbitrary". This warning is repeated time and again by the OECD, EU, WTO, etc.

However, there is abundant empirical evidence that liberalisation prevents governments from promoting sustainability: it leaves them without solutions in the face of external environmental aggressions (acid rain, polluted rivers, climate change, ozone layer hole, etc.); it eliminates border controls; it exerts pressure on governments to relax environmental policies; it makes the application of sustainable product management techniques difficult; it intensifies the flows of materials and their length.

Liberalisation makes it difficult to comply with international trade regulations for ecological, social and sanitary reasons, etc. The result is an acceleration of the rate of the impacts described. In the field of environmental studies there is much information on this phenomenon. The overexploitation of fishing grounds (Sumaila et al. 2007) and the deforestation of non-OECD countries is extensively documented. The IMF's imposition on Indonesia to reduce duties on wood exports from 200 to 10 %, in 2000, led to more than a twofold increase of the previous logging rate (Shimamoto 2008). Many biologists consider that the most pressing environmental problem is the homogenising and degrading impact on the world's biota that trade is causing through "biological pollution", a result of the trade-related transport of species from one country to another. H. French offers abundant information on the subject in her book "Vanishing Borders" and reaches the obvious conclusion: "There can be little doubt that globalisation has accelerated the unprecedented loss of biological riches in recent decades" (French 2002: 28).

The fact that many countries have very relaxed environmental policies, which do not lead to any type of penalty, instead bringing advantages for international trade, exerts deregulatory pressure on the most advanced countries. Globalisation lengthens product chains and makes them more complex. Which leads to a growing difficulty to know the impacts of all types throughout them and, as a result, to assess the problem and adopt corrective measures. Which means that, beyond the intrinsic characteristics of the products, the life-cycle perspective acquires growing relevance: "Globalisation, combined with the continued growing importance of environmental issues, will intensify the conflict between the traditional 'a product is its physical characteristics' perspective, and 'a life-cycle' perspective, which seeks to address environmental externalities wherever they may originate" (OECD 1997a, b, c, d: 49). Also, globalisation leads to more uniform products, technologies and tastes. Uniform products imply difficulties to adapt them to the environmental and cultural characteristics of each location: "Access to foreign technologies may displace existing domestic technologies which are better suited to local environmental conditions" (OECD 1997a, b, c, d: 67). This is particularly evident in the case of uniform agriculture technologies. In the case of tastes, "globalisation might lead to more uniform consumer tastes, influenced by transnational mass media imagery and advertising" (OECD 1997a, b, c, d: 20).

During the last 50 years the world product has increased fivefold, while trade has increased by a factor of 14. In the US, commercial foodstuffs travel 3,000 miles on average. This situation is caused by two phenomena: on the one hand, liberalisation boosts trade; on the other, it produces a distribution throughout the world of the different product chain stages, in such a way that before the international sale of the final product takes place, substantial trade involving its components has been necessary. These factors cause an increased need for transport and its resulting impacts: energy consumption, pollution and the impacts associated with the proliferation of transport infrastructures. The EU Commission (2008b: 14) states that the physical transportation of exported products at the end of their life and of imported raw materials entails significant environmental damage. The OECD (1997a, b, c, d: 56) acknowledges the same idea: "In principle, *any* increase in freight traffic will bring *absolute* increases in environmental damages with it (scale effect)".

3 The Contradictions Between the GATT-WTO System and Sustainability

3.1 The GATT-WTO System

The General Agreement on Tariffs and Trade (GATT) was created in 1947 with the aims of bringing about a reduction and elimination of duties and the elimination of other types of restrictions on trade. Between 1948 and 1994 there were eight negotiation processes (Rounds). The last one, the Uruguay Round, approved the creation of the World Trade Organization (WTO), which was inaugurated in 1994. It adopted the regulations and operating regime of the GATT in their entirety. The trade liberalisation promoted by the GATT-WTO rests on three principles: non-discrimination between exporting countries (Art. I), non-discrimination between domestic and imported products (Art. III) and the prohibition of import or export quotas (Art. XIII). Art. I establishes that there should be no special treatment, be it beneficial or damaging, for any countries. Art. III states that imported products should not be given a different treatment to that of domestic products. Art. XIII prohibits adopting quotas that restrict the volume of trade.

The GATT-WTO has the possibility of regulating trade for environmental reasons based on sections *b* and *g* of article XX. They permit trade restrictions "taken to protect human, animal or plant life or health" (sub-paragraph *b*) or "related to the conservation of exhaustible natural resources" (sub-paragraph *g*). But, as usual, "if they are not discriminatory, arbitrary and do not result in a disguised restriction on international trade" (European Commission 1996: 10, 11). These sections do not reflect the environmental awareness of the GATT because they were written before 1947 and show the concerns of that period (such as the health risks generated by the trade in contaminated foods and live animals). In addition, the countries that wish to apply this article must prove that there is scientific evidence that justifies it (what is called *provisional justification*) and that the measures taken are those that have the least impact on free trade (*final justification*). The provisional justification goes against the precautionary principle (UNEP and IIDS 2001: 32).

The practice of the GATT shows that it never took into account the environmental variable as a justification to regulate trade and the WTO continued along the same path, though intensifying it, despite the fact that in the foundational document's preamble there is support for sustainable development and that the Trade and Environment Committee was created within it. But sustainable development is not a guiding principle, like the three articles mentioned. Its first decision was to force the US to lower the cleanliness standards of imported fuel, established by the Clean Air Act and has supported all the claims against free trade restrictions due to environmental reasons: "After 8 years (. . .) WTO panels have ruled against all food safety regulations under review on the grounds that they restrict trade more than necessary" (Public Citizen 2003:9). This situation explains why the OECD Environmental Outlook (2008a, b, c, d) "concludes that if no new policy actions are taken, within the few next decades we risk irreversibly altering the environmental basis for sustained economic prosperity" (OECD 2008a, b, c, d: 9).

3.2 The Sources of Contradiction

The contradictory areas are many, so we will limit ourselves to an analysis of the policy of the WTO on the Process and Production Methods (PPM), the Agreement on Technical Barriers to Trade (TBT) and the Agreement on Sanitary and Phytosanitary Measures (SPS measures).

3.2.1 Environmental Norms and Process and Production Methods

The regulation of trade due to environmental reasons can be potentially carried out by States on products and on PPMs. When a State wants to protect its environment from the actions of other States, it acts on imported products, either because they are the cause of the problems, or because the PPMs are the cause. But this distinction is becoming less and less clear. Some measures related to PPMs can change features of the product, as happens with agricultural products which use dangerous pesticides in their production processes. The WTO considers that these products and organic products are similar, so there can be no restrictions on the trade of the former. The WTO does not question the right to regulate the trade of products (provided it is carried out within increasingly tighter limits), but if it is not done based on the PPMs.

The protectionist measures of a country motivated by the PPMs of other countries can have two types of causes: the country in question bears direct environmental aggressions as a result of the PPMs, or it seeks to protect the biosphere, because the aggression on it also normally affects it. As there is no international authority that this country can turn to, all it can do is act on the exports of the aggressor country. The GATT established case law on trade restrictions depending on PPMs by means of a ruling on the dispute between the US and Mexico because of a US ban on yellow-fin tuna imports. The origin of the dispute was that the fishing technique used to catch the tuna caused a high mortality rate among dolphins (in the late 1980s, over seven million were estimated to have died and in the area in dispute they were killing 80,000 dolphins a year). In 1988 the US Congress updated the Marine Mammal Protection Act of 1972 to include the case of tuna imports, establishing for 1990 a ratio of dead dolphins for imported tuna 25 % higher than that for domestic fishing: "These provisions do not provide a domestic advantage. In fact, U.S. tuna fishermen must meet standards more stringent than importers, thus putting Americans, if anything, at a competitive disadvantage " (Christensen and Geffin 1991: 577).

But the GATT ruling invalidates the regulation for two reasons:

- PPMs cannot limit trade: "A GATT party can make no distinction between products based upon the manner in which they were produced, even where the importing party has outlawed or limited one method of production because of its environmental destructiveness" (De Belleuve et al. 1994: 55).
- "The Panel considered that if the broad interpretation of Article XX(b) suggested by United States were accepted, a contracting party could unilaterally determine the life or health protection policies from which other contracting parties could not deviate without jeopardizing their Rights under General Agreement" (Christensen and Geffin 1991: 585).

The significance of this ruling is that it invalidates *de facto* the MEA and the capacity of governments to protect the wildlife not only of the planet, but also within their territories, because wildlife knows no borders. And in 1998 the WTO ruled against the US, for the same reasons, due to the ban in this country on the imports of shrimps caught with nets that caused high mortality rates in sea turtles (French 2000: 121).

3.2.2 Agreement on Sanitary and Phytosanitary Measures

Countries have been restricting trade due to sanitary and environmental protection reasons, trying to avoid the hazards derived from pests, diseases and organisms associated with diseases contained in imported goods; from chemicals, fertilisers, pesticides and herbicides, dangerous prescription drugs, etc.

The Uruguay Round approved an Agreement on the Application of Sanitary and Phytosanitary Measures. A key element of this agreement was that restrictions should be based on the FAO Codex Alimentarius, whose standards were reduced after a 4-year campaign carried out by the US. Based on this new Codex Alimentarius, the WTO ruled in 1997 against the EU for its ban on hormonal meat, which it considered a health risk, although it gave the EU time to prove its effects on health. In 1998 the EU set up seventeen research units to study the effects of the six hormones used. In 1999 the WTO authorised the US to impose sanctions on the EU because it had not yet finished its research. In the year 2000 the research results showed that the measure was correct, because one of the hormones was "totally" carcinogenic and the others posed evident risks, in particular for adolescent girls. The WTO has ignored the report, which proves that it does not accept the revision of processes in the event of there being new information. In 2000 it also ruled against an Australian regulation that dated from the 1960s, by which imported salmon is held in quarantine to protect native salmon from imported diseases. It is based, among other reasons, on the fact that Canadian salmon contains some 20 bacteria that are non-existent in Australia. For the same reason, Japan subjects imported foods to a quarantine system. It is revealing that no country that has been reported by virtue of this agreement has received a favourable verdict (Public Citizen 2003).

3.2.3 Precautionary Principle

The precautionary principle establishes the need to act when an important risk is presumed, even if there is no scientific evidence, and it is one of the principles of the EU's environmental policy. The demand that *provisional justification* is based on scientific evidence is impossible to meet (as there is insufficient data) when a product or technology is initially applied. In addition, the scientific community has very little knowledge on how the biosphere works and, as a result, the repercussions our actions have on it. The WTO does not accept the principle because it creates an uncertain framework of action for it.

In short, the non-existence of an international regulatory body for economic activities for environmental reasons has meant that the WTO is the only authority with the capacity to make decisions in this field. This proved to be a capacity that they have used to create very negative case law for the protection of the planet. And this is the logical outcome, as the mission of this organisation is the promotion of free trade. The situation has gone so far that even *The Economist* acknowledges that "it has become a quasi-judicial body, an embryo world government (. . .) it is now being asked to arbitrate on matters that are intensely political. It lacks legitimacy to do so" (French 2000: 125).

4 Contradiction Between WTO Rulings and Multilateral Environmental Agreements

4.1 Multilateral Environmental Agreements (MEA)

The MEA are the biggest challenge to the free trade system. They are international agreements, signed by more than two countries, which seek to solve global and regional environmental and resource-related problems (biodiversity, international fishing, preservation of natural services, eradication of pollution, etc.). Over 1,000 bilateral agreements have been signed, although the UNEP admits that the exact number is not known. Only 20–30, according to different estimates, establish explicit measures that restrict trade (quotas, import and export bans, producer rights not acknowledged by the WTO, etc.), but they are the most important. The UNEP considers that the most relevant are: the Convention on International Trade in Endangered Species (CITES 1975), the Montreal Protocol on Substances that Deplete the Ozone Layer (1987), the Basel Convention on the control of the transboundary movements of hazardous wastes (1992), the Convention on Biological Diversity (1993), the Convention on Climate Change or Kyoto Protocol (1997), the Rotterdam Convention on the Prior Informed Consent Procedure for Certain Hazardous Chemicals and Pesticides in International Trade (1998), and the Cartagena Protocol on Biosafety (2000). All of them take measures to control trade and, although the Kyoto Protocol is an exception, it is the one which has the greatest potential for regulation. These agreements establish a wide scope for controlling trade by the signatory countries, so they can:

- Toughen the agreed regulations individually.
- Establish licensing and permit systems that regulate the trade of certain goods and, even, their transit through their territory.
- Establish sanctioning regimes.
- Discriminate between signatory and non-signatory countries: establishing restrictions on trade with non-signatory countries; applying trade measures against such countries.
- Establish contingent trade.
- Discriminate between similar products based on their PPMs.
- Apply the precautionary principle.
- Promote information transparency and participation.

The UE Economic and Social Committee acknowledges that the framework agreement of the Rio Summit on Climate Change in 1992, the Montreal Protocol and other international agreements drive "a wedge in two fundamental principles of the GATT: the principle of the most favoured nation and the principle of reciprocity" (ESC 1996: 7–8). In addition, they are very dynamic, as they try to adapt depending on new information (level of compliance, new scientific information, etc.) and promote the collaboration of NGOs which, in some cases, such as that of inspection, is decisive. The Convention on Persistent Organic Pollutants (POP) has established the prohibition and control of 12 very hazardous substances. The International Commission for the Conservation of Atlantic Tunas allows member countries to ban imports of tuna and swordfish from countries that do not comply with their regulations. They work on agreements on sustainable fishing, sustainable forest exploitation, etc. (French and Mastny 2001; UNEP and IIDS 2002).

4.2 *Outcomes*

The Basel Convention managed to significantly reduce the amount of waste exported from the OECD to non-OECD countries, but in recent years there has been a spectacular growth in exports of electronic waste (e-waste) to Asia (apparently for recycling) and the convention permits this type of trade. 80 % of the US's e-waste is transported to Asia (Mastny 2003: 27). However, only a small part of it is recycled, and the impacts on the environment and health are huge (UNEP 2009).

Most countries have complied with the Montreal Protocol, which has led to an 85 % reduction in CFC emissions during the 1986–1997 period. The least compliant country is Russia, because it produces them. Funds have been made available to help it comply. On the other hand, China has become the main producer of halons. The result is that, while concentrations of chlorine in the stratosphere fall, concentrations of bromine (from halons) rise (French and Mastny 2001).

The CITES has managed to reduce the trade of some threatened species (such as cheetahs, chimpanzees, crocodiles and elephants), but the illegal trade of many others continues. With the aim of putting a stop to it, agreements on certain species, such as marine species, have been created. In 1995 the Protocol on trans-zone fishing stocks and migrant fish was approved. The International Commission for the Conservation of Atlantic Tunas is taking action (French and Mastny 2001).

The agreements reached in the International Whaling Committee have managed to reduce annual captures of whales from 66,000 in 1961 to 1,100. Oil spills in the sea have been reduced by 60 % since 1981 through regulations of the International Maritime Organization (IMO), although the load transported has doubled. In 1991 a 50-year moratorium on commercial activities in the Antarctic region was approved (French 2000: 146).

These results are limited, though important, as a result of the liberalising trend and of the weakness of the structure and resources of the MEA system. And the weakness comes from the fact that each treaty creates its own institutional machinery, small offices called secretariats and a small team (around 30 people on average). Their function is the supervision of the agreement, which means collecting and analysing data, advising governments, in some cases carrying out on-site inspections, etc. This fragmentation of actions weakens the MEAs and is a source of contradictions among them.

The Montreal Protocol has created a framework for eliminating gases that destroy the ozone layer, but it has not dealt with the possibility that their substitutes are aggravating other problems. This has led, for example, to the development of hydrofluorocarbons (HFC), which are powerful greenhouse effect gases. The Kyoto Protocol encourages the planting of fast-growing trees, without taking into account their effects on biodiversity. On the other hand, it does not include any compensation for the preservation of forests (UNU/GEIC/IAS 1998).

As a result of the MEAS' weaknesses, the agreements reached are often insufficient to solve the problems. Most of them contain few specific goals and calendars, and the mechanisms for monitoring their application are weak when not non-existent (French 2002). However, in recent years there has been a widespread proliferation of networks of global non-governmental organisations (NGOs) that are fulfilling an important role. The Climate Action Network gathers over 250 international and national organisations that work to defend a stable climate. The Pesticide Action Network includes at least 500 consumer, environmentalist, health, trade union and farmer's organisations. The World Forum of Fish Harvesters & Fish Workers brings together small-scale fish worker organisations from six continents. There is an International Network for eliminating POPs. The TRAFFIC network monitors compliance with the CITES and offers recommendations to governments on lists of species to protect (French and Mastny 2001).

The strengthening of NGOs means that, with increasing frequency, they are invited to take part in the work of international bodies and institutions. A very important step in this direction was the invitation to participate in the Sustainable Development Committee carried out in 1993. The opposition of the more conservative members delayed its application by 3 years and in 1996 123 NGOs were accredited. Since then the number of organisations participating in its annual fora has grown to over 700 in 2000. In Johannesburg over 8,000 non-governmental participants were accredited (1,500 at the Rio Summit), which carried out a wide variety of parallel activities, apart from participating in the official fora (French 2002: 172, 173).

5 Government Proposals to Change the WTO's Role in Order to Protect Nature

5.1 Acknowledgment of Principles

Many principles are proposed that are in contradiction with the free trade system. The Rio Declaration establishes, among others, those of Self-Government, the Precautionary Principle, and Common but Differentiated Responsibilities.

- Self-government. Principle 2 states: "States have, in accordance with the Charter of the United Nations and the principles of international law, the sovereign right to exploit their own resources pursuant to their own environmental and developmental policies".

- Precautionary Principle. Principle 15 assumes that: "In order to protect the environment, the precautionary approach shall be widely applied by States according to their capabilities. Where there are threats of serious and irreversible damage, lack of full scientific certainty shall not be used as a reason for postponing cost-effective measures to prevent environmental degradation" Paragraph 6 of the Millennium Declaration makes a generic defence of the Principle: "Prudence must be shown in the management of all living species and natural resources, in accordance with the precepts of sustainable development" (United Nations 2000). The EU's proposal that there be an *ex ante* evaluation of the liberalising agreements, as a prior requirement for their definitive approval, is an implicit acknowledgement of the precautionary principle.
- Common but Differentiated Responsibilities. "States shall cooperate in a spirit of global partnership to conserve, protect and restore the health and integrity of the Earth's ecosystem", but "in view of the different contributions to global environmental degradation" (Principle 7).
- Awareness and participation: "States shall facilitate and encourage public awareness and participation by making information widely available" and provide "effective access to judicial and administrative proceedings" (Principle 10).

Last of all, and as a summary of the principles described, it is necessary to establish the principle of the primacy of nature preservation over trade liberalisation. Up until now there have been many declarations about the need to preserve nature, but in contexts where liberalisation is also defended. An important improvement is that Chap. 1 of the Millennium Declaration (Values and principles) considers as "fundamental values" the following: "Freedom", "Equality", "Solidarity", "Tolerance", "Respect for nature" and "Shared responsibility" (United Nations 2000).

5.2 Regulatory Framework

The EU is fulfilling a leading role in the reform of the WTO's system for environmental reasons, although it has also become the main promoter of free trade. This apparent contradiction is resolved in favour of the second position, because its regulatory proposals have been rejected, which has not changed its leading role in the liberalisation process. In the communication mentioned, which contained its position at the First WTO Ministerial Conference (Singapore 1996), it proposed a revision of article XX of the GATT so that it could contemplate, exceptionally, the legality of the trade restrictions caused by the MEAs, both based on products and on PPMs (European Parliament 1996). In 1999, the European Commission issued a Communication about "The EU Approach to the Millennium Round". In this document, the UE proposed the accommodation of the MEAs in the WTO system: "Consensus should be sought on the accommodation within WTO rules of trade measures taken pursuant to MEAs and the types of multilateral agreements which constitute MEAs". Another proposal was to introduce within the WTO the precautionary principle in particular: "It is necessary to maintain the right of WTO Members to take precautionary action to protect human health, safety and the environment while at the same time avoiding unjustified or disproportionate restrictions" (European Commission 1999b: 15).

The document that the EU presented to the Johannesburg Summit took a further step by abandoning, at least implicitly, the idyllic theory of positive feedback between the two systems: "In order to make globalisation sustainable there is a need for a much better balance between global market forces on the one hand and global governance and political institutions on the other". A proposal that is reinforced by its Chapter 3.1: "Harnessing globalisation: trade for sustainable development" (European Commission 2002a: 4, 7).

As regards agreements, until now two have been contemplated: the evaluation of the environmental effects of trade liberalisation and the establishment of negotiations between the WTO and the MEA system. But neither of them has led to practical results. Canada, the EU and the US announced an agreement to carry out an environmental impact assessment in future negotiations, but these countries broadly differ on their scope: objectives, methods and field of application (Santarius et al. 2004).

As a result of these community proposals, concerning the multiplication and development of MEAs, the WTO has been showing growing concern that its regulations alter the free trade system (UNU/GEIC/IAS 1999), but in the end it has had no choice but to agree to negotiations on the environmental issue. At the World Trade Organization meeting (WTO 2001) in Doha a Ministerial Declaration was approved, which agreed to open up a process of negotiations with the Multilateral Environmental Agreement (MEA) system (2001:10, 11): "With a view to enhancing the mutual supportiveness of trade and environment, we agreed to negotiations, without prejudging their outcome". But their scope was limited "to the applicability of such existing WTO rules as among parties to the MEA in question". The decision was backed by a totally chaotic theoretical framework, because of two frequently cited theories: on the one hand, the positive feedback between free trade and sustainability, and that States have the right and capability to protect their environment, although this right is limited. In the first case, it declared that "We are convinced that the aims of upholding and safeguarding an open and non-discriminatory multilateral trading system, and acting for the protection of the environment and the promotion of sustainable development can and must be mutually supportive". The second theory is expressed in this form: "We recognize that under WTO rules no country should be prevented from taking measures for the protection of human, animal or plant life or health, or of the environment at the levels it considers appropriate, subject to the requirement that they are not applied in a manner which would constitute a (. . .) disguised restriction on international trade" (WTO 2001: 2).

The Implementation Plan, approved at the World Summit on Sustainable Development (UNWSSD 2012) held at Johannesburg, backed the Doha Declaration and dealt with the theoretical chaos in greater depth. Paragraph 92 proposes to: "Promote mutual supportiveness between the multilateral trading system and the multilateral environmental agreements, consistent with sustainable development goals, in support of the work programme agreed through the WTO, while recognizing the importance of maintaining the integrity of both sets of instruments".

There were five MEA-WTO meetings during 2001 and more were scheduled, but since then there has been no news of reaching an agreement. An OECD report acknowledges that "multilateral trade negotiations remain an important focus of discussions for integrating trade and environment objectives" (2008a, b, c, d: 2). Hoffmann, director of Environment and Development at the ECLAC, explains that the difficulty of reaching agreements is due to the fact that "environmental regimes are much more complicated than the international trade regime" (1997). The UNEP backs this conclusion and believes that conflicts between both will worsen in the future. There is "an enormous and complex body of international law". And "as economic globalization proceeds and the global nature of many environmental problems becomes more evident, there is bound to be friction between the multilateral systems of law governing both" (UNEP/IIDS 2005: 2).

5.3 Financing

The provision of sufficient and stable funds for the United Nations system is one of the most debated issues in recent decades. The reason is the current insecurity and randomness of its funding, which is totally vulnerable to the blackmail of donor States, in particular the USA. The secretariats in charge of applying environmental treaties normally have an annual budget that ranges between one and $30 million. These budgets contrast with, for example, the EPA, which in 2000 had a budget of $7.8 billion, and much more with the military budget, over $300 billion, while total military expenses exceeded $750 billion (French 2002).

In 1980 the Brandt Report confirmed the existence of multiple proposals to collect international revenue. Among them there was a tax on international trade, on international investment, on hydrocarbons and exhaustible resources, on energy consumption, on the international oil trade, on international air and sea transport, on the use of 'commons' (fishing and oil and gas extraction from oceans, seabed mining, the use of space orbits and radio and telecommunications frequencies) (Brandt Commission 1981: 277). The WCED (1987: 333) cites the above measures and emphasises some of them: "The Commission particularly considers that the proposals regarding revenue from the use of international commons and natural resources now warrant and should receive serious consideration by governments and the General Assembly" (WCED 1987: 333). However, these proposals were not discussed by the General Assembly nor during the subsequent world summits, due in particular to the opposition of the US; they simply emphasise the need to strengthen the United Nations' system and that adequate funding is essential to achieve this. But no measures have been taken (French 2000: 123).

5.4 *Institutions*

Reforms and proposals of the WTO: The reforms of the WTO refer to the institutionalisation of the assessment of environmental and social impacts of liberalisation measures. At this moment there are two bodies: the Trade Policy Review Mechanism (TPR Mechanism) and the Dispute Settlement Body (DSBody). The former is in charge of assessing the economic impacts of liberalisation measures in each country. The latter is in charge of issuing verdicts on trade disputes between countries. The proposals go in two directions: that the first Body also takes care of assessing the social and environmental repercussions and of the creation of the Strategic Impact Assessment Body (SIA Body). Santarius et al. dismiss the first because the TPR's control function "would always be made only *ex post*, after the ratification of trade agreements". On the contrary, the SIA Body "would investigate the likely ecological and social consequences so that information about these could be input into the decision-making" (Santarius et al. 2004: 44–47).

Environment Protection Institutions: Various types of institutions are proposed, which garner very different levels of adhesion. An International Tribunal for the Environment is proposed, similar to the International Penal Court or the International Tribunal for the Law of the Sea, but very few governments back this proposal (French and Mastny 2001). On the contrary, there is growing support for turning the UNEP into a World Organisation for the Environment, which would integrate all the environmental organisations that exist within the United Nations. This proposal has received widespread support. The European Environmental Council (composed of environment ministers) proposes this measure and although more countries "favour the establishment of a 'UN Environmental Organisation' (. . .) others are not convinced that such an organisation is necessary or desirable" (OECD 2008a, b, c, d: 8). And all this despite the fact that the status of this organisation would be smaller than that of the WTO, as it would still be a body outside the UN's system and controlled, in particular, by the US.

6 The Retreat of Globalization

Economic globalisation has multiplied social problems and the unsustainability of global socioeconomic systems. These problems, and in particular other factors, are starting to produce a retreat of globalisation. We are witnessing the change from the US's hegemony to a multi-polar world, which has been a key factor that has stalled the liberalisation Round started in Doha in 2001 (and which was already delayed 2 years because of the impossibility of starting it in Seattle in 1999 due to the social protests), events that are unprecedented. There are fewer and fewer countries that are committed to a globalised model and, on the contrary, there more and more regional economic cooperation agreements. The European Commission (2003e: 16) confirmed this trend: “In the wake of Cancun, an increasing number of WTO Members appear tempted to pursue RTAs/FTAs as an alternative to multilateral liberalisation”.

World powers systematically fail to comply with the WTO’s regulations. The European Commission (COM(2010)334final) detected in 2008 and 2009 numerous measures that restricted trade and a trend towards increasing these restrictive measures, particularly in the US, Russia, India, China, Brazil, etc. Prominent among them is the limitation by many countries of exports of strategic minerals and metals, the reserves of which they monopolise (Ad-hoc Working Group 2010: 20). The Commission Communication “Trade, Growth and World Affairs” ascertains the slow progress of the Doha Round and the growth of trade-restricting measures. For this reason the Commission (2010: 12) asks “G20 partners to take action to reverse and roll back the restrictive measures”. On the other hand, high oil prices (as a result of its depletion process) reduce international trade, because of the growing transport costs. These topics will be dealt with in Chaps. 10 and 11.

Chapter 5
Sustainable Development in the Brundtland Report and Its Distortion

Keywords Sustainable development • Brundtland's concept • Orthodox concepts • Triple sustainability theory • Dematerialisation theory

1 Introduction

When people first became aware of ecological problems in the 1960s, such issues were seen as discrete and unconnected concerns that each had perfectly identifiable causes (normally related to industry). For this reason, it was thought that such issues could be approached with corrective ad hoc policies, preferably with end-of-pipe technologies. During the 1980s, however, global ocean contamination, ozone layer and forest depletion, and the lack of sanitary drinking water increased the plausible suspicion that we were creating still other problems, such as planetary climate change and the chemical contamination. At that point, environmental issues became seen as systemic and it was held that the economic system, because of its incompatibility with ecological balance, needed to be transformed. *Our Common Future*, the report of the World Commission on Environment and Development, presented in 1987 and better known as the Brundtland Report (BR),[1] is a landmark, not only because it represents the first institutional backing of the concept of sustainable development (SD), but because of its endorsement by the United Nations. The diagnosis of the BR is definitive and categorical: "We are unanimous in our conviction that the security, well-being, and very survival of the planet depend on such changes, now" (WCED 1987: 38).

[1]The report is popularly known by this name because Gro Harlem Brundtland, Norway's Minister of the Environment (she later became Prime Minister of Norway) presided over the commission that drafted the document.

R. Bermejo, *Handbook for a Sustainable Economy*,
DOI 10.1007/978-94-017-8981-3_5,

There is a broad range of interpretations regarding SD, although the BR version currently enjoys the widest degree of acceptance. The present chapter seeks to analyze, on the basis of the BR premises, two approaches to the concept of SD: the theory of the sustainable development triangle and the theory of dematerialization.

2 Developmental Economics

Developmental economics has evolved since its origins in the years following World War II, just as the concept of development itself has. Originally conceived as synonymous with economic development alone (an increase in gross national income), development was reformulated in the 1970s after it was proven that economic growth per se was not able to guarantee an improvement in the actual living conditions of human beings. Later reformulations of the term introduced the social dimension for the express purpose of assuring that economic development was accompanied by social policies that guaranteed a more equitable distribution of income. The incorporation of the environmental dimension following the publication of the BR in the late 1980s was in effect a tacit recognition of the existence of biophysical limits to economic growth. In addition, the report specified that the model of Western development could not be applied to the rest of the world, if the goal was to preserve the earth's natural heritage. Development had to be sustainable. The liberal promise that economic growth would bring with it benefits for the entire planet has not only been demonstrated to be false, given the increase in inequality between the countries of the North and the South, but actually impossible, because if all of the world's countries followed the path of the industrialized countries, six planets would be necessary to provide the natural resources required for, and bear the burden of the waste generated by such "economic progress." The economic development of industrialized countries is thus an asset that depends on their geographical location, since neither the resources nor the capacity of the planet to bear such development are sufficient to allow the reproduction of this model on a global scale. Social inequality and ecological deterioration have without a doubt been the most decisive blows that have been dealt to the era of development initiated during the U.S. presidency of Harry S. Truman.[2]

SD as an economic issue has generated much debate, largely determined by the paradigm of orthodox economics, which generally equates economic growth with an increase in well-being and with full employment, and which therefore sees sustained development as both necessary and good. A high level of economic growth

[2]In his inaugural address on January 20, 1949, Harry S. Truman said: "We must embark on a bold new program for making the benefits of our scientific advances and industrial progress available for the improvement and growth of underdeveloped areas. The old imperialism—exploitation for foreign profit—has no place in our plans. What we envisage is a program of development based on the concepts of democratic fair-dealing."

is thus seen as the proof of any government's successful economic management because it guarantees an improvement of social well-being. This argument is especially relevant in the case of developing countries with high levels of poverty, malnutrition, and infant mortality. It is thus a high priority for the system to defend the compatibility, and even necessity, of unlimited development with ecological balance, the notion behind the hypothesis known as the Kuznets curve (Organization for Economic Co-operation and Development [hereafter OECD] 1997a: 26). This viewpoint stands in contrast with that of those who see the current economic model itself as not only the cause of global environmental deterioration, but as playing a primary role in many of the evils (war, inequality, and poverty) that plague humanity. For its critics, SD is an oxymoron (Sachs 1999, chap. 5). From their standpoint, economic development as it has been conceived and advocated in recent decades is something that itself impedes SD. The debate over SD, along with the subsequent manipulations of that debate, has not only influenced development economics, but has also determined the form that cooperative development policies have taken among the countries of the North, while serving as an obstacle for the adoption of international agreements regarding environmental issues. Such international cooperation is indispensable for addressing global environmental problems such as climate change.

3 Sustainable Development According to the Brundtland Report

During the 1960s the environmental problems resulting from economic development first became clearly evident, and a number of remedies were proposed. In 1972, the Club of Rome published its first report under the title *The Limits to Growth*, commonly known as the Meadows Report, which was presented at the first UN Conference on the Human Environment, held in Stockholm in 1972. Among its proposals, the report made the case for zero growth in developing countries as a response to environmental deterioration and the shortage of planetary resources (Meadows et al. 1972). The work was hugely popular, with millions of copies sold in various languages. Perhaps for this reason, and because the proposal constituted a frontal attack on the philosophical underpinnings of the capitalist economic system, it was bitterly criticized by orthodox economists, who believed capitalism could not survive without unlimited development. In the face of the criticism of this report, the Club of Rome issued a second, more moderate, report in 1974 that defended organic development, understood as involving the kind of limited growth that is intrinsic to all living organisms (Mesarovic and Pestel 1974). Since the publication of *The Limits to Growth*, a considerable number of writers have developed concepts that integrate ecological concerns with economics. Among these concepts are eco-development (Sachs 1981), intensive development (Rizhkov 1986), and various versions of SD.

The debate regarding SD was almost exclusively limited to the scholarly literature until the publication of the Brundtland Report, which defined SD as "development that meets the needs of the present without compromising the ability of future generations to meet their own needs" (54). This interpretation of SD became popular, and was used as a common frame of reference, even though this apparent common ground tended to mask clear differences from other elaborations of the concept, as well as between different versions of SD and the report. David Pearce and Anil Markandya researched differences in the development of the concept, noting that in some cases the ideas developed had nothing to do with sustainability (1989: 43–44).

It is possible that the BR's concept of SD was so widely accepted because it was so generic, although it must be said in its favor that it includes highly important features such as the meeting of the essential needs of the entire global population and environmental protection. Nevertheless, such a truncated definition must necessarily be highly general and, therefore, susceptible to multiple interpretations, even though the most important interpretations (i.e., in terms of those receiving institutional support) are entirely illegitimate.

In an attempt to eliminate the ambiguity of such a generic definition, the BR found it necessary to clarify it with a text that defined SD's two underlying principles. SD, the report said, "contains within it two key concepts: the concept of 'needs', in particular the essential needs of the world's poor, to which overriding priority should be given; and the idea of limitations imposed by the state of technology and social organization on the environment's ability to meet present and future needs" (54).

The first concept explains that, when reference is made to meeting needs, this means essential needs. Because such needs have not been met for the majority of the world's population (mainly those residing in developing countries), those drafting the report found it necessary to add that "Meeting essential needs depends in part on achieving full growth potential" (55). But growth in and of itself is not enough. It is only "part" of the solution—the economic part. The other aspects of the problem have to do with the "state of technology" (insufficient technological development) and the social dimension (the need to address the evils of a "social organization" resulting in an enormously uneven distribution of income). Both of these limitations must be overcome, the report contends, in order to maintain "the environment's ability to meet present and future needs" (54).

In spite of these attempted clarifications, technological obstacles continue to be vaguely defined. It is clear that a deficient social organization (the kind that prevails in most nations of the world, especially in developing nations) results in a highly inequitable income distribution. On the other hand, the clarifying statement might lead one to think that, once problems with respect to "technological development" have been resolved, there will no longer be limits regarding the availability of resources. For this reason, the BR later specifies the scope of technological development when it states that such development cannot be allowed to exceed the limited availability of resources. Such would be the case "for the use of energy, materials, water, and land" (55). In addition, the BR contains other conceptual clarifications and prescriptions. For instance, "sustainability" refers only to the ecological dimension. This premise appears repeatedly and forms part of the conclusion of Chap. 2: "sustainable development requires ... a production system that respects the obligation to preserve the ecological base for development" (74). The report frequently employs the term "development and protection of the environment" and is very clear about what it considers as constituting development: "The satisfaction of human needs and aspirations is the major objective of development" (54). This was the idea expressed at the Rio Summit as well. Principal 4 of the Rio declaration states: "In order to achieve sustainable development, environmental protection shall constitute an integral part of the development process" (United Nations Conference on Environment and Development 1992).

The sustainability requirement is an unavoidable aspect of development and therefore necessitates that a change be made in the regnant model. This is reflected in the BR's dramatic "Call for Action": "Attempts to maintain social and ecological stability through old approaches to development and environmental protection will increase instability. Security must be sought through change .... We are unanimous in our conviction that the security, well-being, and very survival of the planet depend on such changes, now" (37–38). The elimination of dangers that threaten the very survival of the planet makes a change in policy and philosophy an absolute priority. Thus, the environmental dimension of economic development is of vital importance. Finally, in order to transform the development model, it is necessary to implement strategic planning of the required transformations. In the words of the report, "a broad strategic framework for achieving it" is necessary (54).

To summarize, the concept of SD in the BR does not advocate unlimited development, but rather only the development that is necessary to meet essential needs. In addition, development in and of itself is not sufficient to meet essential needs. It is also necessary to guarantee a more equitable distribution of income. Technological development, while by no means a panacea, is also important. Sustainability refers exclusively to the environmental dimension, which is of critical importance because survival itself is at stake. A radical transformation of the current model of production and consumption is vital in order to achieve sustainability, and such a transformation requires strategic planning. The ambiguity in the BR's definition of SD has led to the proliferation of 300 explanatory definitions. These definitions are in fact "the products of conflicting worldviews, differing ideologies, varied disciplinary backgrounds, opposing knowledge traditions, value systems and vested interests" (European Environment Agency 1997, 21). This situation has led writers such as Sharachchandra M. Lele to define SD as "a metaphysics that will unite everybody from the profit-minded industrialist and risk-minimizing subsistence farmer to the equity-seeking social worker, the pollution-concerned or wildlife-loving First Worlder, the growth-maximizing policy maker, the goal-oriented bureaucrat, and therefore, the vote-counting politician" (1991: 613).

4 Orthodox Interpretations of the Concept of Sustainable Development

The phenomena of ecological deterioration and of exhausting natural resources, which have both resulted from the dominant model of economic development, have led to the proliferation of dramatic calls to action that aim at raising the general level of awareness regarding the gravity of a situation in which man is at odds with the natural environment. Such calls have often been accompanied by radical proposals, as was the case with the Meadows Report. But the controversy that ensued following publication of that document rapidly evaporated following the drastic reduction in consumption of resources beginning in 1973. In fact, the 1973 energy crisis by no means invalidated the conclusions of the Meadows Report, given that its main argument was that acute shortages in natural resources would soon emerge as long as the trend of steadily rising consumption (that had begun during the era of postwar economic expansion) continued. The energy crisis disrupted this trend and for a time put the issue on the backburner. But the end of the crisis once again made the shortage of resources evident. This issue has once again gained momentum, and even global institutions have frequently joined the rising chorus of alarm: “The reckless pursuit of economic growth today might leave our children with a larger inheritance of economic assets, but could seriously deplete environmental resources” (United Nations Environment Programme and the International Energy Agency 2002: 5).

4.1 The Theory of the Sustainability Triangle

A large number of international organizations (the European Union [EU], the Organisation for Economic Co-operation and Development [OECD], the World Bank) contend that sustainability refers not only to the environment but to two other dimensions, thus creating a triad of three principal factors that together comprise the notion of sustainability: economic, social, and environmental. This interpretation of SD as an integrated concept that comprises three distinct sustainabilities was presented in World Bank documents beginning in the early 1990s. The European Council declared that one of its guiding principles was to “promote integration of economic, social and environmental considerations so that they are coherent and mutually reinforce each other by making full use of instruments for better regulation” (2006: 5). In the view of the EU, the three dimensions or sustainabilities have an equal weight or scope.

This interpretation of the concept departs in a number of critical respects from SD as defined in the BR. First, while the BR separates the fields of development (economic and social dimensions) from that of sustainability (the ecological dimension), the notion of a sustainability triangle conceives SD as a generic concept applicable to all three dimensions. Secondly, the introduction of economic sustainability allows such a notion to be associated with all of the desiderata of the dominant system: liberalization, globalization, competition, unlimited growth, and so on. Yet it is precisely these characteristics of the system that have made it unsustainable. There is thus a disregard of the BR's admonition "to break out of past patterns. Attempts to maintain social and ecological stability through old approaches to development and environmental protection will increase instability. Security must be sought through change" (WCED 1987: 37). Third, while the BR stresses the importance of the ecological dimension (sustainability), the theory of the sustainability triangle actually deemphasizes the ecological aspect as something of secondary importance. Finally, the BR does not include the premise of indefinite growth—especially in the case of rich nations. In its definition of the concept and throughout the report, the BR continually emphasizes that the goal should be meeting essential needs, with growth seen as a means to this end. The growth of rich countries is only mentioned as a possibility to be considered assuming that there is a consideration on their part for the environment, that they allow other countries access to vital resources, and that they assure the future availability of nonrenewable resources. Yet in the theory of the sustainability triangle, unlimited development has become the most important premise of SD. The theory of the sustainability triangle, by relegating ecological factors to a sphere (the environment) separate from economics, is conducive to a denial that the economy is in large part determined by ecology. The theory thus preserves the traditional status of economics as an independent science that bears no relation to other sciences. In addition, the theory integrates central elements of the most orthodox interpretation of the capitalist system, and thus closely associates itself with this system. It implies that no special attention need be paid to environmental sustainability. In other words, even though there is a pretext of including the environmental dimension within the theory, such a dimension is merely one consideration among the multiple socioeconomic aspects that have traditionally been the central focus of societies. Yet any attempt to define economic or social sustainability independently from biophysical surroundings is condemned to failure. The quest to define sophisticated constellations of socioeconomic sustainability requirements may lead to the attainment of developed societies with a high degree of social integration, yet such societies are doomed to collapse as a result of the deterioration of the biophysical medium within which they exist. This position ignores such pressing problems as the potential short-term collapse resulting from reaching the maximum global level of petroleum extraction. Meanwhile, the notion of the sustainability triangle has been introduced without any prior analysis of what is to be understood by economic and social sustainability, or even on the basis of a consensus regarding the need to introduce these concepts.

The lack of a clear definition of what SD means, and of its repercussions in all of the different spheres, gives rise to a high degree of terminological and conceptual confusion. We now see traditional policies being defended on the basis of a concern for SD. For example, one study of SD in Germany conducted by the OECD showed that the Ministry of Finance had utilized the concept in order to describe the long-term goal of eliminating the budget deficit, while the Ministry of Labor and Social Affairs had used it to define the security of the pension system or the availability of

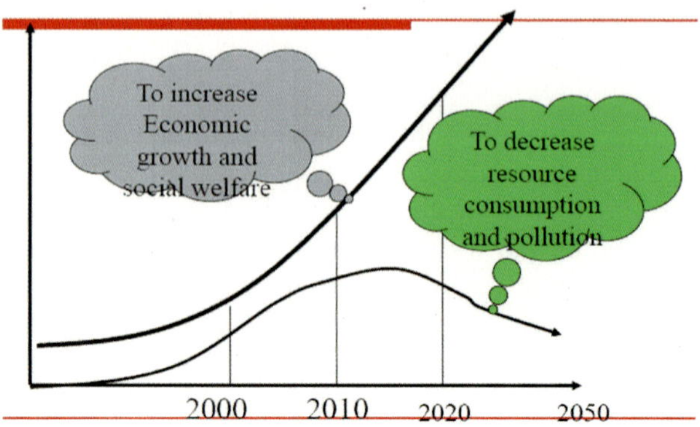

Fig. 5.1 Theory of economic growth with dematerialization (Source: Dajian (2006) Tongji University, www.pmpp.cn)

social capital. It is therefore not surprising that those who conducted the interviews for the study find that "sometimes even public servants who are professionally confronted with sustainability had no clear idea of how to operationalise this concept for concrete policy making" (Janicke et al. 2001: 10).

We can thus see that the fraudulent and abusive use of the concept of sustainability that arises from its application to all kinds of social and economic conditions leads to a term that ceases to be functional, and that the initial goal of integrating the environmental variable has foundered in a sea of "(un)sustainabilities." This situation is the result of a defensive reaction of those in positions of power who, when required to integrate the environmental variable in their decision-making processes, take steps to vitiate its transformational character and thus make its integration conditional on the acceptance of the bases of the dominant economic system.

4.2 *The Theory of Dematerialization*

Public institutions that are reluctant to adopt transformative policies required by such pressing problems as climate change or the exhausting of natural resources have resorted to a theoretical framework that supposedly allows them to respond to such problems without having to renounce unlimited development and free-market economics. The theoretical framework that apparently satisfies such premises is that of the *dematerialization of growth*, or *the decoupling of growth from its physical basis* (Herman et al. 1990). From this perspective, it would be desirable to try to continue to grow in unlimited fashion while at the same time reducing the consumption of resources and any adverse impact on the environment, as shown in Fig. 5.1.

This is a theory that has been embraced by both the OECD and the EU. *Environmental Strategy for the First Decade of the Twenty-first Century*, published by the OECD, establishes decoupling as one of its objectives (OECD 2001). The EU has meanwhile made multiple references to decoupling. Two examples will

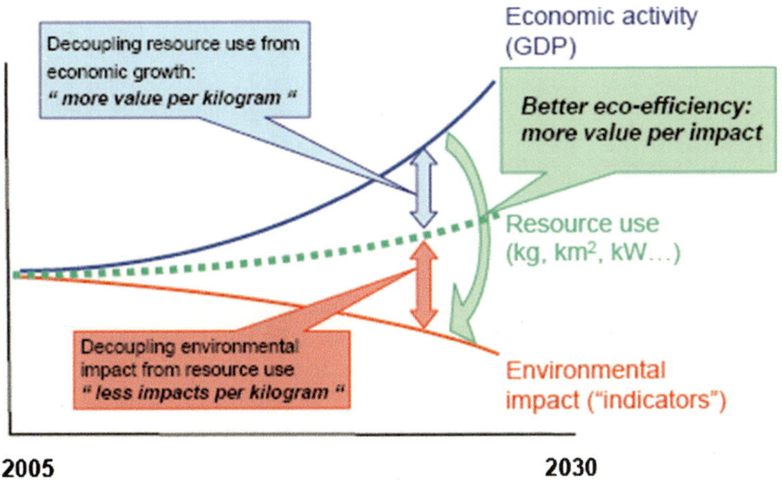

Fig. 5.2 Absolute and relative decoupling (Source: Bringezu (2006, 8))

suffice to give a sense of the typical tenor of EU statements in this regard. The European Commission stated that "Decoupling environmental degradation from economic growth is a central theme of the EU Sustainable Development Strategy" (2003e: 14). Similarly, the first "key objective" of the revised SD strategy is to "prevent and reduce environmental pollution and promote sustainable consumption and production to break the link between economic growth and environmental degradation" (European Council 2006: 3). Yet these EU texts give the impression of a broad-based consensus that does not exist in reality, because they represent two different kinds of dematerialization or decoupling. One of these is (the previously described) "absolute dematerialization." The other is "relative dematerialization" and refers to a more efficient use of resources that aims at increasing the commercial value created by each physical unit of resource that is employed.

This latter version is frequently preferred, because it is more compatible with the dominant paradigm of unlimited development. In fact, statistical data demonstrate a steady advance in efficiency. Yet the importance of such statistics is overestimated because they do not take into account the relocation of a large part of heavy industries from the most industrialized countries to emerging countries. Thus, for example, the energy invested in the manufacture of imported metals does not figure in consumption on the part of the importer. In any case, an increase in efficiency does not absolutely guarantee dematerialization. Figure 5.2 shows that an increase in efficiency does not in and of itself lead to dematerialization, because the consumption of resources grows despite the application of measures aimed at improving efficiency, and such measures are not capable of compensating demands resulting from a higher growth of GDP. On the contrary, the dematerialization curve shows a clear decline in the consumption of resources, and in environmental impact.

Dematerialization of growth (or of absolute decoupling) represents an approach that is conceptually more accurate than that of zero growth proposals.

Dematerialization assumes that unsustainability occurs due to the degradation of the biophysical basis of the economy, and that this problem needs to be addressed directly by drastically reducing environmental impact and conserving resources, while there might be zero growth together with an increase in environmental impact. Nevertheless, there are flagrant contradictions between the theory of dematerialization and the sustainability triangle theory. As is typical, fallacies have been employed to resolve these contradictions. Dematerialization is presented as involving environmental sustainability in a way that would not be inconsistent in principle with the theory of the sustainability triangle. Yet such an approach is incorrect, because the latter involves two different dimensions (the ecological and the economic). Moreover, the former determines the latter, because it is expressed as a necessary condition for growth to be able to continue indefinitely, as is clearly shown in the previous figure. We therefore find ourselves faced with a bidimensional concept (where economic and environmental dimensions are present), but it is the second of these that is decisive. Unlimited growth of an economic product (when it is possible) can only occur when it is accompanied by a decreasing pressure on the environment, at least in the initial phase, so that the pressure can later be stabilized. The theory of dematerialization thus coincides with the BR's sustainability concept, in that it identifies sustainability with the ecological dimension, and in that this is what determines the economic dimension. By contrast, its defense of unlimited growth is not consistent with the BR, nor is its failure to take account of the social dimension (the goal of meeting essential needs).

From the conceptual point of view, the absolute dematerialization of growth, while formally coherent, is not necessarily possible in practice. Those in positions of power defend it on the basis of three arguments: it is necessary to grow in order to be able to invest in environmental protection, the outsourcing of the economy contributes to dematerialization, and the dematerializing power of scientific and technical advances. The theory of a positive feedback loop between economic development and environmental protection is frequently defended, with the mechanism supposedly working as follows: Although environmental degradation is an inevitable consequence of growth in the early phases of economic development, once a certain level of per capita income has been attained, economic growth goes from being the cause to becoming the solution of environmental degradation, given that an increase in a society's wealth allows a greater investment in environmental protection. This theory, known as the environmental Kuznets curve hypothesis, has been defended for some 20 years, and is confirmed on the basis of the correlation observed in some countries between increase in per capita income and a decrease in environmental impact. In many cases, however, it has been shown that behind this supposed dematerialization lie the processes of the outsourcing and relocation of the contamination, or that it applies only in reference to certain contaminants.[3] The theory of the outsourcing process is premised upon an undeniable historical process

[3]Dinda (2004) offers a comprehensive review of distinct viewpoints regarding the environmental Kuznets curve hypothesis.

that has shown that, in its initial phase, industrialization of economic activity moves from agriculture to industry, and thereafter to services, until reaching the current situation in which industrialization is responsible for the vast majority of economic output and employment in the economies of industrialized nations.

As regards the environmental Kuznets curve hypothesis, the argument goes that, while the first movement creates an adverse environmental impact, the second movement reduces this impact. In reality, the increasing importance of the outsourcing sector in the GDP is due to a very debatable distribution of activities among the different economic sectors. Over the course of time, the first and second sectors have been stripped of functions that they historically held and that gave them their identity within the economy. In fact, there has been an externalization of activities that has resulted in the growth of the service sector at the expense of the other sectors. In addition, recent decades have witnessed the phenomenon of more developed countries attempting to reduce costs by relocating certain activities to other countries. Secondly, the dematerialization that has taken place in the service sector is highly debatable. Many studies have demythologized the dematerializing character of the service sector. Jespersen (1999) studies the relative energy consumption of different economic sectors and finds that there was no significant difference in consumption between the private service sector and the secondary sector. Finally, an increase in the consumption of services does not lead to a decrease in disposable goods. Instead, the service sector experiences growth once a high consumption of goods has been attained. Advocates of "environmentalism of the poor" (Martinez Alier 2008) also criticize the environmental Kuznets curve. They emphasize the active role of the most disadvantaged communities with respect to the protection and rational management of the natural resources that are essential to their survival, and prioritize the activities involved in this role over and above economic growth. Some recent examples include the opposition to genetically modified organisms in Latin America, and the movement to protect Yasuni Park in Ecuador.

The dematerialization proposal that allows those who uphold the status quo of the reigning paradigm to defend the notion of a planet of inexhaustible resources depends to a great extent on technology. At first glance, the argument that technological development is a dematerializing force appears to be well founded. Yet on further inspection we see that the dematerializing potential of technology is inherently limited. What can be justly argued is that it is possible to achieve a significant degree of dematerialization if the most efficient technologies that exist in the market are intensively applied and if, at the same time, the patterns of consumption having the highest impact are modified. On the other hand, some scientists (i.e., members of the Factor 10 Club) have reached the conclusion that, on a global level, it is necessary to cut in half both the use of materials (including energy materials) and the present level of adverse impact on the environment. Taking into account the universal right to meet one's basic needs, and starting from the assumption that 20 % of the population is responsible for 80 % of the flow of materials (an assumption that is become increasingly less valid, especially with the recent emergence of China and India), a 90 % reduction in both emissions

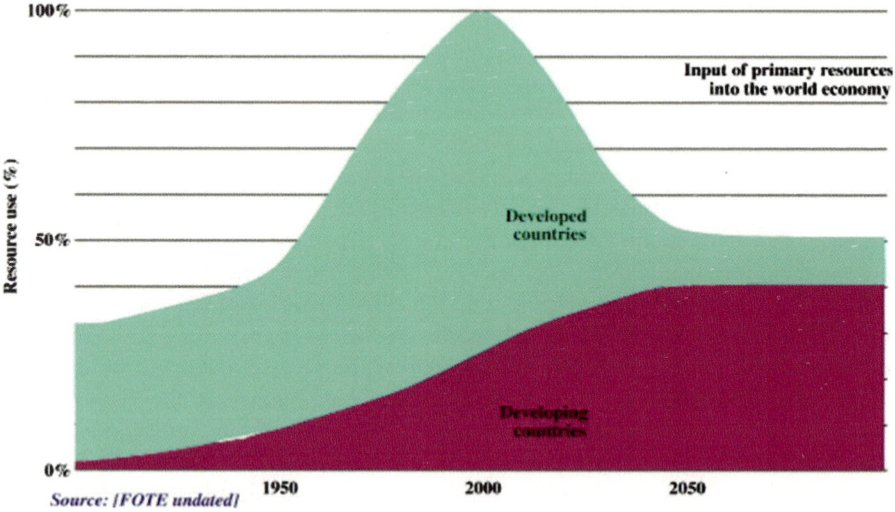

Fig. 5.3 The path to sustainable development (Source: Ayres (2008, 289))

and the consumption of materials and energy on the part of industrialized nations will be required in the future (Fig. 5.3). This novel vision of dematerialization raises the question of the distribution of resources, an issue generally ignored in the contributions of institutions such as the OECD and the EU. Figure 5.3 depicts the evolution of consumption of resources of developed countries and in the rest of the world that would be necessary to attain a factor of 10. The founding manifesto of the Factor 10 Club recommends that governments attempt to achieve a rather less ambitious goal: increasing the productivity of resources by the aforementioned factor in one generation (35 years). This goal would mean a 4.5 % annual increase in the productivity of materials, and a 3 % annual increase in the productivity of energy (Kuhndt and Liedtke 2000). If we estimate an annual economic growth of 3 %, at the end of 35 years, the economic product will more or less triple and, even if the increase in productivity matches what has been indicated above, consumption of materials would only decrease at an annual rate of 1.5 %.

In order to achieve dematerialization within the context of high economic growth, a sharp and sustained increase in the productivity and efficiency of resource use would be necessary. At the outset, the intensive application of the best existing technologies, together with a rapid improvement in consumption patterns, could lead to an absolute decoupling. Speed is of the essence in this regard, because continual development tends to erase the gains made as a result of greater efficiency. For example, an average economic growth of 3.5 % is multiplied by 32 every 100 years. Except for periods of crisis, the world economy has grown at an average annual rate of more than 4 %. The most efficient technologies constitute a technological stock that has been accumulated during the course of many years, and that has not yet been properly exploited. Once these technologies are massively applied, there will remain only the flow of new technologies, which will doubtless be unable to sustain the dematerializing tendency. In addition, the *rebound effect* or Jevons paradox makes any advances in technological development relative. In this regard, it has been demonstrated that when consumers apply more efficient techniques, they reduce expenses on resources and therefore have more disposable income, which they then utilize to increase their levels of general consumption.

This all leads to the conclusion that absolute economic dematerialization is only possible if, following a massive technological transformation and a reduction in levels of consumption, the OECD's favored economy becomes stagnant in physical terms. Other countries would then follow suit to the extent that they can achieve similar levels of economic development.

5 Concluding Remarks

The fraudulent and abusive use of the SD concept by those who have employed orthodox interpretations of the terms is the result of a defensive reaction of the dominant economic system that is conscious of the transformational nature of an acceptance to limits on growth. The current economic model cannot be maintained indefinitely over the course of time, much less applied to the nations of the South. In other words, the opulent lifestyle of the North is oligarchic by its very nature. On the contrary, it is necessary for the North to significantly reduce the environmental burden of the nations of the South, and to make good on an ecological debt that it has accumulated as a result of the overuse of the biosphere's carrying capacity.

Given the fact that the scale of the economy has exceeded the limits of nature, it is necessary for the economy to decelerate in physical terms. What this means practically is that it is necessary to reduce indicators related to the consumption of resources and the generation of waste. Such change is necessary, essentially, because both the economic system and social development are determined by biophysical limits, and sustainability is threatened by the destruction of the planetary ecosystem and the exhausting of the natural resources that sustain our lives.

Yet this in itself is not enough. Consumption on a finite planet is exclusive in nature, in both intragenerational and intergenerational terms. What this means is that an increase in consumption in the developed countries reduces the quantity of resources available for developing countries and for future generations. For this reason, the burden of decreasing in physical terms must necessarily fall on developed societies, so that their decrease allows a simultaneous reduction in the size of the global economy and the development of societies that currently do not meet their basic needs. In addition, reversing the pattern of the present situation would contribute to reducing the ecological debt contracted by the most developed countries, and would allow these resources to serve as a guarantee of the basic needs of the nations of the South (Hoyos 2004).

Within the context of an opulent Western society, physical downsizing should not be seen as implying any reduction in the current state of well-being, but rather as an opportunity to actually enhance well-being. This of course requires understanding "well-being" as a qualitative rather than a quantitative concept (namely, in terms of placing a relatively greater value on leisure time, human relations, equity, justice, and spirituality and not in terms of a never-ending accumulation of material goods). All of this would have to occur within the context of a self-limiting society (a society capable of maintaining itself at an intermediate level of economic output) or, in the words of Wolfgang Sachs, "a society that is able not to want what it would in fact be capable of providing" (1999: 89).

Part II
Principles and Instruments for Sustainability

Chapter 6
Sustainability of Social-Economical Systems

Keywords Complex adaptive systems • Ecosystems main functions • Biomimetic approach to sustainability • Sustainability guiding principles • Epistemological changes of neoclassic economics

1 Introduction

Governmental institutions declare more and more frequently that its policies have the aim of reaching sustainability. But this concept is systematically being manipulated (the sustainable development alike) by the cores of power, with the aim of removing its capacity of transformation, because they want to develop a free market economy and this is the main reason for our process of collision with nature. And, as a result, there is permanent growth in the consumption of materials and energy, the land is being covered over at an increased pace, and biodiversity losses are mounting. For that reason there is a growing need to clearly show where the manipulations are and to determine as precisely as possible what sustainability means. The focal point of the concept of sustainability is the fact that, since the human species are part of the biosphere, its economy has to be a sub-system of a universal economical use of the materials and energy provided by nature. So the socio-economical systems have to imitate nature, and this means that we must deduce the guiding principles of sustainability from the ecosystem's behaviour. There is a growing consensus on the necessity to abandon the materials cycles on which we now depend and to convert to alternatives such as solar energy. But simply, blindly, doing so is not enough. We must also thoroughly understand the parameters that make such aims possible. We must realize that the dynamics of the prevailing economic system run against these aims. We must take a contrary approach. In order to reach those widely heralded aims of sustainability, we must learn how to mimic the many aspects of our ecosystems behaviour, like evolution, diversity, decentralization, self-organization, self-sufficiency and so on.

R. Bermejo, *Handbook for a Sustainable Economy*,
DOI 10.1007/978-94-017-8981-3_6,

This chapter is organized as follows. We begin with a brief description of Complex Adaptive Systems (CAS), because nature is the paramount CAS, and we discuss the main elements of nature's behaviour, that is, the abiotic and biotic functions, and the function of maintenance of its vital constants. Through the lens of these functions, we review the most salient factors that cause the lack of sustainability of the capitalist system. Taking into account that human economy is a subsystem of the general economy of nature, we discuss the possibility of deducing the principles of sustainability from the functions of nature. Finally we argue that this vision is already creating an emergent scientific field, besides a great set of experiences in production, which we will see in several chapters.

2 Sustainability of Complex Adaptive Systems of Nature

A system is a functional aggregate with properties that cannot be deduced from its elements. Complex systems are systems composed of many elements that have diversity, individuality and individual interrelations with each other, but are also influenced by the system itself. Their dynamics are not linear (that is, small changes in critical variables can produce large transformations on the whole system) and, for this reason, many changes are simply not predictable: "the living machinery of Earth has a tendency to catastrophic change with little warning" (MEA-SB 2006: 18). There are many examples of such changes: "disease emergence, abrupt alterations in water quality, the creation of *death zones* in coastal waters, the collapse of fisheries, and shifts in regional climate" (MEA-SDM 2006: 1). Socio-economical and natural systems are self-organized, which means that system development is not pre-determined, but is a result of adapting to its own environment. That is why some authors call them Complex Adaptive Systems (CAS). The principal characteristics of a CAS are: heterogeneity, no linearity, hierarchy and fluxes. By means of interactions, nonlinear CAS organise themselves hierarchically in structures that define them and are reinforced by fluxes among their elements (Callicott 2002: 273, 274; Levin 2000: 12).

Natural CAS show stability, that is, they have the capability of preserving their essential functions after impacts from their environment. Nature is composed of many CAS cells, individuals, ecosystems, the biosphere, etc. Arthur Tansley defined the ecosystem concept at the beginning of the last century. Since then there has been much controversy about its consistency, but it has reached a wide consensus and is even being applied in areas far different from nature. Ecosystems are the first basic unit able to be autonomous: "An ecosystem is a dynamic complex of plants, animals, and microorganism communities and the nonliving environment interacting as a functional unit" (MEA-CF 2006: 29). Ecosystems are composed of biotic and abiotic elements that interact with each other, thus creating complex and multiform structures. They also form a lower level which is complete, that is, which "have all the components required in order to function and survive in the long term" (Odum and Sarmiento 1998: 46). This concept shows clearly the key aspects of the CAS heterogeneity, nonlinearity, hierarchy and fluxes. For this reason Levin (cited by Abel and Stepp 2003) believes that ecosystems are "prototypic examples of complex adaptive systems".

Ecosystems species accomplish the functions required to reach their equilibrium, but their relative importance is variable. While the loss of some species may not affect the basic operation of a given system, the disappearance of others could cause that ecosystem to collapse due to loss of some critical functions. As a result, an ecosystem is hierarchically organized and its highest level of hierarchy is seized by species which perform critical functions. Some ecosystems fluxes are: nutrients, water, toxics, individualistic attributes and information flow. These features give them great stability or resilience because they are able to retain their functions in the face of disruptive impacts (fires, droughts, pests, etc.) (Levin 2000: 14, 15).

Life on the planet is preserved and developed by ecosystems that accomplish three main functions: recycling of material fluxes; use of solar energy; and influence on the inert world that preserves its physico-chemical properties within a range that is suitable to preserve the biotic world.

2.1 *Abiotic Functions*

From the great number of organic simples elements and compounds that there are in the surface of Earth or near it, some are critical for life preservation. They are the so-called biogenic substances or nutrients. Ecosystems tend to hold and recycle the most vital nutrients in greater proportion than the no vital ones (Ring 1997: 241). In nature the organisms "tend to treasure and recycle the vital elements as phosphor, which are scarce in relation to needs" (Odum and Sarmiento 1998: 126). When ecosystems develop (the life cycles become more complex) they tend to increase their level of cooperation and self-sufficiency, the inner recycling of materials and their time of renovation and storage. In essence, they use materials and energy more efficiently (Odum 1992: 196). Although most of the materials are recycled within the trophic chain, the cycles of some others are global, like of cases of carbon, phosphor, nitrogen, sulphur, etc. They are the biogeochemical cycles.

The recycling of materials and, in last instance life, couldn't endure without the permanent flux of solar energy. Plants, as autotrophic organisms, capture solar energy in order to produce biomass during el photosynthetic process. They like decentralized power utilities. The fixed energy passes to other organisms through the trophic chains in a process of cascade like (Ring 1997: 241), and it serves to increase the food available, to create new individuals, and the rest is dissipated in form of low temperature heat. This energy flux has only one direction because entropy grows, that is, the more temperature sinks less useful it is. This is the reason why solar flux of energy must be permanent.

2.2 *Biotic Function*

In order to be able of assuring abiotic functions (and at last to preserve the impressive stability in the evolutionary frame that characterized biosphere) the biotic communities create ecosystems. Mature ecosystems don't grow, they evolve; they are complex or bio-diverse, auto-organized, hierarchically structured, self-sufficient, decentralized, and cooperation out-balances competition. The objective of complex and hierarchical structure is to preserve the stability of natural systems. Goldsmith declares that «continuity and stability have been the most impressive qualities of the world of living beings». The first ecosystems law of Von Bertalanffy is «the organic system tends to preserve itself» (Goldsmith 1996: 239).

2.2.1 Evolution

Organic evolution refers to changes in the organisms during time and as we have seen is non linear. As a rule, it determines a long term development, from the simplest conditions to the most complex or the better ones. Evolution is the result of many processes of interplay. On one side, groups of organisms co-evolve; for example, plants and herbivorous evolution is co-determined. On the other side, the biotic evolution depends on the abiotic medium (Odum 1992: 208). Bio-physic processes unfold in a huge variety of time and space scales; some endure hours or days and happen in very tiny spaces; others unfold during decades, centuries and even millenniums and in regions of thousands of square kilometres. Plants and biogeochemical processes use to be the fastest; animals and abiotic processes have a mean scale; and geomorphologic processes are, normally, the slowest (Holling 1993: 66). For example, the times of generation or regeneration of land and aquifers could be hundreds and thousands of years. The roll that the binomial competition/cooperation plays in evolution is one of the most debatable issues in ecology.

2.2.2 Cooperation/Competition

There are six kinds of principal interplay between two or more species: competition, predation, parasitism, commensalism, cooperation, and mutualism. Competition means that the result of interplay is negative for all the species that take part. Predation is positive for the predator and negative for the prey. Parasitism is negative for the host and positive for the parasite. Commensalism is a simple kind of positive interaction in which one benefits from the other, while the other is not affected either positively or negatively. Cooperation means that species benefit from each other, though this cooperation is not vital for any of them. Mutualism is a vital or very necessary relation for the survival of the species that take part (Odum 1992: 166). On the whole, there are three kinds of negative interactions and another three that are positive. Darwin's theory of natural selection is an oversimplification of reality, because it only takes the negative interactions. (Levin 2000: 20). Plants compete for light and nutrients in a forest and animals compete for food and shelter when these are scarce. On the contrary, "cooperation for the mutual profit is widely spread in nature and it's very important also in the natural selection" (Odum and Sarmiento 1998: 199).

Today many ecologists defend the notion that positive interrelations are more abundant. Negative interplays are risky for the species that take part in the process. For example, when species which have been prey are extinguished or their number is drastically reduced, predatory species are in danger of suffering the same processes. It is agreed in parasitology that virulent parasites tend to decrease in the evolutionary process and that "*commensalism* (peaceful coexistence in which *parasites* no longer harm individual hosts) will be the end result" (Levin 2000: 150, 151). Also negative interactions decrease ecosystem stability because they can produce a strong process of species reduction and, as a result, cause ecosystems to collapse and transform into others of more simple and specialized nature (Rammel y Staudinger 2002). On the contrary, experience shows that biodiversity increases and this means that the selection mechanism is "weak at best" (Levin 2000: 185) or imperfect: "the maintenance of variability and diversity is believed to be mainly a passive process due to imperfection of diversity-reducing principles like selection and competition" (Rammel and Staudinger 2002: 303). Frequently species which aren't strong survive: "many species have survived in the development of ecological eras by more subtle means, as cooperation and camouflage" (Odum and Sarmiento 1998: 235). Often the use of different niches avoids competition. Conclusion: species disappear because they are not able to adapt to quick changes and because niches cease to exist (Matutinovic 2002).

Predatory species were known to have had positive effects on depredated species, events known as "feedback by recompense". This effect can be taken in variable forms. Carnivores limit herbivorous populations and, as a result, these herbivores to not surpass available resources. There is ample empiric evidence that herbivorous populations stimulate plant growth and increase diversity by controlling dominant species (Odum and Sarmiento 1998: 107, 108). Consequently, "no species can exist alone; hence (...) mutualisms and other tight linkages among small numbers of species evolve to provide the participants buffering against environmental vagaries" (Levin 2000: 185, 192). This small group association is known as *small world behaviour* (Matutinovic 2002). Many pairs or major groups of species live together for mutual benefit as competing partners and, furthermore, because of this their ecosystems benefit. From all this evidence, it could be said that mutualism "is extremely generalized and is very important" (Odum and Sarmiento 1998: 187, 201). Species form groups in order to be able to accomplish their vital processes with a degree of autonomy and to defend themselves from other groups, interior collapses and environmental changes. This process is known as modular organization (Levin 2000: 193). A tree can survive only when enough humus exists in the soil. But its capacity to take up nutrients depends on bacteria and enzymes sited around roots, which are fed by the leaves. Concluding, due to modular organization, small world behaviour, niche distribution, subtle means, feedback by recompense, etc., there is a tendency "to diminish the intensity of negative interactions" (and) "it is not an abusive overstatement to say that the trophic chains are in general mutualists" (Odum and Sarmiento 1998: 205).

2.2.3 Biodiversity

Nature has been increasing her diversity for the past 3,500 years, even taking into account that there have been five great extinctions. Biodiversity reinforces ecosystem stability. Odum (1992: 61) says that «*stability of flexibility* (that is, rapid recovery after an impact) is greater when there are many different species in a region». A high diversity gives way to a great rate of redundancy, which generates stability. For example, when there are many species their functions overlap and have many degrees of tolerance for changes (Cantlon and Koening 1999: 109). Ring (1997: 241) declares that a high biodiversity can be considered, besides, an optimization strategy for systems that live in environments that change very little and have an acute shortage of resources, such as the Amazonian forest and coral reefs. But, "we are lowering the resilience of natural systems by simultaneously reducing the variety of species and placing them under unprecedented pressures" (MEA-CF 2006: 18).

2.2.4 Hierarchy

A successful development of a system is determined by two complementary aspects: a growing differentiation and diversity, and its integration in a hierarchical structure of increasing complexity (Schutz 1999: 108–109). Von Bertalanffy's second law of Systems Theory states that a natural system is hierarchically organized. In any given ecosystem, each species accomplishes a useful function for its own survival, although the degree of utility is very different between species. Some member or members of the species implement critical functions that are necessary for the ecosystem's survival. They form the apex of the hierarchical pyramid. They used to be called functional or key species. Control functions (and therefore hierarchy) exerted by species are frequently observed as being stronger as one moves forward through the trophic chain. There are many cases of control: flower pollination by insects; seed scattering by birds; (Jansson y Jansson 1994: 81–82); the role of otters in some marine ecosystems; etc. But there are humble species which implement key functions, such as bacteria and cyanobacteria which fix nitrogen. Normally groups of species, rather than species individually, accomplish critical functions. An ecosystem can have several functional groups, but they do not reach the same level in the hierarchy. And the same applies to species that take part in a functional group (Levin 2000: 10, 11).

In nature the following levels of hierarchy exist in a top-down order: biosfera, biogeographic region, bioma, landscape, ecosystem, biotic community, population, organism, organ and cell. Each hierarchical level influences those standing beneath, usually giving stability, although the lower levels also influence the higher ones by promoting changes (Odum 1992: 29). Nevertheless, «this subordination between levels is always incomplete and each level has its own rules of behaviour and its own specific concerns» (Gowdy 1999: 67). Population dynamics of a species is determined not only by behaviour of other local spices, but also by regional processes, and often very strongly (Lawton 2000: 92). Each level in the hierarchical scale has different timing. For example, fire cycles have devastating effects on some species, but they benefit others that are dominated by the first ones (Gowdy 1999: 74).

2.2.5 Self-organization

As Goldsmith declares: «biological organisms are self-regulating cybernetic systems capable by their own efforts of maintaining their stability» (1996: 153). Ecosystems form a group of interacting elements, which is described as self organization, that is, "structure and processes mutually reinforce each other" (Westley et al. 2002: 106). Ecosystems often collapse due to external shocks; normally they reorganize their structures by a process known as an adaptive cycle. This is the reason why «ecology strives in the holistic study of the parts as the whole» (Odum 1992: 33). During evolution, stabilizing mechanisms of natural systems became more sophisticated and their relations with the environment more stable. In this process the inner mechanisms of control become dominating. On the contrary, in the beginning of the evolution process the controlling mechanisms are external and coarse. Change rhythms of species are very different and faster than in ecosystems. The photosynthesis ratio of a forest is less variable than the ratio of each of its plants. The different rhythms of activity of the elements of an ecosystem act like an equilibrating mechanism (Odum 1992: 33–34).

2.2.6 Self-sufficiency

As ecosystems evolve they become more self-sufficient, reducing their dependency on external forces. This is a strategy to make greater their homeostasis and resilience and, as a result, their stability (Goldsmith 1996: 381). Odum (1992: 15) declares «auto-sustained and auto-maintained are words which characterize the natural landscape» Besides, as ecosystems evolve, trophic chains become more complex and waste is transformed into the principal source of nourishment, reducing furthermore the external abiotic input. In a mature forest, less than 10 % of the net production is consumed in live form (herbs, for example); a great part of the rest is dead material, and the inorganic nutrients become slowly intra-biotics (Goldsmith 1996: 381).

2.2.7 Decentralization

Terrestrial ecosystems are organised in spatial patches: "Nature operating in a decentralised manner, practical, pragmatic and with great intelligence, accumulated after millions of years of experience" (Pauli 1998: 27). Ecosystems have spatial limits due to environmental changes or to self-organization of the systems themselves (for example, a forest structure resembles a mosaic). Spatial limits improve the inner efficiency of a system and show the natural limits to growth (Ring 1997: 242). The shorter the natural cycles are, the greater is the efficiency of materials and energy use. In a forest, leaves carry out the primary production; they fall and become decomposed, roots capture the nutrients, which go up through the circulatory channels of the trunk. In a coral reef, distance between processes is much shorter, because microscopic algae and species which are part of the reef eat them. This could be «one of the explanations for extremely efficient recycling» (Jansson and Jansson 1994: 85).

2.3 Functions of Maintenance of Vital Constants

The biotic world is continually acting over the inert one (*abiotic*) in order to maintain the physico-chemical properties which are necessary to sustain life. That means both environments have evolved over time to form a complex and self-regulating system (Levin 2000: 28). Some of the aspects biologically controlled are: "surface temperature, atmospheric composition of reactive gases, including oxygen, and pH acidity-alkalinity" (Margulis 1998: 121, 123). Our atmosphere is not stable, because some of its gases react between themselves (especially oxygen and methane). Nevertheless, atmospheric composition has remained stable for an extraordinarily long time, because the biosphere has been emitting some gases and sequestering others (especially CO_2) in appropriated quantities in order to maintain equilibrium. On the other hand, land tends to increase its degree of acidity and we do know (as is shown by acid rain phenomena) that the great majority of organisms can only live in an acidity which isn't too high. Nitrifying and denitrifying organisms are able to produce the right quantity of ammonia to stop the acidification process. Nature purifies water and air, detoxifies wastes, controls pests, creates the ozone layer, which protects us from harmful levels of ultraviolet rays, etc. (Odum and Sarmiento 1998: 72; IETC 2003: 21). These functions resemble the immunologic system of organisms.

The Gaia concept aspires to capture this capacity of the biosphere of maintaining stable the physico-chemical properties of the abiotic world. Frequently is conceptualized as «a super-ecosystem (but not a super-organism, because its development isn't controlled genetically)», that is, «interlinking ecosystems which make up a huge ecosystem over the Earth's surface», and « it behaves as a physiological system in limited aspects" (Odum 1992: 62). The dominant vision of pre-capitalistic societies and of a great part of eighteenth century thinking is that there exists a natural order. Nature is seen as a self-organized and self-regulated system (Christensen 2001: 26).

3 Unsustainability of the Capitalist System

Much has been written about the similitude between the natural system and socioeconomic systems. Without doubt there are abundant resemblances, above of all, in the production field. On the contrary, many key capitalistic processes diverge from natural ones, and this divergence is critical to understanding the unsustainability of the socioeconomic systems of capitalism.

A capitalist system is compelled to maintain limitless economic growth, due to its relentless persecution of always greater benefits and money creation by means of credit. But it is impossible to grow ceaselessly in a limited planet. This dynamic can only be aggravated by a demographic explosion. In 150 years the world's population has multiplied by a factor of 6. This behaviour has been compared with that of a carcinogenic cell and we do know the outcome of that type of process.

To defend market forces as the unique and appropriate instrument for organization of economic activity and for social and ecological problem solving, one must deny the systemic character of societies and within it the natural hierarchical order. The result is contrary to the one proposed: instead of solving problems, the market polarizes wealth, disintegrates societies and creates a process of collision with nature. Market primacy means that the lower level of action (microeconomic level) holds the highest hierarchical level instead of the institutions of governance. They are the unique democratic instance and the only one which is able to work for a sustainable welfare. These results are due to the inverted hierarchy: the market acts in the short term, meanwhile social and political institutions act in the half and long term; the natural rhythm is much slower than the market one.

The different scales of time block information fluxes or simply result in unsuitable information for the specific instances in which decisions are required, thus provoking mistaken solutions and the use of erroneous instruments. The market emits signals that change rapidly in the short term and are not suitable to reflect long term structural changes. The opposite occurs when institutions give information about problems that are unfolding in the biosphere (loss of diversity, climate change, population growth, etc.), and the market does not take into account these processes because the temporal scale is not suitable for it (Gowdy 1999: 67, 68). Fast changes of natural product prices often create strong shocks in the environment. When prices of agricultural products rise, land with a high degree of erosion is put into production. When prices come down rapidly, peasants destroy their plants or trees. It has been happening for many years within vineyards, rubber tree stands, coffee plantations, etc. When demand and prices rise again, peasants begin to plant again, but many years may pass by before effective production can begin. For decades the natural gas that appears in petrol wells has been burned in the field due to its low market price. When gas prices rise, this behaviour is less and less frequent, but the burnt gas is gone. The market selects technologies to maximize production and does not take into account the future consequences of this behaviour. For example, the market has selected technologies to increase food production, but these "have reduced the capacity of land and water systems to provide food in the future" (MEA-SB 2006: 62).

In a free market economy, competition is a unique and powerful mechanism of selection that, as a by-product, creates a high degree of specialization and standardization, thus destroying or destabilizing other more diverse economies. On the contrary, "diversity relates to systemic coherence and integrity in terms of adaptations to changing environments, avoidance of head-to-head competition, overall efficient use of resources and energy and the possible range of responses to new selective pressures" (Rammel and Staudinger 2004: 15). It is argued that only competition permits a system to reach maximum efficiency but, in many cases, this is a partial and local efficiency, never a long term and systemic one. The market discards all available technologies to obtain a function except one, even though the others may be better in some, perhaps even most, circumstances. This dynamic reduces the capacity for using natural and human resources existing in local environments and their potential of innovation (Schutz 1999a: 25).

4 Sustainability Principles

As a Millennium Ecosystem Assessment of United Nations declares: "Humans are an integral part of ecosystems" (MEA-SDM 2006: 29). Being that the human species is an integral part of nature, human economics must be a sub-system of the general economics of nature, that is, of its materials and energy. And the principle of natural hierarchy declares that a sub-system's conduct cannot contravene the system's rules. So human economy must imitate nature to be sustainable and this means that we should infer from natural conduct the principles of sustainability. There have been many and important declarations in this sense. Marshall, known as the great synthesizer of the neoclassical school, proposed that biology had "a paradigm suitable for economy" (Christensen 2001: 16). Eminent ecologists (such as E. Odum, Hutchinson, Hall, Margalef, etc.) have proposed the view of imitating nature (Erkman 1998: 22, 41). This vision has been gaining adepts rapidly from the beginning of the last decade, to the point that now it is amply accepted.

Korhonen (2004: 812) states "the economic system is a subsystem of the larger social system. Furthermore, both systems are subsystems of the parent (mother) ecosystem and are totally dependent on it". On the other hand, the former statement implies to this author the conclusion that we have to imitate nature or "learn from nature". In 1995 the University of Maryland held a 3-day meeting of several dozens of authors with the purpose of defining "a detailed, shared vision of a sustainable USA in 2100". The resulting synthesis report declares: "People will recognize that humans are part of nature, one species among many, and must obey the laws imposed by nature" (Farley and Constanza 2002: 247). In the book *Natural Capitalism*, which is north American updated version of the report Factor 4 (Weizsäcker et al. 2004), it is stated that we must redesign "industrial systems on biological lines" in order to enable "the constant reuse of materials in continuous closed cycles" (Hawken et al 1999: 10). Lester R. Brown (founder, former president of the Worldwatch Institute and currently president of the Earth Policy Institute) declares that "if an economy is to sustain progress, it must satisfy the basic principles of ecology. This vision has begun to appear even in the business world, although without giving up the defence of the existing dominant paradigm. If it does not, it will decline and eventually collapse" (Brown 2001: 77). C. Holliday (DuPont) and J. Pepper (Procter & Gamble) state that "the biological designs of nature can be emulated as a role model for sustainability" (Holliday and Pepper 2000: 17). The World Business Council for Sustainable Development (WBCSD 2001: 7) declares: "the biological designs of nature provide a role model for sustainability". But "sustainable development is best achieved through open, competitive, rightly framed international markets". Summarizing, as Benyus (1997) reminds us, "we don't need to invent a sustainable world – that's been done already", it is all around us. "We need only to learn from its success in sustaining the maximum of wealth with the minimum of materials flow" (Hawken et al. 1999: 73).

Many of the above principles are the result of imitation, but some come from our capacity to destroy and create. Nature recycles materials and maintains the physico-chemical equilibrium of our inert environment, capturing, for instance, solar energy. Human economy should adopt this conduct as the basis for guiding principles (they are the abiotic principles), but they alone are not enough to reach sustainability. The human species is currently actively degrading nature, a behaviour that does not exist in other species. So we have to add the principle of stopping the destruction of nature and beginning to recreate it (Keijzers 2002).

But that will not be enough; as we have seen, in order to accomplish abiotic functions in nature, it is indispensable to construct an ecosystemic organization and that means: biodiversity, evolution, self-organization, self-sufficiency, decentralization, hierarchy and cooperation stronger than competition. These elements of the ecosystemic behaviour should be transformed in principles of sustainability of the human societies (they are the biotic principles).

Sustainability principles are equally valid for both developed and underdeveloped countries, the major difference being the departure point. The former have to transform themselves totally towards sustainability. The latter must transform there incipient economies and continue the process of development on a sustainable basis. This vision is shown synthetically in the following figure. It is indispensable in order to reach sustainability to design and implement strategic planning (Fig. 6.1).

But we can learn from nature much more than the principles of sustainability and its ability to adapt. In the measure that our understanding of nature is improving we begin to realize that it also is an inexhaustible source of knowledge. Nature has developed processes to elaborate materials and substances, to capture solar energy, means of communication, etc., that, apart of being sustainable, they are

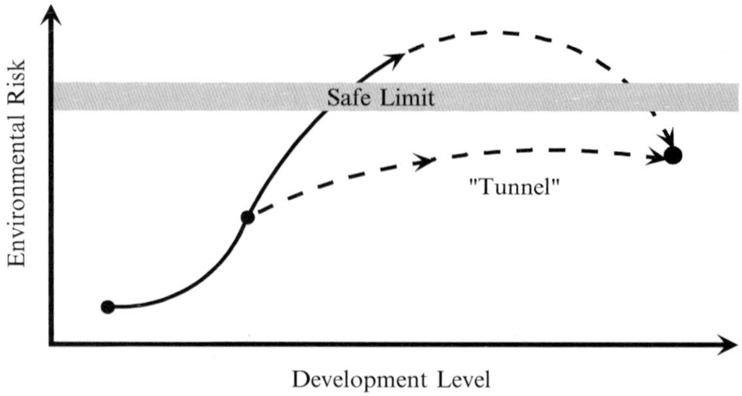

Fig. 6.1 Paths towards sustainability of developed and less developed countries (Source: adapted by the author from Raskin et al. (2002))

much more efficient and simple than those developed by mankind. So in the pursuit of sustainability we shall value nature more and more as a source of knowledge (a living library) than as a provider of resources.

This vision calls for radical changes in the economics model and in the status of economic science. Until now neoclassical economics has strived to be an autonomous field of knowledge, the world of mercantile value. Economics has taken only a few elements from other sciences, largely from physics but, above all, from mathematics, seeking from them a solid and indisputable body of scientific support. Economics being a social science, this is of course impossible. On the contrary, a sustainable economy must not transgress natural limits and the Earth sciences are the only ones capable of fixing these limits. So economic science depends on the former and the monetary world can no longer be autonomous. This is the opinion of Daily et al. (1996: 3) of the Center of Biological Conservation of Stanford University. Scholars of these sciences are those who "*determine the rules of game of the nature* and translate to others (. . .) Experts in ecology, meteorology, toxicology, oceanography, hydrology (and really most of these sciences are necessary to determine the human impact that can be supported by natural systems". Among all sciences, ecology plays a central role. For Barrington Moore (first president of the American Ecological Society) ecology is science "superimposed on the other sciences; a science of synthesis essential to our understanding of structure and function of biosphere" (Goldsmith 1996: 3). So sustainability determines that economics does have to adopt a transdisciplinary approach. It supposes the reduction of its actual high status within the broader sciences, and to become a subordinate field in relation to Earth sciences. Nevertheless, we must recognize that knowledge in Earth sciences is very precarious with respect to performance of large ecosystems and we shall never be able to comprehend integrally the complexity of the life web, because it is incommensurable. Nothing can be isolated in order to be analyzed because all is interrelated; problems extend across many levels of space and time, and data is scant (Funtowicz et al. 1999). And if we cannot know the natural limits, we should act in such a way that we shall never surpass those limits. This precautionary principle is an indispensable premise of strategic planning for sustainability.

In order to accomplish abiotic principles, physical calculation is indispensable. In order to close the life cycles of materials, knowledge of their societal metabolism (by the accounting of materials fluxes) is a necessity. We also have to account for the energy potential of the sun, available land, its allocation among uses, and destruction ratios. Finally, we have to create systems of bio-physical indicators to know if the goals of sustainability have been reached. The aim of monetary economy is to obtain, without harming nature, the greater quantity of goods and services of bio-physical stock. So a sustainable economy is dual, integrating monetary and physical concepts and instruments. This duality is frequent in modern physics, which by this concept can explain such complex phenomena as light performance, using waves and particles (Spangenberg and Lorek 2002).

Industrial Ecology (IE) is the field of applied science in which the idea of imitating nature appears more clearly. During the second half of the twentieth century, IE bases were developed. A wide group of scientists and institutions proposed analogous visions, under the names of industrial ecology, industrial ecosystems, industrial symbiosis, industrial ensembles environmentally balanced, etc. Several States of the soviet block promoted the development of this vision. In Moscow the "Industrial Ecology Department" of the Mendeleev Institute of Chemical Technology has been functioning during the last decades. In Japan an Industry-Ecology work group was created in the beginning of the 1970s, and it has been advising the Ministry of International Trade and Industry (MITI) in its dematerialization strategy for the Japanese economy. This vision has been supported by several international institutions like UNIDO and UNEP, both organisms of the United Nations (Erkman 2002: 27–30). Chapter 17 is dedicated to this issue.

Chapter 7
Transformability

Keywords Adaptive cycle of ecosystems • Usefulness for socioeconomic systems • Resilience concept • Transformability concept • Transformability factors

This chapter has three sections. The first one explains the adaptive cycle of ecosystems; the second one analyses the validity of the concepts useful for studying SESs and the third one explains the concept of transformability and the factors that determine it.

1 Adaptive Cycle of Ecosystems

The evolution of ecosystems after sustaining an impact is interpreted, in general, by means of the adaptive cycle theory. Its study will allow us to obtain tools to analyse the changes in SESs. We will explain it basing ourselves on the concept of resilience and, more specifically, on the factors of potential and connectivity. Later we will describe the importance that a third factor has on the adaptive cycle: Panarchy.

Resilience is normally understood as "the capacity of a system to experience shocks while retaining essentially the same function, structure, feedbacks, and therefore identity" (Walker et al. 2006). The three factors pointed out can be used to infer the level of resilience of a natural system. The potential reflects the richness of an ecosystem and depends on its amount of biomass, nutrients and its physical structure. Biodiversity means a high redundancy of species and groups of fundamental species: "Biodiversity contributes resilience to the functioning of an ecosystem [. . .] Adequate performance of ecosystems function depends on having all the necessary functional groups" (Holling et al. 2002: 406, 407).

Connectivity is determined by the amount and strength of the interior connections of a system and explains whether it controls its own fate or, on the contrary, it

R. Bermejo, *Handbook for a Sustainable Economy*,
DOI 10.1007/978-94-017-8981-3_7,

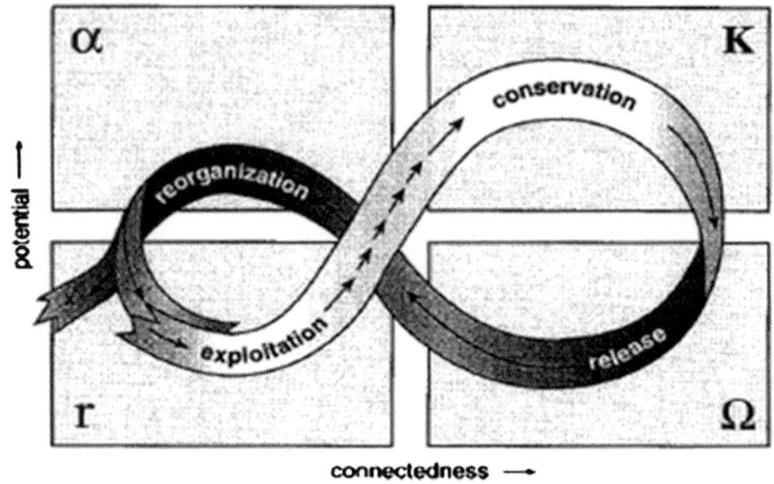

Fig. 7.1 Ecosystems' adaptive cycle (Source: Holling and Gunderson 2002: 34)

is at the mercy of external agents. Low connectivity is a sign of a poor capacity for self-control. This capacity improves with an increase of connectivity, but if it is excessive it is a symptom of rigidity. If the number and the strength of the connections increase, it will reach a threshold from which the system increases its rigidity, losing the flexibility necessary to adapt to the environment and increasing its vulnerability, which will lead to its collapse. When the elements of a system are strongly connected, any internal disruption quickly propagates throughout the entire system. There is, therefore, an optimal threshold of connectivity, which is usually defined as the "window of viability". But the greater the number of species in a system, the less connected they need to be in order to remain within the window of viability (Matutinovic 2002). The existence of a threshold from which the quality of the connectivity changes makes it difficult to determine an adequate level of connectivity.

The adaptive cycle is usually explained based on variations of the potential and connectivity. It has four phases, as shown in Fig. 7.1. The first, that of creative liberation or destruction (omega phase), means that an ecosystem suffers a collapse because the impact received (caused by external disruptive agents: drought, fire, pests, etc.) exceeds its capacity for adaptation and it collapses; there is then a reduction of the level of connectivity, which weakens its self-regulating mechanisms and increases the possibility of renewal (reorganisation). The potential is reduced, as a large part of the biomass is lost. Although it may lose some species, the main problem is a drastic reduction of the population of some species. For example, after a forest fire, few adult trees remain, but it is possible that many saplings and in particular seeds, survive. The loss of connectivity and potential determines the reduction of resilience, which means that the ecosystem has lost its capacity for self-regulation (the feedback processes that allowed it to remain stable). In the alpha

Fig. 7.2 Adaptive cycle in three dimensions (Source: Mearns 2010b)

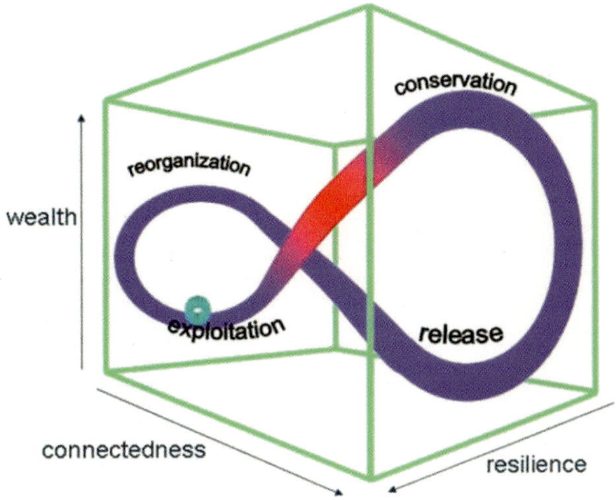

phase the ecosystem is restructured, which will allow the subsequent growth and accumulation of resources. Most of the previously existing species remain and the strong reduction of connectivity allows the addition of external species. There are new associations among them, and the reorganisation may not take place in exactly the same way as in the past. These phases are very rapid (Holling 2009).

The third phase is that of exploitation (r phase), characterised by a rapid colonisation of the degraded areas. The seed banks stored in the past are activated; pioneering species, whose population was previously very low due to a lack of sunlight, start to rapidly develop, occupying the free space. The vegetation that results is a kind of shelter that allows the seeds stored and those from the exterior to germinate. The loss of nutrients tends to fall as the phase progresses. The level of connectivity starts to grow. The greatest level of biodiversity is frequently during this phase. But in the transition to the next it usually falls, as there is strong competition among the pioneering species for the territory and the capture of sunlight. This can no longer be compensated with the introduction of new exotic species, because the connectivity has reached a level sufficient to prevent them from prospering. The fourth and last phase is that of conservation (K phase), where there is a slow accumulation of material and energy. The species create cooperative groups, which are usually vital for their survival. The evolution of the systems is increasingly predictable. Their internal control and that over external variables is strengthened (for example, by creating micro-climates). Potential and connectivity reach their highest peak. The system achieves the highest efficiency in its use of materials and energy (Holling 2009).

Up until now we have used the bi-dimensional diagram of the adaptive cycle, but it is necessary to complete it with the third dimension: resilience. The graph in Fig. 7.2 allows us to understand the variations that occur throughout the adaptive cycle. We see that the exploitation and conservation phases are at the front of the

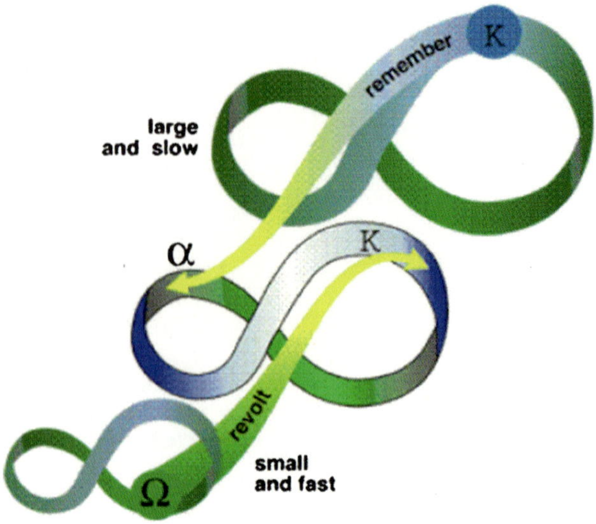

Fig. 7.3 Panarchy (Source: Holling 2004)

cycle, while the other two are at the back. As resilience is greater at the front, the exploitation and conservation phases are those with the highest resilience. On the other hand, the liberation phase is the one with the least resilience and this improves slightly during the reorganisation phase.

Although this theory explains the most typical behaviour of ecosystems, there are cases where there is not a standard succession of phases, so it cannot be used as a tool for predicting, but rather for explaining the changes in most cases (Abel et al. 2006). In addition, these phases are not yet properly understood, particularly growth (r) and conservation (K) (Walker et al. 2006).

So by means of the adaptive cycle an ecosystem proves that it is sustainable: "Sustainability is the capacity to create, test, and maintain adaptive capability" (Holling et al. 2002: 403). These authors, along with Walter and others, offer us a vision of what resilience is (and therefore of the adaptive cycle) focused on the conservationist aspect: an ecosystem preserves its identity through the adaptive cycle. But other authors emphasise its dimension of change, of a process of renewal and revitalisation, an aspect that has attracted less attention: "But, there is also another aspect of resilience that concerns the capacity for renewal, re-organization and development, which has been less in focus but is essential for the sustainability discourse" (Folke et al. 2006: 253).

Panarchy is a new concept that seeks to explain interaction in a hierarchy of systems, meaning that the adaptive cycle can only be understood within the context of interaction. The term comes from Pan (the god of nature in ancient Greece) and hierarchy. It explains that a reference system is subject to the influence of the superior system it is part of and of the inferior systems that form it, generating feedback processes between the different structural levels, as shown in the graph in Fig. 7.3. The larger the ecosystems the slower the rates of change, and with smaller ones the opposite occurs. Larger systems contribute to improving the resilience of the smaller ones, increasing their capacity to recover from collapses because they provide "memory", in the shape of biotic legacies, seed banks and structures. On the other hand, the small ones provide novelty, because their changes are much faster. This function is known as "revolt". So we find a combination of forces that achieve a dynamic balance among the systems that allows them to adapt to the changes in the environment: "Persistence and extinction, growth and constancy, evolution and collapse entwine to form a panarchy of adaptive cycles across scales" (Holling 2009: 4).

2 Applicability to Socio-economic Systems of the Concepts and Factors of the Natural Adaptive Cycle

2.1 The Usefulness of the Adaptive Cycle for Socio-economic Systems

The metaphor of the adaptive cycle is a useful tool to understand the mechanism behind the changes of SESs, because all SCAs show thresholds that lead to a collapse and the phases of the adaptive cycle are a logical sequence of reorganisation and improvement in systems that are sustainable. Once SESs reach sustainability, the adaptive cycle will explain its evolutionary dynamics, adapting to the changing conditions of the environment. Abel et al. (2006: 13) affirm that "the adaptive cycle, as an elaboration of complex adaptive systems theory, is useful in recognizing the change in systems' behaviour during the various phases". But such an idea should not lead to the belief that it is possible to apply it mimetically. There are at least two factors that invalidate the idea. A general one: the human intelligence that allows the prediction of the problems and their avoidance, but which also brings destructive behaviours. Another specific one of many complex civilisations, and in particular of industrial civilisation: unsustainability.

In general, human societies tend to increase their complexity, understood in general terms in SESs as "differentiation in the structure and behaviour, and/or in the level of organisation". More specifically, it means the creation of new institutions, adding more specialists and bureaucracy to what already exists, increasing regulation, the complexity of technologies, etc. (Tainter 2005), which can cause societies to become unsustainable and collapse: "Environmental deterioration is one of the most common explanations for the collapse of ancient civilisations" (Tainter 2005). Tainter (2005: 92) defines a "socio-political collapse as a rapid simplification, the loss of an established level of social, political, or economic complexity".

But although there have been civilisations that have collapsed due to their unsustainability, the pre-industrial world kept to a gentle trend of increasing the population and the consumption of resources—and it was sustainable. On the other

Fig. 7.4 The marginal productivity of increasing complexity (Source: Tainter 2006: 94)

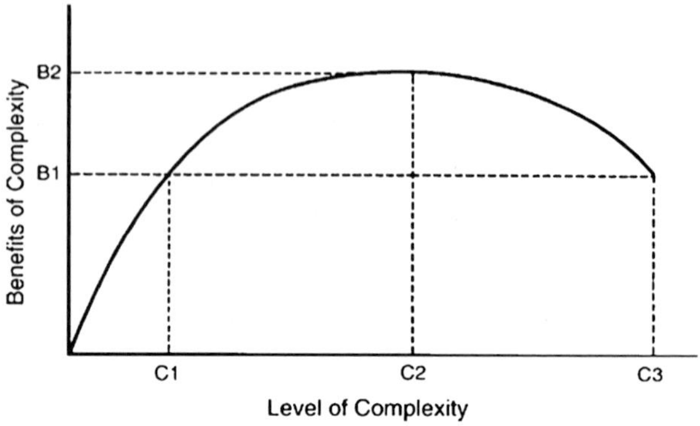

hand, industrialisation started an explosive process of population growth that has been possible thanks to the use of fossil fuels.

Historically, complexity is generated with the intention of solving problems. And greater complexity is not necessarily a negative factor. In the previous chapter we have seen that an increase in biodiversity generates more resilient ecosystems, but they tend to stabilise when they mature. However, SESs tend to grow without limits. There is a widespread agreement that humanity has had over the last 12,000 years "growing populations; greater technical abilities; hierarchy; differentiation and specialisation in social roles; and increasing production and flow of information" (Tainter 2005). The increase in complexity means costs, generally in the form of a greater use of resources.

But continuous growth in complexity is a pathology, non-existent in the natural world. The graph in Fig. 7.4 shows us that in a system that becomes increasingly complex, from a very low level, at the beginning its efficiency improves, but from a certain threshold efficiency is reduced at an increasingly faster rate. The system needs growing amounts of material and energy to function. This dynamic is particularly intense when there is strong competition between different centres of power. The struggle for hegemony among peer agents tends to increase complexity, leading to a situation where no-one is prepared to renounce their goal and the struggle is prolonged, because each one imitates the technological, military and organisational advances of the others. The result is that the performance of their actions decreases, societies are weakened and finally end up collapsing: "Peer policy competition drives increased complexity and resource consumption regardless of costs, human and ecological" (Tainter 2006: 214). This is the case of Easter Island and the Mayan Empire, although there are also many other sustainable societies based on cooperation.

The negative effects of increasing complexity have historically led some societies to reduce their level of complexity in order to survive. The Eastern Roman Empire was much more long-lived than the Western one, because it decided to simplify its organisation, particularly by giving land to its soldiers so that they produced their own food and protected the territory from invasions. Anthropological studies on peasants and fishermen under a regime of subsistence in Russia, Papua New Guinea, Brazil, etc., show a limitation of production to avoid the falling performance of labour. But falling performances are being seen in many areas of industrialised countries: in expenditure on education, R&D, healthcare, etc. (Tainter 2005: 10, 11).

2.2 The Resilience Concept

There is a generalised agreement about what the concept of resilience means when applied to natural systems. K. S. Holling gave a definition of resilience in 1973 that is the reference: "The capacity of an ecosystem to absorb and utilize or even benefit from perturbations and changes that attain it, and so to persist without a qualitative change in the system's structure". This definition coincides with others quoted in the previous chapter and in all of them the concept is positive: the more resilient a system is the greater its capacity to maintain its identity.

When the concept is applied to SESs the agreement disappears, and becomes enormously complicated when it is applied to very different sciences. Among those who study the sustainability of SESs we find expressions of resilience that escape the previous agreement. Folke (2006: 259) states that "it does not imply that resilience is always a good thing. It may prove very difficult to transform a resilient system from the current state into a more desirable one". Walker et al. (2006) declare that "social-ecological systems can sometimes get trapped in very resilient but undesirable regimes in which adaptation is not an option", meaning that in order to transform themselves they require profound internal changes. In these interpretations resilience ceases to always be a positive concept, because it can even be negative (the system becomes rigid). The problem gets complicated when the concept is used by many social sciences such as economics, politics or sociology, and they use it in their planning, to define strong sustainability, when analysing globalisation, in the so-called adaptive management or governance, in international aid, etc. In each one of these applications its meaning changes. So "the specific meaning of resilience becomes diluted and increasingly nuclear. This is due to the use of the concept (a) with many different intentions and (b) with a very wide extension" (Brand and Jax 2007). So it ceases to be operative.

In an attempt to overcome this situation, some authors define two types of resilience: ecological and social. Ecological resilience is that of natural systems and social resilience is the one that has a threshold of change of quality. Folke (2006: 259) defined social resilience "as the ability of human communities to withstand external shocks to their social infrastructure, such as environmental variability or social, economic and political upheaval". By ability we have to understand the capacity to maintain it within a window of viability, like in the case of connectivity. Other authors resort to concepts such as "adaptive environmental management" and "adaptive governance". The first concept is "a formal process for conducting and interpreting ecosystems management". But, "because of the key role of governance in ecosystem management, the researchers have introduced the concept of adaptive governance to study the structures and processes by which humans make decisions" (Carpenter and Forlke 2006). These applications are based on a premise: that SESs can achieve sustainability by adapting without the need for transformation. Therefore, it is incompatible with the premise that we have defended here of the structural unsustainability of the SESs and which has been made clear in Chap. 1.

The terminological confusion (and the differing visions on the level of unsustainability of SESs) when concepts that are used to explain the behaviour of natural systems are applied to the field of SESs, leads many authors to criticise such a mechanical transposition. Anderies et al. (2004) consider that resilience can be difficult to apply to systems where some components are consciously designed". Gallopin (2006: 299) is more emphatic when he states that "the concept of resilience cannot be transferred uncritically from the ecological sciences to social systems", because "using the concept for social systems (. . .) does not really imply that there are no essential differences in behaviour and structure between social and ecological systems". And he adds that "recent formulations of resilience, such as making adaptive capacity and self-organization properties of resilience" are risky. For this reason Joung et al. (2006: 311) defend the separate study of biophysical and social systems: "It is useful to separate biophysical from social systems, and then to consider how differences in responses between these two types of system affect what we can say about responses in SESs".

3 The Transformability of Socio-economic Systems

Having dismissed the concept of resilience because it only makes sense in sustainable systems, it is necessary to find another concept (approach) that enables us to study the dynamics of SESs in the pursuit of sustainability. The pertinent study is that of the capacity for transformation, of transformability. But this study lacks any foundation if we do not define the factors that determine such a capacity.

3.1 The Concept of Transformability

For Walker et al. (2006) "Transformability is the capacity to create a fundamentally new system when the existing system is untenable". In any event, what is noteworthy is that the above authors, despite having the merit of defining the concept of transformability in the case of sustainability, have not gone into it in greater depth (in particular, defining the factors that determine it) and we do not know of any case where they have applied it. The factors that determine the transformability of an SES are the same that explain resilience (potential, connectivity and panarchy), but adapting them (when necessary), though without changing their meaning. And this is due to the fact that the three factors can be used to explain the characteristics of any CAS of their hierarchical relations.

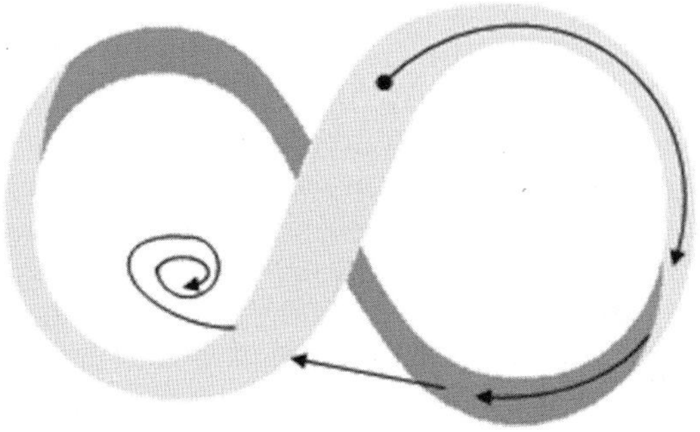

Fig. 7.5 Collapse to chaos (Source: Gallopin 2002: 38)

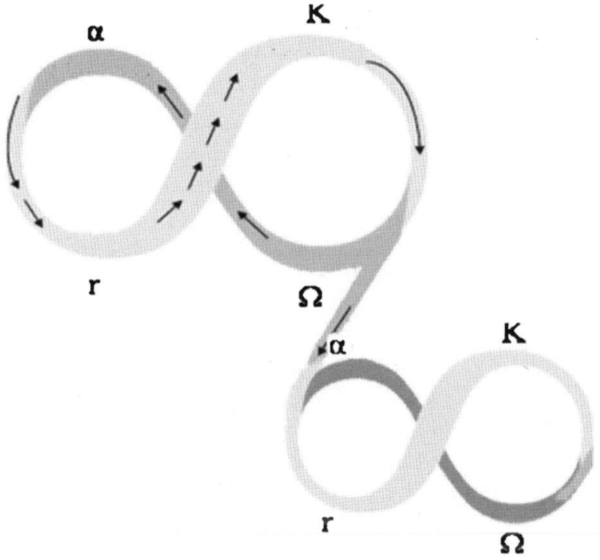

Fig. 7.6 Transformation after collapse (Source: Adapted from Holling 2004)

The characteristics of any SES analysed in light of the principles of sustainability determine the nature and depth of the transformation it must go through. Peak oil and other factors of unsustainability do not leave any temporal margin to carry out a non-traumatic transformation. A more or less severe collapse is a highly probable scenario. Once it occurs, there are two basic options available: to start the transformation or to remain collapsed, because there is an attempt at overcoming it without transformation. Figure 7.5 shows that the CAS that needs to be transformed and is incapable of doing so, when attempting to reorganise itself according to guidelines of the past, cannot begin the r phase and goes back to the omega phase.

On the other hand, the transformation means that the SES abandons, after collapsing, its current structure because it cannot carry out the adaptive cycle; it restructures itself in a totally new way and can complete the growth and conservation phases, as shown in Fig. 7.6. Once the transformation has taken place, the system will have the capacity to evolve with (adapt to) the natural system by means of successive adaptive cycles.

Up to here we have limited the level of analysis to what could happen to SESs when they are isolated from the hierarchy determined by the general system. But if we take it into account, we find that it imposes on the others a series of structures and basic rules of operation. As a result we are facing the threat of the general system collapsing, which means that all individual systems would be affected, although to different extents, depending on whether they are capable of going through deep transformations. Holling (2009) explains this situation inversely; the collapse of one level can cause the collapse of the entire system: "But when the collapse occurs as a consequence of a long effort to freeze the system into a paradigm of development and management, it might involve collapse of a level of panarchy, which in turn threatens all levels". But it is not just a threat—it is a reality. The collapse of the global financial system in the autumn of 2008 (magnified by the sharp rise in oil prices over the 2002–2008 period) is proof of this: "When over-connected, shocks are rapidly transmitted through the system" (Walker et al. 2008).

Once again, we are faced with the confrontation of two models. One designed to succeed in a linear scenario removed from reality (the economic system grows unlimitedly without suffering collapses that require its transformation), of strong competition (which demands constant reductions of prices), without stopping to look at the social and ecological repercussions from such decisions, or, in particular, the capacity to face collapses. The other, designed to maximise the capacity to face impacts, which involves preserving nature's resources and services, strengthening social cohesion and creating diverse productive systems adapted to the social and ecological conditions of its surroundings (Gallopin 2002).

3.2 *Transformability Factors*

We have seen the pertinence and usefulness of applying the concept of transformability to SESs and to our current reality, and that the factors that determine it are: connectedness, potential and panarchy. Although such factors are applicable to any complex system, they require some fine-tuning to be applied to SESs, but without changing their meaning.

3.2.1 Connectedness

It determines the degree to which a system controls its own fate or, on the contrary, it is at the mercy of external agents. In a natural system the connectedness is determined by the amount or intensity of the flow of materials and energy, as well as the hierarchy. The more abundant and the higher the category of the connections, the higher its connectedness will be; that is, its capacity for self-control, but excessive capacity for control generates rigidity and therefore a loss of the capacity to adapt to changing circumstances. So there is a window of viability. The connectedness of SESs is different from natural systems due to their unsustainability and intelligence. This determines its capacity to change its connectedness, so it can transform itself. The first differentiating factor means that the optimal connectedness must be much lower than that of natural systems, because it needs to transform itself, not just adapt to the changes in the environment. It is evident that the general or global SES is characterised by:

- A set of interrelations between different highly structured societies at a global scale, although we cannot speak about a global system.
- A division of labour structured hierarchically among groups of States that make up a centre, a semi-periphery and a periphery. The centre is formed by States that carry out the most sophisticated economic activities. The States in the periphery supply the centre with raw materials or low-quality manufactured products. Semi-peripheral States share elements with the previous groups.
- This division of labour and economic roles determines enormous flows of materials, energy and manufactured goods, in addition to financial and information flows.
- Some of the States at the centre and the GTEs (whose direction is in those States) have up until recently been the basic powers that structured the world order. States impose a political and economic order through control of the world's institutions, the use of force and support for the GTEs so that they can control the world's economic structures (Gotts 2007).
- Some States and GTEs established within them that come the semi-periphery are starting to become part of the world political and economic order. Meaning that a multi-polar world system is being formed.

The existence of political and economic elites, and their control over all types of flows that make up the globalised system, determine a very high connectedness of the global SES. Although there are factors that are starting to weaken it and, probably, they will do so with more intensity as time goes by: the unequal distribution of wealth, the multiple factors of unsustainability, etc.

3.2.2 Potential

In nature the potential is always a positive factor. It grows with an increase in the biomass, in biotic diversity, in redundancy, etc. SESs obtain resources and essential resources from nature. That is, they use the natural potential. In addition, based on natural resources they create an additional potential, the potential or human origin: technological, constructed, institutional, social, etc. The big problem of current SESs is that most of the accumulation of this potential and its use leads to the destruction of the natural potential. We will call it the perpetuation potential (PP) because it is used so that current SESs can function and attempt to perpetuate themselves, but not because we consider that they can actually do it. In addition to the PP, another potential is being built that includes the wealth that is useful to create a transformed, sustainable SES, so we will call it the transformation potential (TP). The accumulation of the PP is a factor of the rigidity against change, because the investment in it limits the capacity to decisively raise the PP.

The reduction of the natural potential carried out until now and the destructive trends that predominate will increasingly limit the capacity of SESs to reach adequate levels of satisfaction of essential needs. It is also evident that the globalisation process is reducing the social potential, by eroding social cohesion, by increasingly promoting individualism, the marginalisation of the weakest and thus creating a breeding ground for separating ideologies. A large part of the technological and constructed potential that is being created is of perpetuation, despite the unquestionable positive tendencies, such as the development of renewable energies. Later we will show that the energy TP grows explosively and that the TP of transport, agriculture, etc. also increases considerably.

3.2.3 Panarchy

We have seen that the concept of panarchy reflects the interrelations between the different levels of a complex system, where the superior systems act as the stabilisers of the rest of the systems, while the lower ones, which are the ones that evolve the quickest, act as the factors of change, of revolt. Panarchy also manifests itself in SESs, although with distinctive features. Superior SESs (powers) are carrying out (particularly during the last two decades) a task of change, strengthening the development of the capitalist system's logic: encouraging the globalisation of the economy through its liberalisation, which intensifies competition and the side effects of such an initiative: the polarisation of wealth and an aggravation of its unsustainability. Liberalisation reduces the level of autonomy of inferior SESs, it condemns them to high specialisation (which subordinates them to the logic of a globalised economic system controlled by the GTE) and, as a result, to high vulnerability and in many cases marginalisation. The exceptions to this rule are cities and regions that are the nodes of the global hierarchical network. As a result of marginalisation, at a scale of cities and regions there is a resistance to the subordinate and vulnerable role. Examples of this defensive reaction are the efforts to restrict or prevent the process of industrial offshoring, incentivising the continuity of businesses and applying dissuasive measures. More examples are the multiplication of incentives to attract new businesses, so that they can maintain their economic diversity and even improve it.

For this reason, some authors state that SESs have reversed natural panarchy. Joung et al. (2006: 310) affirm: "One little known systemic implication of globalization is related to a reversal in the hierarchical structure of large-scale SESs as pointed out by Gallopin (1991)". The superior systems are the driving forces behind the changes and the inferior ones try to soften the impacts they suffer. And no CAS can survive a reversal of panarchy.

But reality proves to be more complex than what has been seen up until now, because there are continual changes made in behaviour of both inferior and superior systems. In the former, we find that an important and rapidly growing number of societies that not only carry out a defence of the economic structure that is being degraded, but which seek to transform themselves to reach higher levels of sustainability involves, in addition, increasing their capacity for self-government. And by doing this they carry out the function of revolt. On the other hand, we find that the hierarchical pinnacle is going from a unipolar stage (The US after the collapse of the USSR) to a multi-polar one due to the emergence of new powers. This phenomenon is being translated into a restraint on economic globalisation and, on the other hand, a reinforcement of regional organisations. The failure of the WTO's Millennium Round is a clear sign of this process. Therefore the reversal of panarchy is being weakened. This leads us to wonder about the adequate panarchy for the current period of energy emergency. The answer can only be general. The inferior systems must carry out the same function as their peers in nature: to revolt. And the superior systems must carry out two types of functions: to extend and coordinate the task of transformation, which means a prior transformation of their structures and objectives.

4 Conclusion

The global SES shows a very low level of capacity for transformation because its connectedness is very high, the TP is low and the panarchy is basically reversed. The positive side is that the trend is towards weakening the first and third factor and strengthening the second.

Chapter 8
Instruments for Sustainability: Strategic Planning and Ecological Tax Reform

Keywords Strategic planning for sustainability • Premises of strategic planning • Environmental tax reforms • Ecological tax reform • Perverse subsidies • Escalating taxes

The implementation of a sustainable economy requires carrying out an in-depth transformation of the economic system in force. Many governments have been developing National Sustainable Development Strategies (NSDS), and for this reason it is pertinent to evaluate their capacity for transformation. An essential requirement of the evaluation is to analyse whether these governments carry out an ecological tax reform, because the current tax system subsidises the destruction of the planet.

1 Genesis and Premises of the Strategic Planning of Sustainability

Chapter 8 of Agenda 21 (Rio Summit of 1992) urges governments to design sustainability strategies. From that moment an important number of States started to fulfil the mandate. In 2001 the EU approved the so-called SDS, but it only contains generic reflections and proposals, meaning that it lacks any effectiveness. The generalised criticisms of this document led to the approval in 2006 of the SDS Revised, which continues to lack any appreciable potential for transformation. Most of the 30 countries of the OECD have prepared SDSs. Of them, 23 have designed formal strategies or plans (OECD 2006: 10). All members of the EU are obliged to fulfil them.

In order for the NSDS to have a clear potential for transformation, they need to fulfil a series of requirements, which are common sense. The OECD and the UN have reached a high level of coincidence regarding these requirements, which

R. Bermejo, *Handbook for a Sustainable Economy*,
DOI 10.1007/978-94-017-8981-3_8,

can be summarised in the following premises: integration of the three dimensions; achieving an agreement on a vision of long-term sustainable development; basing strategies on rigorous diagnoses; involving the highest institutional levels; establishing commitments at all regional levels; a high level of participation; including objectives that are accompanied by adequate budgets; carrying out a follow-up of the strategy and correcting it if there are deviations (OECD 2006: 13).

A Statement by the European Environment and Sustainable Development Advisory Councils (EEAC), which is supported by twelve national councils and one region, declares that the "present patterns of governance in the EU and in member states (. . .) rarely takes all aspects of sustainability fully into account. The outcome is that European governance inadvertently acts to increase inequality, social disruption, and ecological dislocation"; for these reasons the difficulties of reaching sustainable development "are enormous" (EEAC 2008: 2, 4). The outcome of these facts is a "huge uncertainty as to what could happen to the fabric of the planet, the habitability of many populated regions, and even the viability of humanity itself". For this uncertainty "is an almost impossible set of predictions", and, although some changes "will be incremental, building on experience, courage and experimentation, others may be convulsive, abrupt and ill-coordinated, driven by desperation or panic as conditions for human well-being deteriorate" (EEAC 2008: 3). The Statement estimates that there is limited time for action: "At best, there is a 25 year window for a genuinely significant response" (EEAC 2008: 3).

Finally, the Statement invites the Commission to institute a 1–5 years "series of visionary exercises", conducted at an EU level. With the support of the EEAC: "These visions should look into the issues of managing within nature's limits, creating a fair, just and resilient society, generating sustainable livelihoods, designing robust and appropriate technology, and laying out space for prolonged sustainability with the emphasis on rural viability, city regions and effective community mobilization" (EEAC 2008: 6). Last of all, it makes it very clear that the markets cannot carry out these transformations: "Markets cannot handle this (sustainability). For one reason, there is too much uncertainty, too little reliable measurement [. . .] This is why a robust framework for markets is needed in which they can operate" (EEAC 2008: 5).

But it is doubtful, as we have seen in previous chapters and will see in many subsequent chapters, that we have 25 years for the transformation to occur. Nor is there any time to lose carrying out prolonged "visionary exercises".

1.1 Analysis of the Fulfilment of the Premises in the Case of the NSDS

1.1.1 Preparation of a Diagnosis That Conforms to Reality

The Statement affirms that "effective and organised knowledge about long-term system dynamics and potential crises plays a key role for preparing for the future"

(EEAC 2008: 4). Except in the case of the Dutch government (which carried it out before preparing its first plan, the NEPP1), the known cases of NSDS are not based on the preparation of a diagnosis. The European Commission (2004b: 4) reached the conclusion that "most documents analysed do not reveal how policy choices were made". Another very notable exception was the SDS proposal by the European Commission (2001c), which we will comment on later.

1.1.2 Definition of Measurable Objectives Coherent with the Diagnosis

The objectives must be measurable, coherent with the diagnoses and must reflect priorities. But without a diagnosis there can be no coherence. In addition, many of the strategic objectives of the NSDSs are not measurable (for example, the fight against climate change and improving the efficiency of resource usage), although they do normally appear in the plans (5-year). Out of 19 countries analysed by UNEP (2004: 13), only seven "employed measurable and/or time-bound objectives, although the degree of specificity varied considerably". The German SDS is prominent because its most important long-term objectives and fulfilment periods are:

- Climate change: to comply with the Kyoto commitment (21 % reduction of equivalent CO_2 emissions).
- Energy: the fastest dismantling of nuclear energy; doubling of the energy intensity (GDP units created per energy unit) during the 1990–2020 period; doubling of the percentage of contribution of renewable energies to consumption during the 2000–2010 period.
- Land: reduction of the land sealed by urban construction and transport infrastructures from the 129 ha/day today to 30 in 2020.
- Pollution: a 70 % reduction during the 1990–2010 period.
- Agriculture: 20 % of agricultural land dedicated to ecological farming by 2010.
- Transport: a reduction of the intensity of goods transport by 5 % and that of passenger transport by 20 % for the 1999–2020 period (Federal Environment Ministry 2002: 10 and ss.).

1.1.3 Clear and Adequate Definition of the Responsible Institutions

It is evident that in order for the SDS to be fulfilled, it is necessary that the responsibility of their fulfilment be located at the top of the political hierarchy. The OECD (2006: 21) considers that the achievement of the goals "depends a great deal on high-level political commitment, well-functioning government institutions and overcoming co-ordination failures in public policies". However, the responsibility usually falls on the Environment Ministry, which is normally the weakest. On the other hand, "a good practice is to assign overall co-ordination to a Prime Minister's office or the equivalent", because it has the power to obtain the necessary financial backing and to settle disputes (OECD 2006: 21).

1.1.4 Adaptation, Integration and Prioritisation

Normally, the actions planned have been clearly insufficient to achieve the chosen goals: "To what extent NSDS remain declarations of intent or actually have contributed substantially to changing their policy measures and the way they are made, in many cases remains to be seen" (European Commission 2004a, b, c: 20). On the other hand, it is not often that general goals are broken down into sectoral goals in order to define the contribution of each government department to achieve the former: "In many cases interdependencies or spill-overs between sectors are not fully taken into account, so policies in different sectors pull in opposite directions" (European Commission 2004a, b, c: 6). There is usually no vertical integration either, that is, when the general goals are divided into others that lower-level government institutions must achieve (regions, municipalities, etc.). There is no connection between the German strategy and that of the landers. These facts explain that "only a few OECD governments have attempted to catalyse and fully co-ordinate their activities with the sustainable development efforts in effect at sub-national governments levels" (OECD 2006: 23).

In addition, there is normally no definition of what actions are priorities: "A lack of prioritisation can be noted in many NSDS, and reflects the difficulties that countries face in designing NSDS with concrete, realistic and credible intermediate targets and measures". Integration and coherence between measures frequently does not appear: "The objectives and measures contained in the NSDS are often a mixed bag or assembly of individual actions. Therefore they are not always integrated into a broad framework (. . .) Many decisions that are contrary to the aims of the NSDS also prevail" (European Commission 2004a, b, c: 20).

1.1.5 Maximum Legal Backing

It is not at all frequent that sustainability is a mandate contemplated in national Constitutions. Among the cases studied, only Sweden and Switzerland have Constitutions that place sustainable development as a national goal. The new Swiss Constitution establishes, in addition, the obligation of sustainability in all government actions, both at home and abroad. On the other hand, legal backing for the Strategic Planning for Sustainability (SPS) is a revealing sign of the level of commitment of governments to them. Greater commitment is achieved when they are approved by law. Only Canada, Mexico and Madagascar have fulfilled this requirement (IISD et al. 2004). So "a lack of a legally binding basis means that NSDS mainly rely on the political commitment of the government in place and the engagement of the different stakeholders for its implementation" (European Commission 2004a, b, c: 11, 18).

1.1.6 Adequate Financial Resources

The application of the NSDS means very important transformations that require substantial funds. And these can only come from a profound budget reassignment. The EEAC (2008: 6) proposes "a change in view on public management where strategic goals are translated into an output oriented budget based on best available knowledge", and "a budget and a budget process that is focussed on European values and strategic goals". But most national strategies lack provisions to systematically assess the costs and benefits of alternative actions (OECD 2006: 19).

1.1.7 Promotion of Participation, Information and Education

Participation is vital to creating a feeling of ownership of the NSDS among the population, essential to fulfil its goals. There is increasing insistence on this, but there is still a long way to go. The OECD (2006: 25) affirms that an "active stakeholder participation (e.g., business, trade unions, non-governmental organizations, indigenous peoples) in the development and implementation of national strategies should be an inherent feature". However, in the OECD countries "the extent to which the stakeholders are involved in policy processes reflects national institutional settings and preferences". Belgium, Finland, France, Germany, Hungary, Ireland, Luxembourg and Netherlands, Poland, Portugal and Slovenia have established Sustainable Development Advisory Councils which form the EEAC Network (2008). They are composed of representatives of social and economic organisations and of experts.

1.1.8 Follow-Up and Adaptation Process

Strategies must be subjected to periodic assessment to correct their deficiencies and to incorporate new knowledge, international commitments, community Directives, etc. The OECD backs this point of view: "National strategies for sustainable development are not meant to be static plans. Rather, they should evolve as more information becomes available", and this requires "a process to monitor strategy implementation" (OECD 2006: 29). However, "for many countries, effective monitoring of progress remains a difficulty" (European Commission 2004a, b, c: 21).

1.1.9 Promotion of International Cooperation

Most ecological problems are of a global dimension. Meaning that it is necessary to cooperate in order to implement and improve the existing Multilateral Environmental Agreements, to approve new ones, to help make the development of less developed countries sustainable, to create environmental institutions at a global level. At a state level worth mentioning is the development aid in Sweden which in for guiding and reporting on long-term broad developments and promoting forward-looking reflection on sustainability (. . .) However, merging cross-cutting strategies does not seem feasible". Within this context, it seems ironic that the Commission concludes that "there may still be room for further clarification of the specific role of the EU SDS in relation to other strategies" (European Commission 2009d: 13, 14).

However, "Europe 2020: A strategy for smart, sustainable and inclusive growth" (COM(2010)2020 final), which represents the vision of the economic area on sustainability, ignores the conceptual balances that the area carries out to integrate the EU SDS and the economic strategies. It becomes the number one strategy, which includes elements from the environmental field. Smart growth means "an economy based on knowledge and innovation". Sustainable growth is achieved by "promoting a more resource efficient, greener and more competitive economy". So the concept of sustainable development is replaced by that of sustainable growth, and this represents an efficient use of resources and more competitiveness. Besides, the Commission put forward seven flagship initiatives, though only one refers to sustainability: "Resource efficient Europe to help decouple economic growth from the use of resources, support the shift towards a low carbon economy, increase the use of renewable energy sources, modernise our transport sector and promote energy efficiency" (European Commission 2010d: 5, 6).

2 Environmental Tax Reforms and Ecological Tax Reform

2.1 Environmental Tax Reforms

There are many terms associated with environmental tax reforms, because on some occasions the same terms have different meanings, and on others different terms mean the same thing. The terms Environmental Tax Shift (ETS) or Green Tax Shift (GTS) mean raising taxes on things that are harmful to the environment in order to cut taxes on labour. But when the term 'fiscal' appears instead of 'tax' it means a deeper reform, although the scope of the reform could be broad. The terms Environmental Fiscal Reform (EFR), Green Fiscal Reform (GTR), or Ecological Fiscal Reform (EFR) are often used. The European Environment Agency (EEA: 84) establishes clear differences between the two sets of definitions:

"Environmental tax reform (ETR) is a reform of the national tax system where there is a shift of the burden of taxation from conventional taxes, for example on labour, to environmentally damaging activities".

"Environmental (or ecological) fiscal reform is a broader approach, which focuses not just on shifting taxes and tax burdens, but also on reforming economically motivated subsidies".

The Green Paper of the European Commission on the relaunch of environmental fiscal policy coincides with what has been pointed out: "An ETR that shifts the tax burden from negative taxes for welfare (e.g. on labour) to positive taxes for welfare (e.g. on environmentally damaging activities, such as the use of resources or pollution) can be a policy that greatly benefits both employment and the environment" (COM(2007)140 final). This double positive effect is called the "double dividend". But the EEA (2005) adds the requirement of fiscal neutrality. An ETR is the "shift of the fiscal burden on conventional activities to those related to the environment, but maintaining the neutral fiscal burden". But it is convenient to point out that such a policy cannot be the only instrument for environmental improvement. These instruments "do not substitute for but complement and strengthen regulatory and other approaches to fiscal and environmental management" (OECD 2005: 24).

In this paragraph the practise of tax reform is analysed in relation to the environment, and in the next paragraph we will go into more depth in the ecological fiscal reform.

Some European countries started to apply environmental taxes from 1990 and focused on energy fiscality. But in 1993 they only collected 1.5 % of the total of the EU-15 (EEA 1996: 8). In the OECD countries it barely reached 2 % of GDP at the start of the twenty-first century. In 1995, energy taxes amounted to 90 % of environmental taxes, the rest came from taxes on emissions, waste generation, chemical products, etc. In 1999 they reached 7.6 % of the tax collection of the EU-15, thanks to the special effort of Denmark and the Netherlands (OECD 2002a). In the EU-27 environmental fiscality accounted for 3 % of GDP in 2005. Denmark stands out with 6 % of GDP, although it is a sharp drop from the 9.5 % it reached in the year 2000. It is followed by the Netherlands with 4 %. At the other extreme we have four countries: Estonia, Romania, Spain and Lithuania with percentages slightly higher than 2 %. It is usually considered [as S. Speck does (2009)] that Denmark and the Netherlands, along with Finland, Sweden, Germany, the United Kingdom, Switzerland and the Czech Republic, are carrying out environmental fiscal reforms. But the latter occupied intermediate positions in the EU-27, meaning that such a treatment is not justified. The tax reforms in the EU-15 only managed to slightly reduce energy intensity and, much less, labour costs (EEA 2006: 30).

Environmental taxes are divided into four categories: transport, pollution, energy and other resources. Taxes on energy represented 3/4 of environmental tax revenue and around one twentieth of all fiscal revenue. Transport taxes represented 1/4 of environmental revenue (EEA 2008a: 22).

2.2 Concepts of Fiscal Reforms in Relation to Sustainability

The ETR seen in previous paragraphs has a scarce potential for transformation. As we have seen, for the EEA "environmental (or ecological) fiscal reform is a broader approach, which focuses not just on shifting taxes and tax burdens, but also on reforming economically motivated subsidies" and for these reasons it "offers more opportunities for progress" than the ETR (EEA 2005: 84).

But leading authors propose a more radical ETR. They defend the goal of transforming the economic system in order to bring it closer to sustainability. This transformation means that the most unsustainable activities are drastically reduced or even eliminated, and that on the other hand others that are now weaker or non-existent have been intensely developed. The ETR uses two instruments: it eliminates the subsidies that are perverse for sustainability and establishes escalating taxes, that is, they are increased annually until the goals pursued are achieved. This way the monopolistic conception of the monetary universe is broken and is placed at the service of achieving a set of physical sustainability goals. The escalating taxes are maintained until a solar and materially circular economy, sustainable transport, sustainable consumption, etc. are achieved. So it regulates the market, as the main function attributed to it, that of defining prices, would be deeply altered. Here the taxes will determine the main component of prices.

In addition, the ETR seeks to eliminate the tax categories that do not consider the use of resources as an activity that should be taxed: "In the European context, tax revenues represent over 40 % of gross domestic product. In the vast majority they are a derived form of taxation of human effort and only a small fraction is derived from environmental taxes (. . .) and particularly of the Value Added Tax (VAT), which is one of the worst taxes from an environmental and social point of view" (Paleocrassas 2002). This means that the use of raw materials is tax free and only labour and profits are taxed. Both the EDSs of the EU and the Sixth Environmental Action Programme proposed a "reform of the subsidies with a considerable negative effect on the environment" (European Parliament and Council 2002: 5). This proposal disappeared in the ERDS (European Council 2006: 5).

The ETR explained is an essential requirement to achieve sustainability, but it does not, as traditional economists think, come about because environmental taxes are the most efficient tool, but because the current tax system encourages the destruction of the planet. But the ETR can only be successful if it is part of a planned transformation strategy to make the productive system more sustainable, requiring the integrated application of multiple policies. It will fail if adequate alternatives to the activities subject to growing fiscal pressure are not offered: if business owners do not have alternative production methods at affordable costs, if car owners do not have an efficient collective transport system, etc.

2.2.1 Subsidies

The amount of perverse subsidies is huge and for this reason countries must reduce them as an urgent matter: "The reduction of perverse subsidies is that of primary importance" (Beherens 2004: 16). The IEA defines energy subsidies in general terms as "any government action that concerns primarily the energy sector that lowers the cost of energy production, raises the price received by energy producers or lowers the price paid by energy consumers" (UNEP/IEA 2002: 9). The OECD defines subsidies in a broader perspective: "any measure that keeps prices for consumers below market levels or for producers above market levels or that reduces costs for consumers and producers". This definition includes activities such as direct payments from public budgets towards the maintenance of certain sectors, exemptions or tax reductions, as well as rules that benefit certain economic players: preferential introduction to certain markets, limited liability (EEA 2007: 11).

Subsidies via budgets are usually the most important. In the EU over two thirds of transport subsidies have this origin, of which another two thirds go to roads. Among the most usual fiscal policies are: cancellation of taxes; tax reduction. Among those derived from regulations are: assuming most of the damage insurance costs in some types of private companies, such as in the case of nuclear plants; the sale of public resource extraction permits at low prices; restrictions on access to the electrical grid or competing companies (EEA 2007: 11, 12).

In the energy sector fossil fuels and nuclear energy received the most subsidies. Subsidies are given to prospection, extraction, consumption, research, the construction of plants, etc. Therefore, for example, in most OECD countries VAT on fossil fuels is only applied to private homes. International agreements eliminate taxes on hydrocarbons for aeroplanes and ferries. The countries that are introducing high taxes on energy do not apply it, or do so with strong reductions, to energy-intensive companies (OECD 2002a). The average electricity tariff for companies in the EU is less than twice as low as that applied to private homes (EEB 2002: 5). One of the most decisive subsidies is access to cheap capital. For example, in the 1980s loans for the construction of nuclear plants had an interest of 5 % in the US, when the official interest was 10–11 %. With this interest electricity prices should have risen by about 40 % (Roodman 1998: 82). A comparative study of subsidies for nuclear and wind power in the US, during the 15 years of development of the former, reaches the conclusion that nuclear energy received 30 times more than wind energy per kWh generated (EEA 2004: 16).

The OECD states that only 10 % of state R&D budgets go to renewable energies; the rest goes to conventional energies (European Commission 2004a, b, c: 38). The EEA (2002b: 18) states that in the EU, investment in R&D in the energy sector dropped during the 1990–1998 period, but that "it is still focused on nuclear energy". According to the AIE, wind energy has historically received 10 % of R&D investment, as opposed to the 60 % for nuclear. Although this inequality is changing, it is still enormous. A study carried out by the GAO, a US administration body, found that, during the 2002–2007 period, nuclear energy amounted to 54 %, fossil fuels 27 % and renewable energies 12 % (Levesque 2007). For the World Council of Renewable Energy (WCRE 2000) the annual subsidies to conventional energies reached in the world around $300–350 billion and the authors of the following table have reduced it to $240 billion. However, the AIE estimates that in 2008 subsidies to fossil fuels reached $557 billion. And a study by Bloomberg estimates the subsidies for renewable energies at $43–46 billion, of which one-half are biofuels (BNEF 2010). That is, the former are greater than the latter by a factor greater than 13 and, in addition, most of the subsidies for renewable energies go to the promotion of fuels, whose impact on the environment, society and on the production of food is increasingly evident. Chapter 13 deals with this issue.

D. Pearce (2002) estimates that agriculture subsidies amounts to $400 billion per year. The most important subsidies take place in fuels, phytosanitary products, chemical fertilisers, subsidies for exports, etc. In transport, roads receive a large majority of the subsidies. The free or subsidised use of roads and subsidies for goods transport fuels make up most of them. In 22 countries of the OECD the total subsidies for industry during the 1986–1989 period reached $66 billion (OECD 1997a: 20), to which we would have to add the $30 billion in subsidies for mining (Pearce 2002). The EEA (2007: 7) estimates that in the EU road transport subsidies reach $125 billion, of which $110 billion are for infrastructures. In addition, it estimates at 30 billion the generic subsidies for transport. The IEA (2012) estimates fossil fuels subsidies amounted to $523 billion in 2011, up almost 30 % on 2010 and six times more than subsidies to renewable. Taking into account that nuclear energy subsidies are 50–100 billion (IISD 2010), we can estimate that global total subsidies would be at the least in the region of $1. 5 trillion.

The first step to suppress perverse subsidies is to record them, something that has only been partially done to date due to opposition of the privileged sectors. And this happens despite the fact that the OECD states that worldwide elimination of direct subsidies for fossil fuels would mean a 6 % reduction in CO_2 emissions and a 0.1 % increase in economic growth (UNEP/IEA 2002: 16). The Directorate-General for Transport and Energy published in 2002 an inventory of the public aid given to different types of energy. Although "the abolition of such subsidies has long been demanded" (Beherens 2004: 16), until now no government has designed a broad plan to do it. Only some countries have been reducing them in some cases (pesticides, coal . . .). The RSSD of the EU mandated the Commission to carry out in 2008 an analysis of the subsidies and o ways to abolish perverse subsidies. In 2010 it had not yet published any results (Hontelez 2009).

2.2.2 Escalating Taxes

It is important to distinguish between escalating and progressive taxes. An escalating tax annually raises the tax burden on an activity until it is eliminated or drastically reduced. A progressive tax is one that is raised depending on the volume of consumption of a resource. Many municipalities raise the price of water as consumption increases. The United Kingdom is the only country that has used escalating taxes systematically. It applied them to landfills (in operation from 1998 to 2002) and in particular to fossil fuels. The Conservative Party implemented this in 1993 and it was intensified by the Labour Party until it was frozen in the year 2000. The initial escalating tax was 3 %, it was then raised to 5 % and in 1997 to 6 %. The revenue was used to reduce the social burdens of labour and to finance the health system. It also brought environmental improvement. But public opinion rejected it due to the growing cost of petrol and the government was forced to abolish the system when lorry drivers brought the country to a standstill by blocking the refineries (Green Tax Commission 2009).

2.2.3 Some Visions of Ecological Tax Reforms

Ernest U. von Weizsäcker, considered a prime reference in the defence of the ETR, proposes a simultaneous double process to reduce subsidies and to gradually raise the fiscal burden on unsustainable activities. These taxes would grow at an annual rate of 5 % during approximately 40 years, until they reached a revenue of 5–10 % of GDP, which would mean contributing 12.5–25 % to tax collection in the EU (bearing in mind that in the EU tax revenue reaches 40 % of GDP). This author seeks the transformation of many economic sectors (mining, the construction of electrical energy plants, nuclear energy, the manufacture of lorries, basic chemistry, cement plants and the manufacture of metals, etc.) through taxes on emissions, materials and products. As a result of this fiscal pressure, "long-distance freight transport, coalmining, non-recyclable container manufacturers, the construction of roads and part of the chemical industry are expected to experience an economic crisis" (Weizsäcker 1994).

The European Environmental Bureau (EEB) has been developing a campaign throughout the previous decade in favour of an RFE that has the following objectives: to abolish anti-ecological subsidies; to increase tax collection for environmental reasons by 10 %; and to destine it (mostly) to reducing the fiscal load on labour (it is estimated that it could lead to a 26 % reduction) based on distributing 50 % of the reduction between the employers' and workers' contribution and the rest to incentivising nature protection activities (Maro 2007). Some authors propose taxing quantities of total material requirements (a Material Input Tax- MIT), but only raw materials, where "macroeconomic models for Germany have shown that dematerialisation by means of MIT is possible without frictions in the system" (Beherens 2004: 19).

The increased mobility of people and goods and the growing use of means of transport with a higher impact make transport the sector that most urgently needs to be transformed. Establishing high taxes on land would slow down the current trend of proliferation of transport infrastructures. Escalating taxes on fossil fuels would lead to a reduction of mobility and to a shift towards more energy-efficient forms of transport.

2.3 *Overview of the Development of an ETR*

2.3.1 General Reforms

Denmark and Germany have both made attempts to advance towards an ETR very focused on energy (although the governments called them ETRs), and the results have coincided. In Denmark environmental tax reached 9.5 % of total revenue in the year 2000. Apart from high taxes on fossil fuels, it contains very high taxes on the purchase of cars, solid waste, wastewater, plastic bags and hazardous chemical products. The revenue allowed it to reduce taxes on labour from 27 to 25.5 % during the period from 1994 to 2000. The reaction of the public was negative, because they only perceived the rise in taxes. The conservative parties took advantage of the discontent, won the elections and kept the experience on hold (2001–2008). In 2005 revenue fell to 6 % and in 2009 it was 5 %. The new government plans to continue with the reform (Danish Ecological Council 2002). In Germany the reform took place between 1999 and 2009. The results have been: an increase in the demand for collective transport; sales of efficient cars greatly increased; the consumption of fuel dropped by 17 %; fossil fuel imports dropped by 13 %; industry reduced energy costs by 1 billion euro; etc. Public opinion also expressed its opposition to the reform (Cottrell 2009).

2.3.2 Land Tax

Classical economists Adam Smith, David Ricardo and John Stuart Mill agreed that, of the three factors of production (work, capital and land), land should be the most heavily taxed. The following text by Adam Smith is enlightening: "Both ground-rents, and the ordinary rent of land, are a species of revenue which the owner, in many cases, enjoys without any care or attention of his own. Though a part of this revenue should be taken from him in order to defray the expenses of the state, no discouragement will thereby be given to any sort of industry. The annual produce of the land and labour of the society, the real wealth and revenue of the great body of the people, might be the same after such a tax as before. Ground-rents, and the ordinary rent of land, are therefore, perhaps, the species of revenue which can best bear to have a singular tax imposed upon them. Ground-rents seem, in this respect, a more proper subject of singular taxation than even the ordinary rent of land (. . .) Nothing can be more reasonable than that a fund owing its existence to the good government of the state, should be subject to a special tax, or should contribute something more than the greater part of other funds, towards the support of that government." (2005: 693).

In the eighteenth, and in particular the nineteenth and early twentieth centuries, politicians who declared themselves physiocrats or who were simply influenced by Smith's ideas (and in the US by Henry George, a nineteenth century thinker) decided to turn land tax into the main financial resource of their governments. In some cases (Denmark, Japan—after the Second World War—, Taiwan, etc.) this tax was applied at the same time as the distribution of the land of big landowners among peasants took place. In France, the Republic that emerged after the monarchy was abolished obtained 80 % of its income from land tax. In 1830 it was still 30 %, and in 1980 it provided 13 % of income. In the early twentieth century, it was applied in States that were the product of colonisation (Australia, New Zealand, South Africa, etc.) and also in South America in States that were part of the Spanish Empire (Argentina, Colombia, Uruguay, etc.). Today this fiscal policy is maintained in many cases (especially in cities) and there are some signs of its resurgence. Between 1919 and 1986 half of the local governments of the metropolitan area of Melbourne abolished taxes on buildings and increased land taxes. The population density in these municipalities is 50 % higher than in the rest and even higher in the municipalities that started earlier (Roodman 1998: 122–123).

The state of Pennsylvania allows (based on a law from 1951) separate taxes on sites and buildings. Fifteen cities, including the capital, Harrisburg, have two separate taxes on sites and on buildings, with the former being taxed 4–16 times more than the latter. The capital and another town started the experience in 1913. Between 1979 and 1980 Pittsburgh increased even more the tax on land, with the aim of achieving the commercial revitalisation of the city centre. Washington D.C. and many municipalities of the state of Virginia started to apply this policy in the 1990s. All Australian states, except Victoria, tax land, although the weight of this policy is being reduced. In 1991, 90 % of the municipalities of New Zealand taxed land. In 1990 the Estonian government decided that the easiest tax system to apply was that based on income from the rental of land, whose public property it had inherited from the Soviet regime, reaching up to 95 % of its income (Roodman 1998: 124; Smith and Nelson 1998).

The defenders of this fiscal policy claim that it has many advantages: it improves the use of the land, it increases economic activity, it generates a high density urbanism, it improves the sustainability and improves the redistribution of income. Land tax does away with speculation with land by discouraging land grabbing. There are many studies that prove that, when this policy is applied, economic activity and employment increase. It is considered that the application land tax has allowed turning Johannesburg (which was a mining town) into South Africa's economic activity hub, displacing Cape Town, despite the fact that its port is one of the best positioned in the world. It promotes the creation of denser cities and encourages the modernisation of buildings, which involves less use of resources and greater labour intensity as opposed to new construction. The best distribution of wealth takes place particularly when the price of housing is made cheaper. Studies on the towns of Pennsylvania that apply this policy showed that 75 % of the population pays less than before (they are the groups with the least income), 20 % remains the same and 5 % pays more. Taxes on agricultural land allow a more intensive use of it. The ageing of the farming population, faced with the refusal of the new generation to continue with the activity of their parents leads to abandonment of the farms. Taxes on land forces them to sell it or rent it, allowing young people who want to carry out this activity to do so. Last of all, factors that improve sustainability are: reduced use of materials in the modernisation of housing them in new constructions, the creation of dense urban planning and pressure to abandon unproductive land, which allows the restoration (through support policies) of the habitats that existed in these places. Despite these advantages, the power of the landowning and construction oligarchies frequently manage to abolish or minimise this fiscal policy (Smith and Nelson 1998: 10–12).

3 Forecast Evolution of the ETR

In view of the analysis carried out it seems very unlikely that States will apply an integral ETR, although it is possible that some partial measures will be implemented: a reduction of some perverse subsidies and an increase of benefactors, which is what is happening with renewable energies; the slow extension of land tax; an increase of tax on some scarce resources or those with large environmental impacts associated to their use, as is the case of coal; an increase in taxes on road transport; etc. This forecast is based on experience. The fact that the relative weight of environmental tax has decreased in the last decade in Europe, and that the same has happened with land tax at a global scale (except for some cases of progress), indicates the difficulty of advancing in the ETR. In the EU, the requirement of unanimity to approve any fiscal change is an insurmountable obstacle for the progress of an ETR.

However, very important elements of an ETR have started to be produced by means of escalating prices of scarce resources. It has already happened with fossil fuels (in particular with oil), with a large part of strategic materials (materials that are essential for the most important technologies) and food, reflecting the growing scarcity of land and water. The scarcities indicated will determine new price hikes in the future. This is the result of a study by the Green Fiscal Commission (2010), which suggests three scenarios. Scenario B1 suggests that the escalation of oil prices (which in 2008 reached $147 per barrel of oil) continues very gently until reaching $170 in 2020. B2 suggests a fall in the price until reaching $70 in 2020. And B3 is one in which the rise in oil prices continues rapidly until reaching $500 in 2020. In scenarios B1 and B2 the energy goals described in Directive 20/20/20 are not achieved, meaning that an ETR characterised by a displacement of taxes is proposed. The study concludes that the results are very positive in the cases of fiscal reforms and a very negative in scenario B3, due to inflation, the enormous sums of money transferred to oil exporting countries and the ensuing crisis.

Chapter 9
Science and Technology for Sustainability

Keywords Dominant paradigm of science and technology • Rebound effect • Science for sustainability • Sustainable technological systems • Bio-mimetic technologies

In this chapter we will briefly describe some of the key aspects of a new system of science and technology. First of all we will offer a critical vision of the dominant paradigm, then define the bases of the paradigm of sustainability, and finish with an analysis of the technological concepts and models that are presented as contributions to sustainability.

1 Criticism of the Paradigm of the Current Dominant Science and Technological Model

In recent centuries there has been an enormous development of science and technology that has allowed us to know much more about the birth and development of life on Earth, how nature works, genetic codes, the universe and its evolution, subatomic physics, etc. Many technologies have brought about significant advances in the well-being of societies, such as electricity, telecommunications, transport, some productive technologies, etc. But there are many patent social and environmental impacts that cause an evident reduction in our quality of life (atmospheric and acoustic contamination, hazardous waste, the deterioration of nature, etc.), in addition to the depletion of resources. These impacts show the unsustainability of the dominant science and technology system.

In Chap. 1 we saw that this civilisation is the first that we could consider secular, but it has enthroned a new god: the science-technology system, though in reality the gods would be the scientists and technologists that develop it. The dominant science paradigm establishes that the facts on which science acts are indisputable

R. Bermejo, *Handbook for a Sustainable Economy*,
DOI 10.1007/978-94-017-8981-3_9,

evidences and that, in order to solve the problems that arise, science applies a rigorous methodology that determines appropriate solutions, meaning that scientific development is free from any ethical or regulatory aspect. Different fields of science are perfectly delimited and each science progresses without the need to co-operate with other sciences. The development is accelerating of four scientific-technical fields (nanotechnology, biotechnology, renewable energy capture technologies and information and communication technologies). But the first two pose a serious threat to life on the planet.

In addition, biotechnology also has profound repercussions on the dominant paradigm. J. Benyus (2009) believes that "with the advent of genetic engineering, some of us have come to fancy ourselves as gods, riding a juggernaut of technology that will grant us independence from the natural world". An opinion confirmed by Allenby, though not as criticism, but as the advent of a new civilisation full of promise. Nanotechnology, biotechnology, ICT, and applied cognitive science "in some ways are the logical end of the chapter of history that began with the Greeks 2,500 years ago" (Allenby 2009: 173). This author thinks that "increasingly biodiversity becomes a product of design choice and industrial and political imperatives (security issues, e.g.) rather than evolutionary pressures", to the point that as "the behaviour of biological systems increasingly becomes a function of human dynamics and systems the more it requires an understanding of the relevant human systems". The outcome of this thinking is that "biotechnology makes us 'lords of the biosphere'" (Allenby 2009: 169, 174).

The OECD and the US National Intelligence Council present a technocratic vision, less ideological, but very optimistic too. Like in the 1950s and 1960s, when we were told, under the slogan Atoms for Peace, that nuclear energy would provide us with unlimited, cheap and even free energy, now a report by the US National Intelligence Council announces "significant improvements in human quality of life and life span [...] with accompanying prosperity and reduced tension" (Anton et al. 2001: XI, XII). The OECD report "The Bioeconomy to 2030" declares: "Biotechnology can increase the supply and environmental sustainability of food, feed and fibre production, improve water quality, provide renewable energy, improve the health of animals and people, and help maintain biodiversity by detecting invasive species" (2009b: 8).

This optimistic vision is not tarnished, in the opinion of the authors, by their admission that such a revolution will generate wider gaps between classes, reduce privacy, generate cultural threats, the danger of an enormous ecological catastrophe due to genetically modified organisms, etc. (Anton et al. 2001: xviii). Allenby (2009: 173) acknowledges that technological evolution is nearly impossible to stop, and difficult to moderate: "technological evolution is unlikely, if not impossible to stop, despite the efforts of many in the environmental and sustainability communities", and "whether and how technological systems can be moderated in the age of global elites become important research questions".

The development process of these technologies follows a pattern similar to that of genetically modified organisms (GMO), agro-fuels or many developments in nanotechnology. Companies identify technologies that maximise their profits, they start to make strong investments in R&D, then they carry out massive campaigns to convince society and in some cases producers (such as farmers). Last of all, and when social opposition grows, they put pressure on governments to establish a regulatory framework that exempts them from the collateral effects that societies will suffer (Wakeford 2003). On the other hand, the UCS (2009: 33, 34) declares in an analysis on the capacity of transgenic varieties to increase production that, "genetic engineering has delivered only minimal gains in operational yield (...) Most yield gains are attributable to non-genetic engineering approaches", despite "the tremendous resources being devoted".

Nanotechnology is the manipulation of particles (nanoparticles) (NP) smaller than a micrometre: <0.000001 m. It is expected to have a massive application in computers, materials science, medicine, the production and storage of energy, etc. It is already being applied in cosmetics, clothing and sports material without any information or monitoring. The health and Consumer Protection Directorate General organised the "Mapping out Nano Risks" workshop which reached the following conclusions; among others it:

- "highlighted that some engineered nanoparticles produced via nanotechnology may have the potential to pose serious concerns";
- "revealed that panel experts were of the unanimous opinion that the adverse effects of NPs cannot be predicted (or derived) from the known toxicity of bulk material";
- "exposed the limits that preclude a complete risk assessment today, in particular, the present scarcity of dose-response and exposure data" (European Commission 2004c: 11).

Allenby and Rejeski (2008: 269) state that, "despite early calls for a life-cycle approach to nanotechnology development, proactive management of emerging risks, and the greening of the product infrastructure, little has happened".

Despite this, the official discourse is that we are heading towards a nature-friendly post-industrial society thanks to the dematerialising potential of technologies such as, for example, microelectronics and nanotechnology, and the high efficiency reached in the OECD is boasted about. Respect for nature is in contradiction with being 'lords of the biosphere'. This self-praise about the OECD's efficiency does not take into account two factors: the dismantling of most of the heavy industry within the OECD and that reality has not shown the dematerialising role of the electronic industry. As for the first factor, we must take into account that the OECD imports the products that it used to produce itself (steel, aluminium, ethylene, etc.) and does not take into consideration the enormous amount of energy that has been invested in their manufacture. The OECD acknowledges that the effects of these structural changes are more important than the environmental policies themselves: "In particular, the structural shifts toward services in the OECD countries, and toward industry in developing ones, seem to have been more important than changes in environmental policies themselves" (OECD 1997c: 37). On the other hand, the predictions concerning the dematerialising capacity of new technologies have not only failed to become a reality, but have caused an increase in the consumption of resources. Although there are cases of intensive dematerialisation (25 kg of optical fibre provide the same service in telecommunications as a ton of copper cabling), it is subject to cycles, as a result of which materials that at one moment are in much less demand than in the past, are later in demand again. Copper is increasingly used due to the promotion of electrical technologies in transport, for example. It was suggested that new telecommunications and software technologies would drastically reduce the use of paper, and the opposite has happened. Nor has the written press been replaced by online information and electronic trade generates more energy consumption and packaging than conventional trade.

We are also given the example of miniaturisation in microelectronics, but the manufacture of increasingly powerful semiconductors requires making increasingly pure silicon wafers. This means the need for increasingly bigger treatment and ventilation installations and that an increasingly higher amount of waste is generated. In order to manufacture a 32 MB DRAM chip (that is now considered outdated) that weighs 2 g we need 1,700 g of resources. And it is considered that the new ones (VLSI) need more resources. This industry is one of the ones that uses the most technical products. They use hundreds of tons of chlorine, acids, solvents, etc., and generate enormous amounts of waste. In what is known as "Silicon Valley" (Santa Clara Valley, California) there are more dumping sites for hazardous waste than in any other county in the US, and the aquifers have been contaminated (Korowicz 2010: 25).

The big corporations create technological monocultures that are formed by a dominant technological system in each field, of a planetary nature (and, therefore, centralising and not adapted to local circumstances), capital-intensive with few labour requirements, and unsustainable. We have an example in the case of detergents. Companies design a single type that works in the case of water quality that is the least suited for washing (producing the highest environmental impact possible), instead of making detergents adapted to the water quality of each area and taking into account the specific environmental impacts (Curtis 2003). The need to reach or maintain a scientific-technical lead involves promoting rapid developments, often with large environmental and social impacts. The new technological systems are introduced without defining the safety limits, without any information whatsoever about the resulting products. Achieving technological leadership is put before all these aspects. Governments usually support this behaviour in companies: "When we wrap up the national flag around any technology in a global race to the top, we can quickly kill the kind of analysis that is critical to the early warning of future social and environmental problems and systems failures" (Allenby and Rejeski 2008).

The arrogance of the scientists who defend the dominant paradigm is negated by sciences such as physical cosmology, quantum physics, evolutionary biology and the new field of research into conscience, which have discovered unexplainable phenomena that undermine the solidity of the theories in vogue. Cosmology cannot explain the amazing precision and constance of the parameters shown by the universe as it spreads through space and time. To the point that if these parameters had been even slightly different, the universe would not have existed. Quantum physics cannot explain why the subatomic world shows the same characteristics as the universe, in addition to "entanglement", which prevents a separate analysis, and which leads to phenomena such as when a particle, faced with an insurmountable obstacle, disappears in front of it and reappears behind. Biology has found an extraordinary coherence, that it is unable to explain, between the parts of a living organism and between the latter and its environment, enabling it to adapt to it. Last of all, it remains unexplained that certain native tribes are able to communicate using telepathy and that some images, ideas and archetypes appear recurrently in civilisations that have not had any communication among them (Laszlo 2007: 19–33).

Faced with these phenomena, scientists established within the dominant paradigm claim that they are anomalies that do not question the basic make-up of the scientific system. Others state that they can only be explained from a new paradigm of physical and life sciences. What Ervin Laszlo (2007) defines as "An Integral Theory of Everything" (which is what Einstein sought without success). This author describes two approaches in the words of two scientists. Cosmologist Steven Weinberg states: "I believe that what we have found so far (an impersonal universe which is not particularly directed to human beings) is what we are going to continue to find. And that when we find the ultimate laws of nature they will have a chilling, cold, impersonal quality about them".

On the other hand, physicist David Peat states that leading researchers accept the challenge of finding the meaning of life through science: "We are constantly confronted by a series of overwhelming questions: What is the nature of the universe and what is our position in it? What does the universe mean? What is its purpose? Who are we and what is the meaning of our lives?" (Laszlo 2007: 2, 3). Scientists like him defend risky positions by exploring other alternatives, exposing themselves to discredit and academic marginalisation, though, as they further their knowledge, their theories are becoming increasingly solid and the general approach is being strengthened (Laszlo 2007: 16).

The development of science and technology is to a large extent determined by the interests of the centres of power. The Bush period is characterised by taking this policy to the extreme. A report by the Union of Concerned Scientists (UCS 2008: 26, 33) in the US makes this clear in a chapter entitled "Patterns of Abuse", the thematic index of which speaks for itself: falsification of data and fabrication of results; selective publication of documents and the creation of false uncertainties; rigging of scientific procedures; intimidation of and pressure on scientists; censorship and expulsion of scientists; failure to fulfil legal mandates on the use of science when making decisions in some cases; the creation of conflicts of interests by placing people from private companies in charge of organisms who have interests in how these organisms position themselves, corrupting scientific advisory commissions. The same can be said about the behaviour of corporations. The UCS (2012: 13–30) accuse them of "corrupting the Science", "Shaping the Effectiveness of Federal Agencies", "Influencing Congress", and "Exploiting Judicial Pathways".

But the EU and the OECD also carry out manipulative practices. The EU has ignored the many reports by its Joint Research Centres about the enormous impacts of the production of agro-fuels. The OECD defends the goodness of so-called biotechnology, without the slightest reference to the problems that some lines of

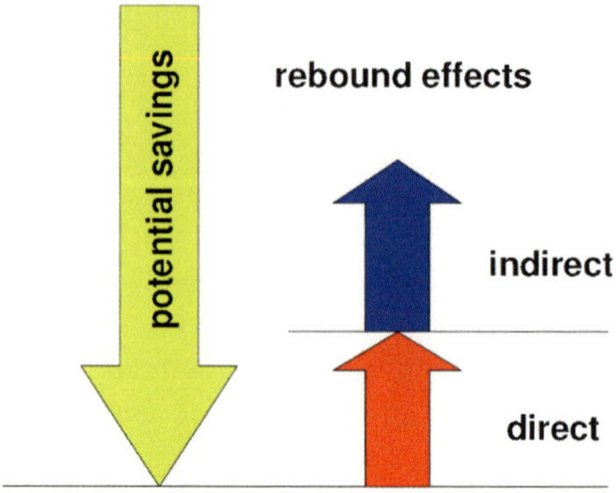

Fig. 9.1 The rebound effect (Source: Schettkat 2009)

research generate (OECD 2008a). But its Environmental Directorate acknowledges that globalisation inflicts pressure on the environment: "The growing extent, intensity, and velocity of global interactions can be associated with their deepening impact such that the effects of distant events can be highly significant elsewhere" (OECD 2008b: 20). Though it may seem contradictory that it considers that a reversal of globalisation will not solve the environmental problems.

To finish this section, we will focus on two aspects of technological development that show it can only be, in the best of cases, part of the solution: the rebound effect and planned obsolescence. Two aspects of the rebound effect are normally distinguished, which are defined, on the one hand, as the direct or micro-rebound, and on the other, the indirect or macro rebound. Figure 9.1 shows both effects.

The micro-rebound or direct rebound refers to the behaviours derived from a reduction in consumption of a resource due to the higher efficiency brought about by improved technology, which causes an increase in the consumption of this same resource. For example, the reduction of the energy costs of a house, due to the increased efficiency of household appliances, sets off collateral effects that reduce its effectiveness, such as raising the average temperature of the household heating or the number of light bulbs that are turned on. The OECD (1998: 52) states: "If consumers spend less on these (eco-efficient) products and services, they are likely to spend more on other goods and services, again leading to a net economic stimulus". It has been calculated that a 10 % improvement in the efficiency of vehicles stimulates an increase in traffic of 1–4 %. The increased efficiency brought by mobile telephony as opposed to classic land lines is compensated by the massive use of mobiles. The macro-rebound or indirect rebound refers to the general and vague effect whereby a reduction in expenditure leads to increased consumption in other areas of the economy. Some authors estimate that the rebound effects could reach up to 40 % of the energy saved. According to other authors, the effect of the macro-rebound is at least 10 % and often exceeds 50 %, and the micro-rebound is probably less than 30 % in the heating and air conditioning of houses and around 10 % in transport. All this can be observed in OECD countries (Schettkat 2009).

So this author estimates that the rebound effect reduces by more than 50 % the savings in resources produced by improved efficiency.

Last of all, the rapid change of products (and often of the associated technology) is achieved by designing products with rapid obsolescence, which is a structural feature of waste, of unsustainability. There are three types of obsolescence:

- Technical obsolescence is understood as the design of products so that they have an increasingly shorter lifetime. Another collateral element that has the same result is the business decision to cease to produce components, making repairs impossible.
- Functional obsolescence happens when a new product appears on the market that does the job better than the old one. In almost all cases, the functional improvements are not very relevant or only affect a very small number of highly specialised users. Most computer users are forced to frequently upgrade them even though their use of is modest and stable. The reason is that new software packages include more functions and require increasingly powerful computers to run. So the change of basic technologies produces a rapid obsolescence of all the technologies of each technological trajectory.
- Psychological obsolescence happens when fashions change. In this case functionality and duration become irrelevant. This is the main cause of obsolescence in many products: clothing, footwear, household furniture, decoration, personal aesthetics, etc.

2 Towards a Science for Sustainability

The science of sustainability rejects the dominant paradigm and is based on another that changes the vision of the position of the human species on Earth and in the universe. We must get down off the pedestal from which we think we rule over nature, which is supposedly at our service, to see ourselves as just another species: "The core idea is that nature, imaginative by necessity, has already solved human problems (. . .) The conscious emulation of life's genius is a survival strategy for the human race, a path to a sustainable future" (Benyus 2009). The development process of the science and technology system for sustainability requires the fulfilment of the following premises: informative transparency, democratic decision-making, geared towards the universal satisfaction of essential needs in a sustainable way. Informative transparency means that the population receives all the relevant and summarised information that exists in society about the problems of unsustainability, its levels of seriousness and about the most adequate actions to solve them. The population participates in the decision-making enriched with the flow of information. In order to evaluate the sustainability of innovations, new production processes and new productive systems it is necessary to contrast them with the natural model, because nature is the measure of sustainability (Benyus 2009). First of all we will analyse the requirements so that science is at the service of society, to then end by studying the science of sustainability.

The International Council for Science (ICSU), which is a non-governmental organisation that represents 103 national organisations and 27 international organisations outlines in the report "Science and Society: Rights and Responsibilities" its vision of scientific activities in five aspects: flows of science and scientists; the production of scientific knowledge; the speed of innovation; the government of science and technology; the relationships between science and society. The ICSU defends the need to guarantee "the mobility of scientists and the free flow of science", because "traditional threats to mobility (. . .) continue in many areas of the world in the form of state discrimination against scientists and repression of research and communications". On the other hand, "the growth of research in (or sponsored by) the private sector raises particular questions about the ethics of conducting and communicating science in industry". These problems and the fact that "science can no longer be seen as intrinsically pure (if indeed it ever was so) but rather as serving many masters" generate a greater concern about the fact that "peer review (. . .) needs to be supplemented by additional processes that secure more open communication and effective criticism of scientific results bearing on public health, safety and welfare" (ICSU 2005: 12–15).

The higher speed of innovation increases its own risks. Some scientists have reached the conclusion that we are living in a "'risk society', in which everyone (. . .) is always to some degree at risk" (. . .) The growing distance between producers and users (. . .) have rendered inadequate the system of regulatory checkpoints through which national authorities historically control risks (ICSU 2005: 11–16). This problem is particularly serious in the case of transnational risks: "This implies that there is a need for International institutions to participate in identification, analysis and management of cross-border risks (. . .) resulting from innovations in science and technology" (ICSU 2005: 16).

One scientific school of thought defines the alternative paradigm as post-normal science, following Funtowics and Ravetz (1993). It is based on various premises: we are immersed in complex systems; there are large knowledge gaps; the scientific establishment is not the only source of knowledge; and scientific-technical development is conditioned by ideological factors, and economic and political interests. There are huge knowledge gaps, particularly regarding how nature works, and a high level of uncertainty concerning the effects of our actions on the natural system, on health, on future generations, on social cohesion, on the economy, etc. In societies there is a broad plurality of values, of concerns, of desirable goals, of different ways to achieve them and of criteria to justify actions. In addition, science no longer appears as the sole source of knowledge. Many groups of humans (aboriginal, peasant, fishing populations, etc.) have developed knowledge and ways of acting (in medicine, sustainable production, biotic materials, the ability to predict changes in the environment, etc.) that constitute a very enriching scientific heritage. For this reason, science cannot have a monopoly on the definition of changes, but rather all social agents are experts in different ways and in relation to the different aspects of each problem, and are legitimised to participate in the definition of solutions and in their application: "Citizens (as well as scientists) become both critics and creators, providers and recipients in the knowledge production process" (Spangenberg and O'Connor 2003: 5). Sustainable development requires "the incorporation of knowledge generated endogenously in particular places of the world. This represents a great opportunity to use inputs from other forms of knowledge, by exploring the practical, political an epistemological value of traditional, local, empirical or indigenous knowledge" (Modvar and Gallopin 2005: 8). This need is recognised at an international level: "Numerous International treaties and agreements have recognised the need to include holders of local, indigenous and traditional knowledge in treaty implementation" (ICSU 2005: 17).

NGOs have a growing influence on the decisions of governments and international bodies. Their work has led the World Bank to reconsider its investment policy. The protests of those affected by the system of dams of the Narmada river (India) and the confirmation of its enormous environmental impacts led the WB to withdraw its funding and to the creation of the World Commission on Dams. They were also decisive to make Monsanto desist from selling its "Terminator" seeds, whose degradation prevented farmers from reusing them. For these reasons "bioethics has emerged as a recognized field of study and substantial research programs on the ethical, legal and social implications of the human genome project have been established in many countries" (ICSU 2005: 19 and 20).

It is necessary to move away from the sealed compartments that scientific fields have been locked into. Anyone who does research realises that in order to have an understanding about multiple aspects of each scientific field it is necessary to look to other sciences. This realisation leads (apart from the incursions of each researcher into knowledge on the aspects of other sciences that affect their own field) to the creation of transdisciplinary research groups, because only they can offer alternatives to the complex systems that integrate social, economic, ecological and other aspects. But always bearing in mind that societies must be consulted. A democratic society is, among other aspects, one that is capable of deciding what fields it considers to be more beneficial (Spangenberg and O'Connor 2003). The CEPAL reaches the conclusion that what is needed is "the involvement of scientists and technologists in broad processes of consultation and dialogue with relevant stakeholders", and states that "the change of direction was officially recognized at the United Nations Conference on Environment and Development (...) and reconfirmed at the World Summit on Sustainable development in Johannesburg" (Modvar and Gallopin 2005: 9, 23). Many groups have been created to analyse the consequences of genetic engineering and nanotechnology. Some governments have been sensitive to such demands. In 2003 the British government organised a public debate to evaluate the risks and advantages of transgenic foods; it created three groups formed by experts and representatives of the public. As a result of the debate, the government adopted a much more cautious policy. Denmark has very successfully institutionalised the "Consensus Conferences"; participants represent the Danish social fabric in a balanced way and receive information from experts on the subject dealt with. Based on this they reach the broadest consensus possible, which has a big repercussion on society and has even led to laws being passed (ICSU 2005: 17, 18).

Most of all, the approach of post-normal science must be complemented with the biomimetic vision of scientific-technical development. Nature has been developing for over 3 billion years and has been capable of sustaining 30 to 100 million species by designing a variety of much more efficient techniques than ours, that are also perfectly sustainable. The Biomimicry Institute, created by Janine Benyus, is the theoretical and practical reference of the vision of imitating nature. In 1998 it popularised its theories with the book Biomimicry. Innovation Inspired by Nature, J. Benyus states that no species that appropriates all the available resources can survive and that this is what the human species is doing: "The most irrevocable of these laws (the ecological laws) says that a species cannot occupy a niche that appropriates all resources-there has to be sharing. Any species that ignores this law winds up destroying its community to support its own expansion. Tragically, this has been our path" (1997: 5). Biomimicry is a paradigm shift: "Biomimicry introduces an era based not on what we can extract from organisms and their ecosystems, but on what we can learn from them. This approach differs greatly from bio-utilisation, which entails harvesting a product (. . .) It is also distinctly different to bio-assisted technologies, which involve domesticating an organism to accomplish a function (. . .) they are inspired by one idea". Instead of using the Earth as a source of raw materials, we should see it as a permanent and inexhaustible source of knowledge: "Unlike the Industrial Revolution, the Biomimicry Revolution introduces an era based not on what we can extract from nature, but on what can learn from her". Many civilisations have disappeared after exhausting their resources: "The real survivors are the Earth inhabitants that have lived millions of years without consuming their ecological capital" (Benyus 1997: 2, 9, 2009).

J. Benyus (2010: 3) describes three levels of biomimicry. The first level is "the mimicking of natural form. For instance, you may mimic the hooks and barbules in an owl's feather to create a fabric". The second level is "the mimicking of natural processes. The owl feather self-assembles at body temperature". The third level is "the mimicking of natural ecosystems".

The imitation of mature ecosystems has profound implications: "nature would be model, measure, and mentor". This means:

"Nature as model. We would manufacture the way animals and plants do".

"Nature as measure (. . .) we would look to nature as a standard against which to judge the rightness of our innovations"

"In the end, I think biomimicry's greatest legacy will be more than a stronger fibre or a new drug. It will be gratitude, and from this, an ardent desire to protect the genius that surrounds us" (2010: 7).

For this reason it is important that the ICSU and the IGFA (International Group of Funding Agencies for Global Change Research, which is a forum of national agencies with fields of research of mutual interest) have agreed to finance the Earth System Science Partnership, "which brings together researchers from diverse fields, and from across the globe, to undertake an integrated study of the Earth system: its structure and functioning, the changes occurring to the system, and the implications of those changes for global and regional sustainability" (ICSU and IGFA 2008: 5).

All this means a radical change in the scientific paradigm and in the status of sciences. Earth sciences must fulfil a central role when defining the problems of unsustainability and to validate the solutions depending on their contribution to solve the problems. And, ultimately, ecology (as the science of synthesis) should give the final validation based on comparing the innovations with nature.

3 Towards Sustainable Technological Systems

The process of technological change is usually divided into three stages: invention (the development of something new), innovation (the development of a product capable of being commercialised) and diffusion (the commercialisation of the innovation). Although the importance of basic research (the first stage) cannot be ignored, the main obstacles are in the other two, particularly in diffusion. Even if the innovation is very good, its diffusion cannot be taken for granted. And the question is that, among other reasons for failure, individual technologies act within a technological trajectory, so that diffusion can only be the product of a systemic change.

When approaching technological transformation we must bear in mind that we are not talking about isolated technologies, but about integrated technological systems. Each technological system includes knowledge, productive equipment, products and services as well as organisation and management (UNEP 2003). But in these systems there is also a hierarchy and, in it, the highest position corresponds to the main technologies (core technologies), that is, technologies that can be incorporated to many products and processes (microelectronics, new materials, new energy technologies, etc.). By technological trajectory we mean a system that contains a vision of the world or paradigm, relationships between producers and users, productive technologies, transport, repair, waste management, institutions with standardised regulations, a financial regime, insurance policies, etc. (Hall and Kerr 2003). Microprocessors, for example, are not used in isolation, but within a system of communication, information and data processing that includes hardware and software, cable and optic fibre networks, telecommunications satellites, the Internet, etc. For this reason, a technological policy focused on sustainability must direct research towards the development of new technological trajectories and incorporate them as a whole (integrated transport, energy and other systems) and facilitate their rapid diffusion.

Despite the fact that technological optimism is prevalent in our society, it is very frequent that from government institutions (and in particular from the environmental department) they insist that the prevailing system of production and consumption is unsustainable: “Decoupling environmental degradation and resource consumption from economic and social development requires a major reorientation of public and private investment towards new, environmentally-friendly technologies. The sustainable development strategy should be a catalyst for policy-makers and public opinion in the coming years and become a driving force for institutional reform, and for changes in corporate and consumer behaviour" (European Commission 2001d: 2, 3). Chapter 3 of the Implementation Plan adopted by the World Summit in 2002 is dedicated to "Changing unsustainable patterns of consumption and production", which acknowledges that "fundamental changes in the way societies produce and consume are indispensable for achieving global sustainable development" (United Nations 2000: 7). The requirements of a sustainable technological system are even often defined with notable precision. A Report by the Expert Group on Sustainable Production commissioned by the European Commission (2001a, b, c, d, e: 7) defines it as one which has the following characteristics: "sustainable use of renewable resources and renewable energy"; "management of non-renewable resources, for example in closed material loop systems"; "maintenance or restoration of ecological and environmental systems"; "minimisation of transportation needs". But this is a typical text of the general documents of the EU and the more specific they are, the less relevance sustainability has (Spangenberg and O'Connor 2003). In addition, it is still considered that technology is capable of solving by itself the problems of unsustainability. The European Commission "Report of the Technologies Action Plan" (2007a, b, c, d, e: 3) states: "Through appropriate measures, e. g., financial support or regulation we can support eco-innovation and steer market forces towards a world-leading economy that is both competitive and green".

3.1 Overview of Some Concepts and Technologies Which Are Placed in the Field of the Dominant Paradigm

We think that the concepts and technological groups that we analyse here are within the dominant paradigm because they are focused on achieving incremental improvements: reducing contaminant emissions, increasing efficiency in the use of raw materials, increasing the use of recycled materials, etc. In addition, everything is put at the service of economic growth.

3.1.1 Concepts

Eco-innovation

The OECD (2009c: 11) defines eco-innovation as the "production, assimilation or exploitation of a novelty in products, production processes, services or in management and business methods, which aims, throughout its life cycle, to prevent or substantially reduce environmental risk, pollution and other negative impacts of resources use (including energy use)". The definition by Reid and Miedzinski is similar, but includes competitiveness and the satisfaction of human needs: "The creation of novel and competitively priced goods, processes, systems, services, and procedures designed to satisfy human needs and provide a better quality of life for everyone with a whole-life-cycle minimal use of natural resources (materials including energy and surface area) per unit output, and minimal release of toxic substances" (Wuppertal Institute et al. 2008: 15).

Eco-innovation is an incremental, divided-up and non-systemic concept: environmental improvements are sought in the productive system, but without questioning it, as a step prior to its transformation; and it only covers the field of the so-called environmental technologies. Despite this, the EU has launched quite a large number of eco-innovation programmes. The most important are the Eco-design directive, the Competitiveness and Innovation Framework Programme, the Environmental Technologies Action Plan (ETAP), and the Directive on the energy performance of buildings. The Eco-design directive is focused on increasing the energy efficiency of products. The Competitiveness and Innovation Framework Programme (CIP) is aimed at companies and has a variety of programs, among which is eco-innovation, representing one fifth of the Programme's funding, while the sub-programme of Entrepreneurship and Innovation obtains around 60 % of the funding. It is estimated that the 7th programme (the most important, by far, as regards funding) assigns around 30 % of its budget to eco-innovation (around ten billion euro), but includes actions (agro-fuels, agro-refineries and CO2 sequestration) that do not contribute to sustainability. The ETAP "aims to harness their full potential to reduce pressures on our natural resources, improve the quality of life of European citizens and stimulate economic growth" (European Commission 2004a, b, c: 3).

But the Report by the Environmental Technologies Action Plan (2005–2006) acknowledges that there has been little improvement, and that in order to make a significant improvement "much greater levels of deployment and take-up of environmental technologies are required at EU and global level (. . .) There is no time for complacency" (European Commission 2007e: 5). This text is significant, above all, because it defends the view that eco-innovation must be the rule in all aspects of the economy, but there is no analysis of the causes that prevent such a goal from being met.

One reason behind the poor result is the lack of coherence in the EU's policies. We see that the 7th Programme is far from making eco-innovation the dominant driving force. The Directive on energy efficiency only establishes minimum efficiency requirements and does not contemplate strong measures to improve the energy efficiency of already existing buildings, despite the fact that they consume 40 % of the EU's final energy (Wuppertal Institute et al. 2008).

Cleaner Production

The UNEP created in 1989 the cleaner production concept. It was looking for a more general and integrated approach to the changes in productive processes that would overcome the restrictions of the various techniques that had been used until then, such as "waste minimisation", "pollution prevention", "reduction at source", etc. The UNEP states that the concept "offers an approach more systemic and holistic" than eco-efficiency. And that "cleaner production is the continuous application of an integrated preventive environmental strategy applied to processes, products and services to increase eco-efficiency and reduce risks for humans and the environment". Cleaner production includes the preservation of raw materials and energy, the elimination of toxic or raw materials and the reduction of the amount and impact capacity of all the emissions and waste before they are withdrawn from the process. So we are still in the field of incremental improvements.

Eco-efficiency

The term 'eco-efficiency' was coined by the World Business Council for Sustainable Development (WBCSD) in its 1992 publication 'Changing Course'. It is based on the concept of creating more goods and services while using fewer resources and creating less waste and pollution (www.wbcsd.org). It soon came into vogue in the 1990s and is now very used in relation to the theory of the dematerialisation of growth or when speaking about integrating the environmental variable in industry. Today it is studied in universities and the UDEP and the OECD hold conferences on the subject. It is being discussed at numerous international fora (SDC of the UN, ISO, UDEP, OECD, WTO, etc.). But it is interpreted in very different ways, which generates much confusion.

The WBCSD states that "eco-efficiency is achieved by the delivery of competitively priced goods and services that satisfy human needs and bring quality of life, while progressively reducing ecological impacts and resource intensity throughout the life-cycle to a level at least in line with the Earth's estimated carrying capacity" (WBCSD 2000). The OECD report on "Eco-efficiency" states that this term "can be considered as a ratio of an output divided by an input: the 'output' being the value of products and services produced by a firm, a sector, or the economy as a whole, and the 'input' being the sum of environmental pressures generated by a firm, sector or economy" (OECD 1998: 15). But there are other meanings: "Some companies and governments use the term so that it is virtually synonymous with 'cleaner production' (. . .) Other companies have used eco-efficiency in a way that brings new meaning, in particular emphasising the dynamics of innovation in technology and organisation" (OECD 1998: 69).

Therefore, eco-efficiency is an indicator, a concept, a strategy for ongoing improvement, a ratio, a management philosophy, a dynamic of innovation and a goal. So it is a flexible concept.

But despite the conceptual confusion, eco-efficiency has various basic features: an incremental improvement is sought in existing products, technologies and productive systems; it is applied separately, that is, to each product or productive process, without taking into account the existing interrelations in the productive process; the scope of application is each company, so the interrelations that condition individual actions are not contemplated: "Plainly put, eco-efficiency only works to make the old, destructive system a bit less so" (McDonough and Braungart 2002: 62). This general approach is not a substantial improvement in relation to the traditional behaviour of industry, because companies have always sought to improve efficiency as a means of reducing costs. This was the aim of the American mechanical engineer F. Taylor, who designed the "Principles of Scientific Management" of work around 1900. His theories were applied by Henry Ford in the automotive industry in the early twentieth century. He developed a management creed: "You must get the most out of the power, out of material, and out of the time" (McDonough and Braungart 2002: 51). The unique difference is "that eco-efficiency has adopted a broader conception of the value of services and a wider range of natural resources" (Hukkinen 2003).

3.1.2 Technologies

Environmental Technologies

For the European Commission they are "all technologies whose use is less environmentally harmful than relevant alternatives". And there are three types: "technologies and processes to manage pollution (. . .) less polluting and less resource-intensive products and services, and ways to manage resources more efficiently" (European Commission 2004a, b, c: 2). It is evident that a gradual reduction of contamination is sought (in particular through the application of technologies to the existing productive systems, that is, end-of-pipe technologies). Only a third of technological investment is destined to integrated technologies (those that are integrated in productive processes to minimise contaminant emissions and the use of resources), as opposed to the two thirds dedicated to end-of-pipe technologies (European Commission 2002a, b).

Environmentally Sound Technologies

It seems that the UNEP is reluctant to define what are sustainable technologies and limits itself to describing environmentally sound technologies following the criteria established in Chapter 34 of Agenda 21: they protect the environment, they are less contaminant, they use resources sustainably, they recycle most of the waste and products and manage all the waste in a more environmentally acceptable way than the technologies they replace. It summarises it with the following definition: "Environmentally sound technologies cover the full spectrum of production and consumption technologies that are more environmentally sound than the technologies for which they are substitutes". We are still in the incremental field. They are the stage prior to that of sustainable technologies in the process of transition towards them. The UNEP considers that it is difficult to establish rules that define, a priori, what technologies are sound, because this will depend on the places of application, on the available infrastructures, on the technical training of the workers, on the forms of use, management capacity, etc. For this reason, their qualification can only be given after their application (UNEP 2003: 16, 17).

3.2 Analysis of Transforming Concepts and Technologies

3.2.1 Appropriate Technologies

The appropriate technology concept was designed and applied by Schumacher during the mid-twentieth century. In 1978 he published his most important work, Small is Beautiful, which has been very influential, particularly in the context of a technological transfer to less developed countries, Schumacher (1978). The UNEP states that "the technology and associated equipment should be relatively simple and understandable, as well as suitable for local maintenance and repair". They are simple, they do not require advanced training to use them, they do not depend on complex materials that do not exist in the community (in such a way that they could be controlled by it) and they do not produce harmful effects. They are heterogeneous and adapted to the local culture, economy, communities and environment. In short, they improve their socio-economic conditions and are sustainable (UNEP 2003: 18, 19).

3.2.2 Clean Technologies

This is a concept that is used in North America and defined as "a diverse range of products, services, and processes that harness renewable materials and energy resources, dramatically reduce the use of natural resources, and cut or eliminate emissions and waste" (Makower and Pernick 2001: 2). The CleanEdge consultancy states that clean technologies reduce child illness and mortality and improve abilities to create "meaningful jobs". In addition, it defines four main areas of application to goods and services: transportation, energy, materials and water. The table shows the most important technological groups in each area and represents a positive evolution in relation to previous reports. For instance, sustainable bio-fuels appears instead of bio-fuels. In some cases the technological categories are broader and in others they are more precise, to show the best option. Instead of describing different renewable energies there is the category of "Renewable Energies". Instead of fuel cells we have "Hydrogen Fuel Cells". Last of all, there are new categories, such as "Smart Grid Devices and Networks", "Electric Rail", "Cradle to Cradle Systems" or "Reuse and Recycling" (Makower and Pernick 2001: 2; Permick et al. 2010:4) (Table 9.1).

3.2.3 Biomimetic Technologies

As we learn more about how nature works, the satisfaction that permeates the centres of power and broad sections of the population concerning developed technologies is increasingly incomprehensible, because they cannot stand up to a comparison with similar ones of nature: "Heat, beat, and treat has become the de facto slogan of our industrial age". This is what happens with Kevlar, the strongest fibre we have

Table 9.1 Top clean-tech job sectors

Energy	Transportation
Renewable energy (e.g., Solar, Wind) Energy storage Energy conservation and efficiency Smart grid devices and networks Electric transmission and grid infrastructure Biomass and sustainable biofuels	Hybrid-electric vehicles All-electric vehicles Electric rail Hydrogen fuel cells for transport Advanced transportation infrastructure Advanced batteries for vehicles
Water	**Materials**
Energy-efficient desalination UV and Reverse-Osmosis filtration Membranes Drip & smart irrigation systems Automated metering and controls Water recovery and capture	Biomimicry Bio-based materials Reuse and recycling Green building materials Cradle-to-cradle systems Green chemistry

Source: Permick et al. (2010: 4)

ever made: it is manufactured based on oil derivatives, it is placed in a pressurised chamber that contains concentrated sulphuric acid and high temperatures are applied to it. A large amount of energy is consumed and highly toxic waste is generated. However, spiders make a silk with water and insects, at ambient temperature and pressure, that is much stronger and more flexible than Kevlar and five times more resistant than steel. But, nature "can afford to follow this strategy (. . .)nature manufactures its materials under life-friendly conditions – in water, at room temperature, without harsh chemicals or high pressures" (Benyus 1997: 6, 97). It is amazing how trees transport cubic metres of water to a height of dozens of metres through transpiration fuelled by solar energy, mangroves desalinate seawater, leaves capture solar energy with great efficiency, termites thermo-regulate their dwellings through structural design, some species produce ceramics that are as hard as pearls, the fact that bones are so light and resistant, etc. The cybernetic system of a living being is incomparably faster and more capable of processing large amounts of information than the most advanced human communications system. Our radars bear no comparison with the multi-frequency system of bats. Turtles, butterflies, birds, etc., navigate without maps. Many species build long-lasting structures with a minimal amount of material. Bees, for example, build the cells of their honeycombs in a way that maximises the available space, but with resistant and long-lasting walls (Benyus 1997: 265).

As we have seen, nature "is model, measure and mentor" (Benyus 2010). To be a mentor means being a source of inspiration. It is a model to imitate, in products, in production processes and in environmental organisation, and the measure of what is and what is not sustainable. For this reason, when faced with any innovation we must question whether it promotes life and whether it has precedents in nature. An example of how nature is the measure of sustainability is the fact that synthetic biology (introducing genes from certain species in others) cannot be sustainable, because it does something that nature does not.

In truth, humanity has never ceased imitating nature. For example, in the ancient world renewable energies were used to transport and to move devices in artisan and manufacturing production. The industrial civilisation has also innovated by imitating it: in air flight navigation, in sea travel, in the capturing methods of renewable energies, etc. However, starting in the twenty-first century, biomimetic research is becoming more intense. In 2005 the Biomimicry Institute was founded and has developed educational programmes: the Biomimicry professional ("a 2-year master's equivalent programme") and the Biomimicry Specialist ("an 8-month program (. . .) to complement the busy schedules of active professionals") (Casey 2012). In addition, many institutes (The Land Institute, Sandia National Lab, etc.), universities (Arizona, California Santa Barbara, Oxford, Manchester, Harvard, Cornell, etc.) and many companies are working on the development of products inspired by nature.

There is an enormous amount of biomimetic products that are being commercialised or which are in the prior stages. The most active fields are: surfaces; packaging; aerodynamics; robotics; energy and efficiency in the use of resources; medical and pharmaceutical applications; architecture; etc. But only a few products reach mass markets, except in the case of capturing renewable energies. Various factors can explain, at least in part, such a result: the dominant paradigm; innovation that imitates nature needs long maturing processes, thus maintaining or reaching the lead in the competitive race requires fast innovations (Mueller 2008).

There are also some notable cases of productive processes and systemic approaches that are inspired by nature. Green chemistry imitates nature when it uses, as nature does, few materials and many reactive agents, the opposite of traditional chemistry. There is significant activity in the construction of biomimetic systems. Architecture is introducing an increasing number of biomimetic elements (energy efficiency, solar energy capture, water recycling, etc.) but, in addition, work is being done on a construction that imitates the natural system. The complex of the Isthmus of Santa Catalina in Las Palmas de Gran Canaria seeks to be self-sufficient in energy and water (through the desalination of water using a biomimetic process). Philips is developing a self-sufficient building, whose main feature is that the façade will be built with sensitive materials (like a membrane) that capture air, light and energy and channel them towards the interior. The city of Masdar, which is being built in Abu Dhabi, will use solar energy exclusively, will recycle materials and water and transport will be electric. On the other hand, permaculture (permanent agriculture) creates highly efficient and self-sustained productive systems by imitating ecosystems (ECOS 2007). The field of biomimetic technology is growing at a rapid pace. A report by the Fermmanian Business and Economic Institute in San Diego says: "While the field is just emerging, in 15 years biomimicry could represent $300bn of US GDP and (. . .) Globally, could represent about $1,000bn of GDP in 15 years" (P. Miles, Financial Times, August 12, 2011). In subsequent chapters we will analyse renewable energies and industrial ecosystems.

Last of all, nature develops extremely diverse technological systems to adapt to the conditions of each environment and to each circumstance. This diversity gives it resilience in the face of rapid changes. This reason we must create infrastructures and platforms that can be used by a wide array of technologies. Systems that, although based on universal technologies, can be adapted to each society: "The existence of diversity and the considerable depth of knowledge about many alternative technical options is a potential source of systemic self-renewal and adjustment to new circumstances" (UNEP 2003: 6).

Part III
Sustainable Production and Consumption

Chapter 10
The Limits of Fossil Fuels

Keywords Fossil fuels resources • Oil scarcity • Peak oil • Energy balance • Extra-geological limiting factors

Energy has determined the development and the survival or death of civilisations. The industrial civilisation has been built thanks to fossil fuels, and in particular, to oil, due to its high energy density and the ease with which it is extracted, handled and transported, in addition to being the raw material for an immensely wide range of products. The use of fossil fuels meant a quantum leap in the availability of energy. It is estimated that a barrel of oil (159 l) contains energy equivalent to 25,000 h of human labour. Price estimated in 1995 that the energy used equalled around 50 slaves per person (Gowdy 2006). According to the International Energy Agency (IEA 2008), the consumption of fossil fuels account for 82 % of all global primary energy, broken down as follows: oil (35 %), natural gas (21 %), coal (26 %). Oil is used in transport (70 %), in electricity production (10 %) and the rest in petrochemicals. Natural gas is used especially in electricity production, heating and in industry. Coal is used, above all, for electricity production and ironworks. The increases in average annual consumption have been 2.2 % (gas), 1.8 % and coal that, after being stagnant in the 1990s, during the next decade grew at a rate of 4.8 %.

In this chapter I analyse the factors that determine the limits of fossil fuels, the current and future development of their respective offers, focusing in particular on oil and to a lesser extent on natural gas. Lastly, I briefly analyse the effect of a scarcity of fossil fuels on climate change. And although in the graphs the term production appears, out of respect to their authors, in the texts I will apply the term extraction, as it is the closest to reality.

R. Bermejo, *Handbook for a Sustainable Economy*,
DOI 10.1007/978-94-017-8981-3_10,

1 Peak Oil

1.1 Reserves

Included are: crude oil (which is oil obtained directly and of which there are very different qualities); condensate (oil which is the result of the spontaneous condensation of raw natural gas associated with oil); liquid natural gas, which is obtained at raw gas separating plants and is formed by propane, butane, pentanes, etc.; extra-heavy crude oil (particularly in Venezuela) from which the lighter elements of crude oil do not exist, though intermediate and a large amount of heavy components do; bitumen from bituminous sands (mainly in Canada) and bituminous shale; fuels obtained from coal; agro fuels; etc. Oil reserves have been classified as conventional and non-conventional. The former is light, with scarce sulphur content and is extracted from easily accessible areas. Non-conventional oil is that which has, at least, some of the following features: high density, high content in sulphur and/or heavy metals and difficult access. However, the IEA only includes in the non-conventional chapter the fuels that are not oil: liquid natural gas; bitumen; fuels based on coal; agro fuels, etc. Last of all, an important number of authors is in favour of counting the different types of fuels as “all liquid fuels”. In this text I will use the classification of the IEA.

Despite the lack of reliable data, most studies reach the conclusion that the reserves of conventional oil in place are in the region of 2000 Giga barrels (Gb). The report “Global 2000”, published in 1980 on the orders of president Carter, which is the most exhaustive one to be carried out, estimates the oil in place at 2,100 Gb. Another study called World Oil Supply 1929–2050 and carried out by Petroconsultants in 1995 (which analysed the existing 10,000 oilfields) totally coincides with the previous report by the Association for the Study of Peak Oil (ASPO), an international organisation dedicated to studying peak oil and to raising awareness in governments and societies about the problem, and makes the lowest estimate: 1,900 Gb. The average estimate of 65 consultants, oil companies and other entities is slightly less than 2,000 Gb. On the other hand, a study carried out in the year 2000 by the US Geological Survey (USGS) estimates the conventional oil in place at 3,300 Gb, which, along with other also excessive estimates of non-conventional oil, allows it to state that there will be no supply problems until after 2030. The difference between both estimates is more important than what may seem at first sight because, as we have already consumed around 1,000 Gb, the second estimate is over twice the remaining amount than the first. This study is being discredited because the current rate of discoveries is one-quarter of that predicted. As a result, the USGS reduced its estimates by 500 Gb in 2007 (ASPO Newsletter 2003, December; Zittel and Schindler 2003, 2004; IEA 2008).

Figure 10.1 shows the decline in new discoveries of conventional oil within a context of increasing consumption. The curve of new discoveries reached its peak in 1964, as shown by the graph, and now has a downward trend of around 5 % per year. Since the late 1970s (period during which the North Sea oil province and

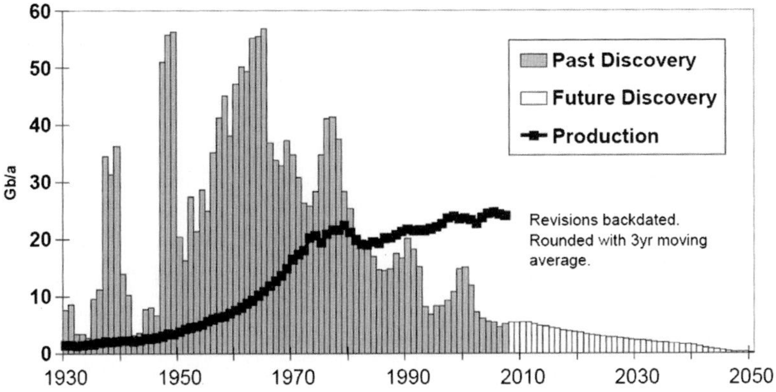

Fig. 10.1 Oil growing gap between discoveries and production (Source: N. Hagens 2011)

the massive oilfields in the Bay of Prudhoe in Alaska and the Cantarell in Mexico were discovered) no oilfields of this type have been discovered and the massive ones found tend to zero.

The new main sources of oil come from: deepwater sources (>500 m.); bituminous sands; the Arctic Pole; and extra-heavy crude oil. The extraction of very heavy crude oil is more similar to mining than to the typical extraction of crude oil. Its extraction is very slow and it later goes through a complex industrial process. Only half of the bitumen from bituminous sands is processed and turned into synthetic oil (synfuel). In the process around 10 % of the original product is lost (Schindler et al. 2008: 7 and 8). Deepwater oil is the most important new resource (estimated at 7–8 % of global reserves), but it will have a short lifetime (once the peak is reached, the oilfields lose capacity at an annual rate of 6–12 %). And due to the enormous risk of accidents in ultra-deepwater drilling (>1,500 m.), due to its extreme conditions, it should be banned by governments. ASPO estimates at 525 Gb the amount of non-conventional oil in place. It breaks it down as follows: heavy (226 Gb), including heavy crude oil and bituminous sands; deepwater (89 Gb); polar (52 Gb); and liquefied gas (156 Gb). And rounds it up with 2 Gb to reach the total. But it approaches another limitation: net energy. An increasing amount of energy is required for the extraction process (ASPO Newsletter, 2009 April; Bukold 2010).

1.2 The Main Agents

1.2.1 Private Companies

Up until the 1960s, seven companies (until the United States antitrust legislation forced the division of American monopolies) controlled oil supplies through the International Petroleum Cartel (IPC). They shared the market among themselves using a quota system. They later ceased controlling prices, due to the nationalisation of oil in the Persian Gulf (which took place in the 1970s) and to the disappearance of the Cartel. In addition, they were forced to search for oil in increasingly inhospitable areas, although their situation was relieved with the extraction of oil from the North Sea, which started that same decade. But the decline of this oil from the year 2000 and the nationalisations that are taking place, particularly in Russia and South America, mean that their situation is increasingly precarious. It is estimated that private companies control less than 20 % of global oil reserves (with Exxon, the largest private company, in 17th place) and this percentage is falling (Baker Institute Policy Report, 2008, Number 37).

Given the decline in reserves, companies increase their investments destined to buying other businesses (BP absorbed Amoco and Arco, Exxon took over Mobil, Chevron purchased Texaco and Unocal, etc.) and to increasing their reserves of natural gas (Shell's strategy focuses on this). As their reserves dwindle, companies multiply their ploys to hide the real data. The estimation of companies' reserves does not distinguish between oil discovered and that of the companies that are absorbed. And as their typical behaviour is to absorb other companies, most new reserves come from this policy. In addition, they declare their estimates of new oilfields as an increase in reserves, even if they do not yet know whether these new oilfields will be profitable. Last of all, they provide joint statistics of crude oil and natural gas reserves in barrel of oil equivalents (Rodrigue 2008).

1.2.2 OPEC

In 1960 Venezuela, Iran, Iraq, Saudi Arabia (SA) and Kuwait created the Organisation of the Petroleum Exporting Countries (OPEC) to increase profits, the distribution of which was imposed by the above mentioned IPC But in the 1970s many countries started a nationalisation process (Libya, 1971; Iraq, 1972; Iran, 1973; Venezuela, 1975; Saudi Arabia (SA), 1979). From then on the OPEC sought to fix prices by complementing the supply of private companies and that of non-OPEC exporting countries. Member countries had supply quotas depending on the reserves they were estimated to have and the OPEC increased or reduced them proportionately to maintain the prices it deemed adequate. And during the period from 1974 to 1978 the OPEC managed to control prices. But it later lost this capacity and the usual situation is of a lack of control, due to: the war between Iran and Iraq; internal disagreements about quotas; Saudi Arabia has been exceeding its quota, based on an agreement with the US to maintain low prices in exchange for military protection. But since 2004 it started to extract at full capacity to satisfy a growing demand, but as it was not successful, there was an escalation of prices between 2002 and 2008. For this reason it ceased controlling prices. Last of all, the OPEC is formed by the exporting countries from the Persian Gulf, Algeria, Libya, Angola, Nigeria, Venezuela and Ecuador (Rodrigue 2008).

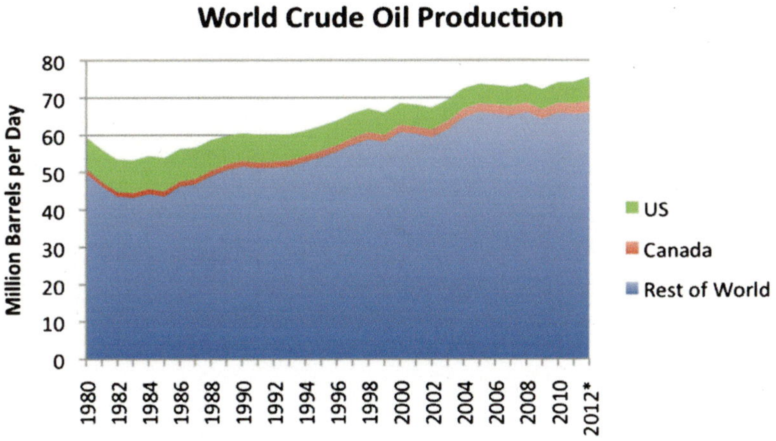

Fig. 10.2 Total liquid fuels (Source: Tverberg 2013)

1.3 Supply

The gigantic oilfields that have satisfied most of the demand are running out and the new ones are incapable of replacing them. 70 % of the oil is from oilfields that are over 30 years old. Around 120 oilfields, with a pumping capacity higher than 100,000 b/day each, supply 50 % of the oil. 70,000 oilfields complete the other half of the demand. This indicates that the new oilfields are increasingly smaller and are depleted faster. The 20 biggest oilfields provide 27 % of the global supply and have an average life of 50 years. Of these, four are at the maximum pumping capacity stage and the rest at different subsequent stages (with extraction rates between 85 % of the maximum and less than 50 %). Figure 10.2 shows the evolution of the supply of all liquid fuels during the 1997–2011 period. There were the following phases: intense growth until 2005; stagnation until 2007; a small and brief rise in early 2008; a reduction of 3–4 Mb/day of more condensate crude oil (done by the OPEC to adjust it to demand); recovery in 2010 of 2, 3 Mb/day; and, last of all, a rise in the offer of all liquid fuels. Despite this, the supply has spent 7 years on an oscillating plateau (+/−5 %) (Foucher 2011; IEA 2010a).

However, the previous oscillation is much lower, because the graph is built accumulating different fuels with unequal energy power. The energy power of liquid natural gas (which is the main non-conventional contribution) and of agro-fuels, etc., is just 65 % of the crude oil. Even applying a potential of 70 %, as is the case in Fig. 10.3, results in an almost flat supply curve (Staniford 2012).

In the next section we will see that exports fall, due to the growing domestic consumption of exporting countries. But there is another factor in the reduction of available oil: the net energy obtained, that is, the energy obtainedafter subtracting

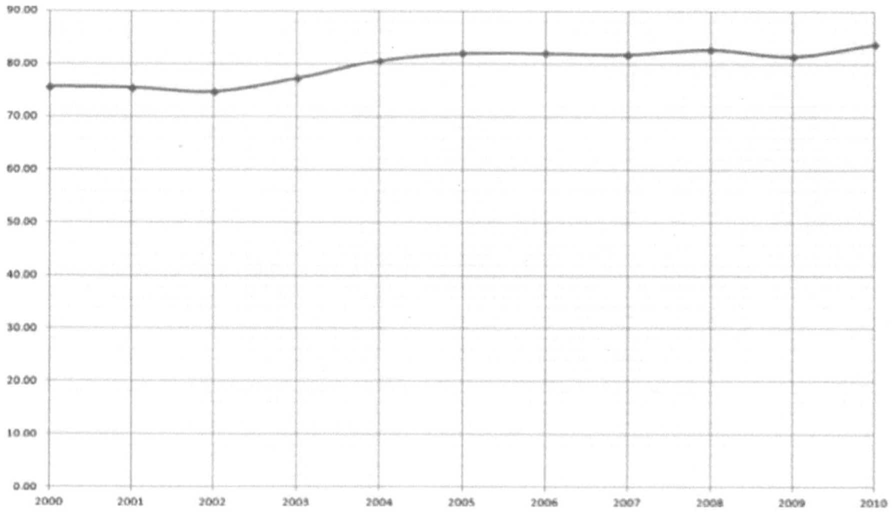

Fig. 10.3 Supply in millions of barrels of oil equivalent (Source: Staniford 2012)

the energy invested. In 1950, they could obtain 100 barrels investing one. Now they only obtain 15 barrels and in the case of bituminous sands the ratio is 5/1 (Hirsch et al. 2010: 141).

1.4 Demand

Average global demand during the 27 years before 2008 grew by 2.0 % a year. But in 2008 it dropped by 0.3 Mb/day and in 2009 by 2.5 Mb/day. The IEA predicts that demand will increase at an annual rate of 1.2 Mb/day, as will the increase in supply. But this demand will only be 50 % of the total demand for fuel in 2050, where the rest will be hydrogen, agro-fuels, etc. This forecast is due to the fact that oil exporting countries and emerging countries are greatly increasing their consumption, due to their high economic growth and to subsidised petrol. In the Persian Gulf countries there are additional factors: the proliferation of energy-intensive industries and a high birth rate. China has been responsible for over 50 % of the increase in demand during the 2003–2008 period. The IEA forecasts that 93 % of the increase in future demand will take place in Non-OECD countries (except in former USSR countries) and that China will absorb 37 % of all the increase in demand, going from a market share of 17 % in 2010 to 22 %. The Persian Gulf countries will have the highest increase in demand (IEA 2010a, b, c).

On the other hand, the drop in demand in the OECD started in 2005 and worsened in 2008 (−1.87 Mb/day, due to the crisis and the energy policies of the

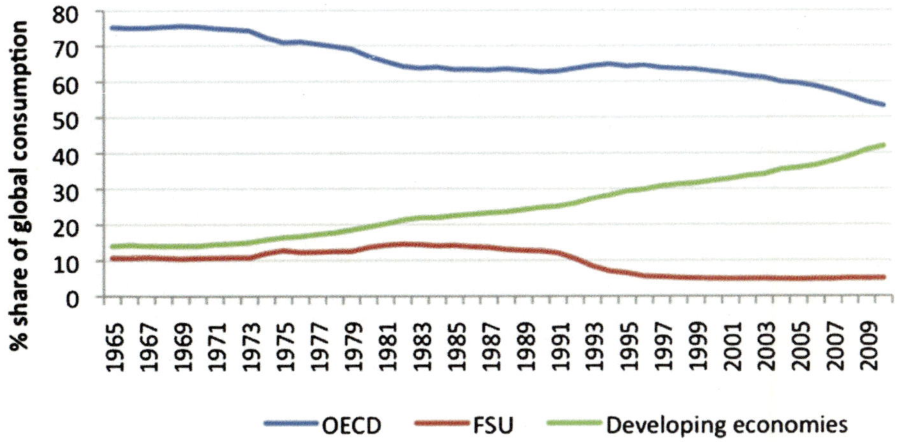

Fig. 10.4 Global oil consumption trends (Source: Mearns 2011a, www.theoildrum.com/node/8483)

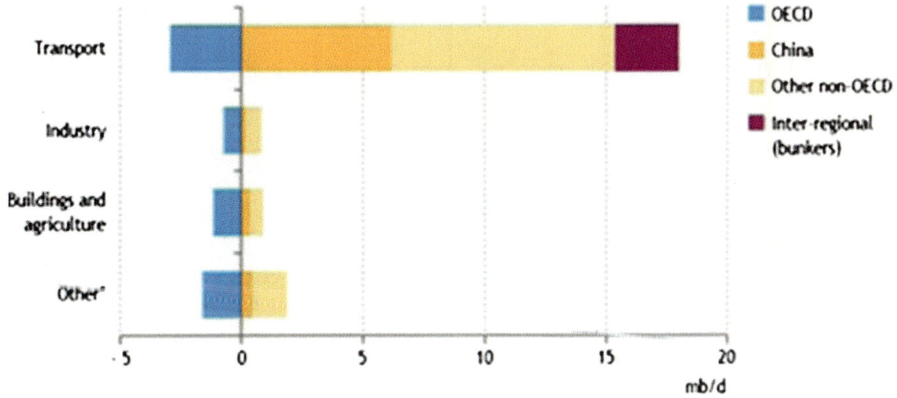

Fig. 10.5 Consumption of primary energy by sector and region forecast by the IEA, 2009–2035 (Source: IEA 2010a, b, c, New Policy Scenario)

EU, Japan, etc.). In 2009 the strong downward trend continued, although this trend has later slowed down a little. Figure 10.4 shows the diverging consumption trends indicated.

The IEA (2010a, b, c) forecasts that the demand of the OECD will fall by 6 Mb/day between 2010 and 2035 (which means an annual reduction rate of 0.6 %, where transport is the sector with the biggest drop), as shown in Fig. 10.5. The reduction rates among OECD countries vary substantially. In Japan it would be 1.3 % (going from a consumption of 4.1 Mb/day in 2009 to 2.9 Mb/day in 2035); in Europe it would be 0.9 % (going from 12.7 to 10.4 Mb/day); and in the US 0.8 %. The last chapter of consumption corresponds to interregional transport.

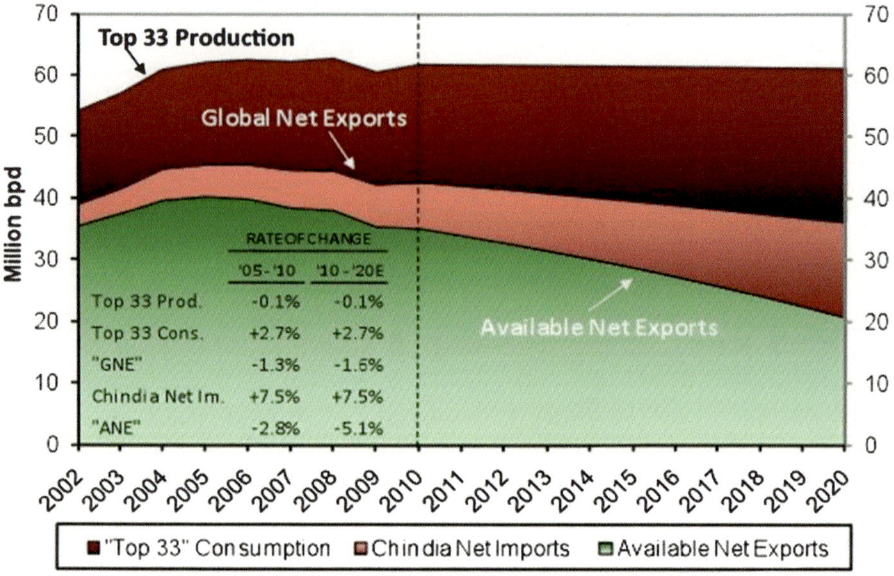

Fig. 10.6 Evolution of exports of top 33 Net Exporters (Source: Callahan 2011)

The stagnation of the offer worsens the supply problem, because the domestic consumption of the exporters grows considerably and as a result the amount of oil exported falls, as indicated in Fig. 10.6. The exports of the 33 countries that pump over 100,000 b/day have increased consumption from 16 % of the global total to 17.5 % in the last 5 years (Brown 2010). The falling amount of oil exported (around 3 Mb/day since 2005) has not prevented emerging importing countries from escalating their consumption, above all due to the reduced consumption of the OECD countries mentioned. The Fig. 10.6 shows the drop in exports until 2010 and the future trend forecast, taking into account the main exporting countries. The author extrapolates the past trend until 2020 and, in addition, establishes the hypothesis that the supply of oil will fall at an annual rate of 1 % and that China and India will continue to increase their imports. The result is that the rest of the world will go from importing around 35 Mb/day in 2010 to little over 20 Mb/day in 2020.

1.5 Prices

The setting of oil prices in the Persian Gulf went through three phases, which have been decisive in the evolution of the world: the period of cartelisation and colonial concessions (1901–1950); the transition period (1950–1972); and the globalisation phase (since the mid-1970s). In the first phase, Great Britain distributed concessions (long-term, 50 years or more, in uniform terms) among the seven companies

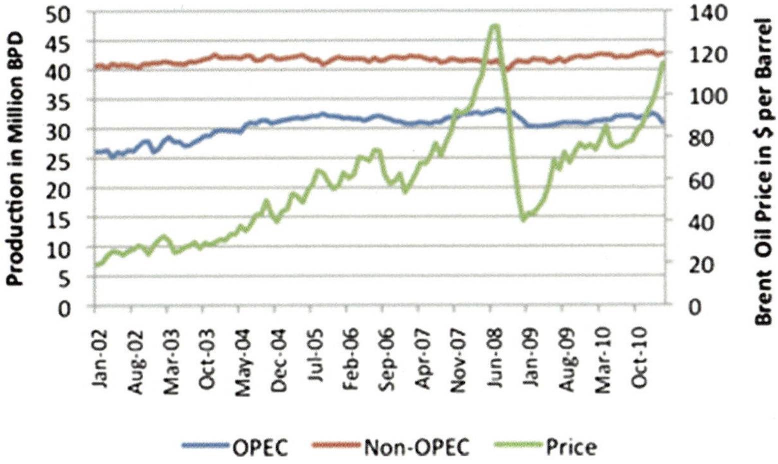

Fig. 10.7 World oil supply (crude and condensate) and prices (Source: Tverberg 2011, www.theoildrum.com/node/8268)

that dominated the sector. These companies defined the prices and distributed the market among themselves by means of quotas through the IPC mentioned. During the second part the decolonisation process took place and the IPC imposed the distribution of profits among companies and States at 50 %. In 1960 the OPEC was created with the aim of fighting against this distribution. The third period meant the nationalisation of oil in the Persian Gulf, the conversion of the OPEC into a potential supply oligopoly and the disappearance of the IPC (Bina 2009).

From 2002 there was an escalation in prices that worsened during the 2004–2008 period, in a context of high volatility. That is due to the fact that there are many factors that make prices fluctuate: economic growth forecasts; stock variations; seasonal consumption variations; reduced extraction capacity due to political reasons (wars, sabotage, strikes, etc.) or to the effect of climate phenomena; etc. But only the growing discrepancy between supply and demand explains the exceptional upward trend of prices during the 2004–2008 period, which ended in July 2008 with a price of 147$ per barrel (159 l) at the New York stock exchange. Figure 10.7 shows a clear correlation between the stagnation of the supply (OPEC and non-OPEC countries) from 2005 and the escalation of prices, despite the collapse of prices from late 2008 to early 2009. Goldman Sachs and J.P. Morgan were right in 2010 when they predicted that in 2011 the 100$/b threshold would be surpassed (it was the year with the highest average price since 1864) and that the escalation would continue in 2012 (www.bloomberg.com; Financial Times, November 27).

Regarding the future, there is a very widespread agreement among experts that oil prices will continue to rise. F. Birol, Chief Economist at the IEA, states that "we must be prepared to see very turbulent markets and high prices" (Schneider 2009).

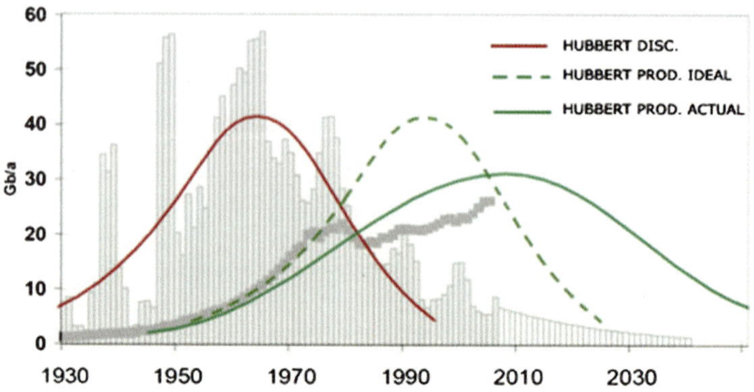

Fig. 10.8 Hubbert's theoretical curves of discoveries and the extraction of oil (Source: J. McManus (2011) in comments on S. Foucher's article, www.theoildrum.com/node/7785)

1.6 Peak

The exhaustion of any resource depends on two factors: the existing reserves and the rate of consumption. But the geology of fossil fuels (and in particular oil) adds an additional factor: the rate of extraction. There are many factors that obstruct the extraction of oil: it sits in the crevices in rocks, impregnating sands and porous rocks; oilfields lose the initial pressure, meaning the oil has to be forced out, injecting water or gases; an increasingly heavy remnant is left; etc. These difficulties determine: that the rate of extraction reaches a peak and that, from that moment on, it irreversibly declines (Fig. 10.8).

The studies by M. K. Hubbert (petroleum geologist from the mid-twentieth century) and decades of studying the behaviour of oil-rich countries show that the curves of oilfield discoveries and extraction rates are bell-shaped. And a few decades after the first curve reaches its peak the second one does too (and corresponds approximately to a consumption of half of the reserves) and pumping starts to decline until the oilfield is depleted. The peak is reached because a decreasing rate of new discoveries and rising consumption inexorably lead to the point where the oil discovered is less than that consumed. So consumption increasingly depends on reserves. The global peak of conventional oil discoveries happened in 1964 and the disparity between the oil discovered and that consumed started in 1981. Hubbert correctly predicted in the mid 1950s that in the US the peak would be reached in 1970 (Hemmingsen 2010).

A variety of methods are used to calculate the moment the peak is reached, but here I will only explain the two most frequent ones: by estimating the future extraction capacity of the main oil-rich countries; and by estimating the rate of exhaustion of all the oilfields being exploited, the existing projects for new oilfields and those predicted to be found (Hemmingsen 2010). Following the first method, I make a separate estimate of the future supply of the main OPEC and non-OPEC countries. The pumping capacity of most OPEC countries is in slow decline or stagnant (Iran, Kuwait, UAE, Libya, Venezuela, Ecuador, etc.). Only a few have a significant capacity to increase extraction (Angola, Nigeria and Iraq). The latter is the only country in the area that has the potential to greatly increase its extraction. Some state that it has an extraction potential of 6 Mb/day, but more reliable analysts place it at 3–4 Mb/day. But the IEA expects Iraq to maintain its current capacity until 2015, due to political instability. Nigeria pumps 2.0 Mb/day of oil, but it could reach 2.5 Mb/day, without civil war. It is considered to have a future potential of 3.5 Mb/day, but for a very short period of time. Angola extracts 2.0 Mb/day and can reach 3.0 Mb/day, maintaining this level for quite a long time. There is widespread debate about the capacity of Saudi Arabia (SA). In recent years it has pumped around 10.5 Mb/day and the official discourse is that it will raise it to 12 Mb/day, although it could reach 15 Mb/day. But many analysts consider that it can only maintain the first level, and during a short period of time. The global result of the OPEC is that, after reaching its highest level at the end of the previous decade (34 Mb/day), it will maintain it for a few years. It will then decline slowly before accelerating after 2020 (Koppelaar 2010a April; Eriksen 2009).

The scenario for non-OPEC countries is worse. Extractions from the North Sea are falling at a rapid rate. Norway is going from 3.4 Mb/day in 2004 to 2.2 Mb/day in 2010. Mexico went from 3.9 Mb/day in 2004 to 2.9 Mb/day in 2008, due to the rapid decline of its Cantarell oilfield. Brazil raised its capacity of 1.5 in 2005 to 2.0 Mb/day in 2010, an insufficient amount to satisfy its domestic demand. But it is discovering important oilfields at enormous depth in the sea, although their exploitation is being delayed due to the big technical problems they face. Canada maintained the same capacity between 2007 and 2010, compensating the decline of conventional oil with that from tar sands. In 2006, the latter yielded 1.2 Mb/day and some people estimate that they could reach 5 Mb/day. But other authors reduce that amount to 2 Mb/day. Last of all, of particular importance is the behaviour of Russia, because between 2000 and 2008 it increased its pumping from 6 to 10 Mb/day and satisfied 40 % of the new global demand. In 2010 it pumped 10.4 Mb/day, though this figure is expected to drop soon. The total supply of these countries is stagnant, but it is expected to decline soon (Koppelaar 2010b; Chaykovskaya 2010).

In the second method the future supply capacity is calculated, based on an estimate of new projects and the rate of depletion of existing oilfields. But there is no agreement on this rate. A referential analysis of peak oil is the Wiki Oil Megaprojects (WOM) study, which is a collaboration of The Oil Drum and Wikipedia to offer information on large oil projects and to contrast the supply that they will provide with the depletion rate. Its premises are: depletion rate of 3.6 %, which the authors themselves consider to be conservative (the IEA places it at 4.4 %), because "there is great uncertainty about this rate, as it represents the average of tens of thousands of small oilfields and several hundred gigantic oilfields" (Foucher and Border 2011); the study's projection is short: 2015. The authors consider that longer-term studies are not reliable. The graph in Fig. 10.9 shows the oil supplies of the megaprojects from the OPEC(minus Iraq), from Iraq

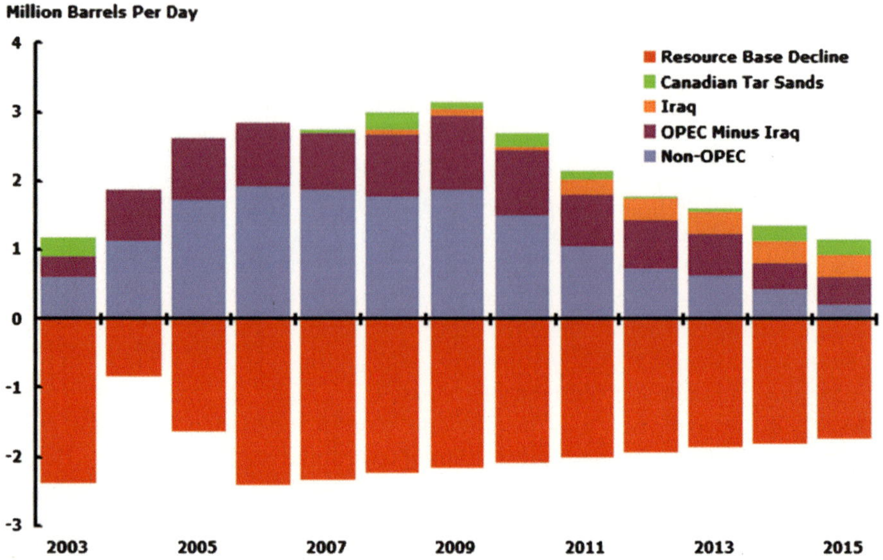

Fig. 10.9 Gross new capacity additions per annum from conventional oil megaprojects (Source: Foucher 2011, www.theoildrum.com/node/7785)

and from the tar sands in Canada. However, deepwater oil, which is the main factor for the increase in extractions, in particular in the Non-OPEC group, does not figure explicitly. The result is that from 2011 the net supply of oil decreases (Foucher and Border 2011).

Based on the different methods, a numerous and growing group of experts estimate that the peak has already happened or will happen during the first half of this decade. When the supply plateau started in 2005 some analysts identified it with the peak. Among them are al Husseini, Bakhtiari (both presidents of exploration and development until their retirement at ARAMCO and NIOC, national companies of SA and Iran), the World Energy Group, etc. But others considered that the rise (temporary) to 85.5 Mb/day in 2008 was the peak. As the plateau continues to oscillate, there is an ever-growing agreement (as shown by the ASPO USA 2010 conference) that the plateau is the peak and that the decline is soon to come. A study carried out by the German army states that we have possibly arrived at the peak. Although other experts consider that the peak will be reached in the next few years. Two reports by the Pentagon (2008, 2010: 28,29) predict that "a severe energy crunch is inevitable without a massive expansion of production", because "by 2012, surplus oil production capacity could entirely disappear, and as early as 2015, the shortfall in output could reach nearly 10 Mb/day" (Eriksen 2009; Petroleum Review, October 2009; CGAT 2010).

There are also other estimates that put back the date of peak oil. C. de Margerie (president of Total) has expressed doubts that 90 Mb/day can be reached (ASPO Newsletter, 2008, March). In 2008 Chatham House, a reputed British think tank, published a study that predicted a shortage in oil in 5–10 years. A group of British transnational companies issued a report in which they predicted the peak by 2011–2012 and in 2010 another where they slightly put back the date of peak oil (due to the crisis): between 2012 and 2015. The BGR (the German Federal Institute for Geosciences and Natural Resources) and the World Energy Council (WEC) put the peak at 90–100 Mb/day (Petroleum Review, October 2009; IPTOES 2008, 2010a).

Last of all, the IEA constantly contradicts itself. It publishes reports in which peak oil appears in the medium to long-term. But with each annual report it comes closer to reality. The WEO2004 put the peak at 121 Mb/day. The WEO2008 lowered it to 104 Mb/day and the WEO2010 puts it at 99 Mb/day. But these data do not coincide with the dramatic tone of the reports. The WEO2008 states: "The world's energy system is at a crossroads. Current global trends in energy supply and consumption are patently unsustainable (. . .) The time to act is now". The WEO2010 states that "the energy world faces unprecedented uncertainty" and wonders whether peak oil will "be a guest or the spectre at the feast". The reason behind the contradictions is that it has a clear idea of the seriousness of the situation, but must issue optimistic reports to avoid upsetting the US (Macalister 2009; IEA 2008, 2010a, b, c).

1.7 Extra-Geological Limiting Factors

Up until now I have focused on the geology of oil, but the extraction capacity is limited by many extra-geological factors: economic, environmental, technological and political, which can reduce exploitable reserves, increase them or delay their exploitation. The high costs of extracting and transporting oil or gas from a deposit can prevent its development. There is a close correlation between the rise in sea temperatures and the severity of hurricanes (which has duplicated since the 1970s). Most of the hurricanes that have damaged (sometimes irreversibly) oil exploitation sites in the Gulf of Mexico have occurred during this century. Hurricanes Katrina and Rita permanently reduced extraction capacity by 0.255 Mb/day (Rubin and Buchanan 2008a). On the other hand, global warming will make it possible to explore the Arctic region as it melts. But there is growing alarm concerning the predictable environmental impacts of exploiting the Arctic region and very deep waters. The Horizon rig oil spill (from the Macondo field) in the Gulf of Mexico caused the biggest environmental disaster ever to happen in the US and has shown the enormous problems brought by this type of installation: the inability of the US to carry out acceptable supervision; the use of untested new technologies; etc. S. Bukold (2010) requests that installations greater than 1,500 m. are forbidden, because the spills cannot be controlled. But we know that "Macondo will increase deepwater drilling and development costs and delay projects, as yet there is no way to quantify this" (IPTOES 2010b).

As tension mounts concerning the distribution of oil, the frequency of armed conflicts increases. These have a limiting effect on reserves. When the US attacked Iraqi troops in Kuwait, they retreated and set fire to many oilfields, which burned for months. Attacks against oil pipelines are another example of the destruction of resources. There are systematic attacks in Iraq, Mexico and Nigeria. Lacking the capacity to invest due to wars (Iraq and Nigeria) or being subjected to an economic embargo (Iran) decreases the exploitation capacity (Bermejo 2008: 115–125).

On the other hand, the oil nationalisation process, which is unstoppable, leads to a national debate on what the extraction reduction rate should be to lengthen the lifespan of resources. A phenomenon known as resource nationalism. This is particularly relevant in countries with a small population and plentiful resources (such as Canada or Norway). And the conclusion is usually to lengthen their lifespan. The Norwegian parliament refuses to increase the rate of gas extraction. The king of SA has made many declarations concerning the need to limit extraction to satisfy future needs (www.peakoil.net; www.thegulfonline.com). Technological development has offered the possibility of exploiting reserves that were initially unrecoverable, as is the case with deepwater reservoirs. No-one disputes this fact. But there is a debate about whether technological development is allowing an increase in the amount of recoverable oil in oilfields being exploited. Although many state this, the most important geologists consider that all they manage is to extract the same amount of oil, but at a faster rate, thus accelerating the depletion of resources. Others consider that they increase the rate of extraction, but only during the final extraction phase. But in most new oilfield projects there are long delays before their commercial exploitation. Thunder Horse (Gulf of Mexico) started production in 2008, 4 years behind schedule. It was planned that Kashagan (Kazakhstan) would be operative in 2005, but it is being delayed and the latest prediction is that it will start operating in 2012–2013. Of the 32 new oilfield or existing oilfield expansion projects that were supposed to begin operating in 2006, only 12 managed to do so. The main reason for the delays is that the technology is not up to the new challenges (Konyrova 2008).

2 Peak Natural Gas

Like in the case of oil, we also find that there is conventional gas and unconventional gas (from coal and slate deposits, etc.). Conventional natural gas is found in deposits on its own or in combination with oil. It is considered that 9 % is consumed in the extraction, treatment and transport. In addition, a lot is wasted. When oil reservoirs lose pressure, they emit light hydrocarbons, formed by natural and liquid gas (liquid natural gas). These are recovered (condensate) in plants and the natural gas is injected into the reservoir to maintain pressure or it is burned at the site. This happens in particular in Nigeria and in Russian oilfields. The World Bank estimates that 150,000 m^3 of gas are burned annually around the world, the equivalent of 30 % of Europe's consumption and 5.5 % of global consumption. This squandering of gas

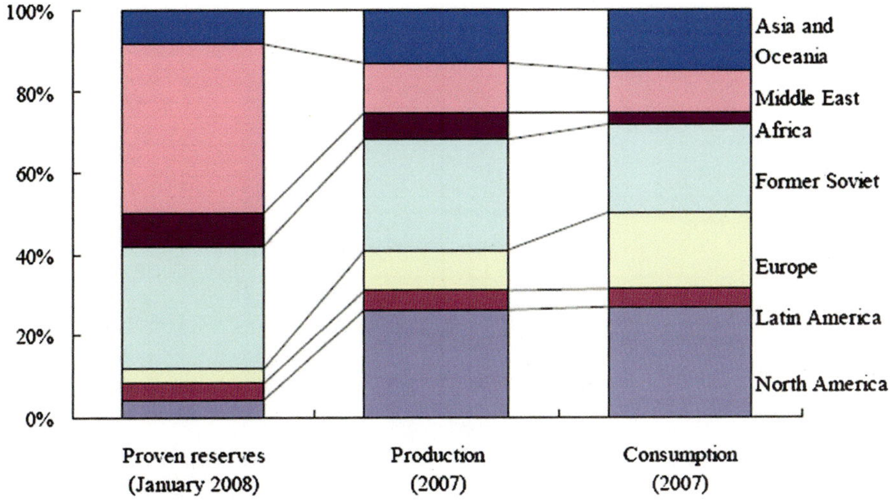

Fig. 10.10 Proven reserves, production and consumption by regions (Source: Kobayashi 2010: 3)

is because oil is the priority as it is more expensive. In addition, gas is used to extract ultra-heavy and tar sand oil. Conventional gas reserves are concentrated in a small number of countries and deposits. The three main gas producing countries (Russia, Iran and Qatar) have 56 % of the world's conventional reserves. A few massive deposits hold a large portion of the reserves. The largest gas deposit contains 15 % (North Dome, shared by Qatar and Iran) of the total reserves. Ghawar, the largest oil reservoir, only provides 8 % of the total. Figure 10.10 highlights the enormous discrepancy between reserves and consumption in North America and Europe (Gobierno Federal de México 2008: 24; Tverberg 2009a).

Like in the case of oil, we find a peak in new discoveries and the extraction of gas. In Fig. 10.11 we see that new discoveries of conventional gas took place in the 1960s and 1970s, the latter being the decade when the discoveries peaked. From the peak onwards the number of new discoveries plummeted. And since the year 2000, the amount of gas consumed exceeds that discovered and the difference continues to grow. Due to the high fluidity of natural gas, the extraction curve of a typical reservoir presents upon starting production a very steep curve, then a plateau, the length of which is determined by the transport limitation imposed by the diameter of the oil pipelines and the reserves of the reservoir. But when the plateau ends extraction rates plummet. Governments never plan for the phase of decline, which generates supply problems, because building the necessary infrastructures (gas pipelines or re-gassing plants) for importing gas is expensive and takes a long time (Bermejo 2008: 97).

Global consumption of gas has been growing by 2.2 % a year, but due to the crisis in 2009, it fell by 3 %. It is mainly used for electricity generation, heating buildings and in industry. Europe-Eurasia, North America and the Asia-Pacific

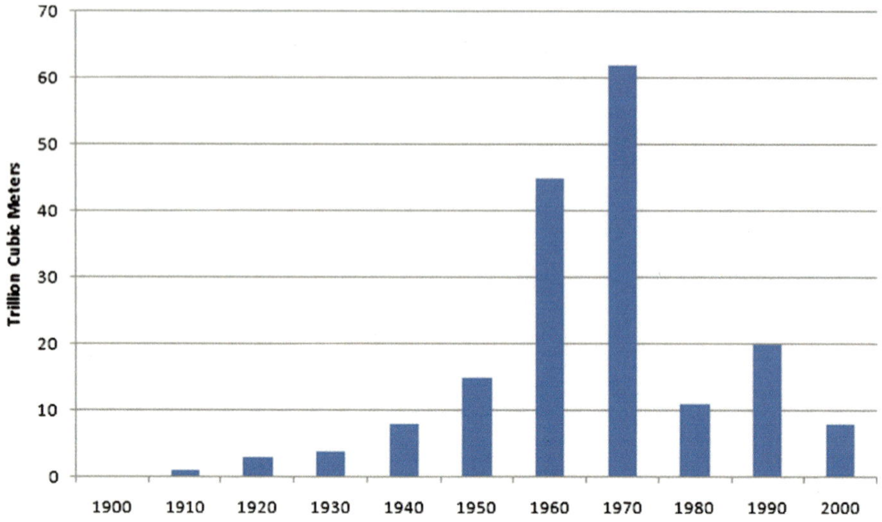

Fig. 10.11 New discoveries in billions of cubic metres (Source: http://en.wikipedia.org/Wiki/Peak_natural_gas)

region consume 82.3 % of the total gas. The US consumes 26 % and Russia 15 % approximately. In its annual Natural Gas Market Review reports prior to the crisis, the IEA had been alerting of the inability to satisfy demand in the near future. The cause put forward, as with oil, is that investment in the development of new reservoirs is far lower than what is necessary to satisfy the demand. But it has changed its tone by sharing the optimism that the US government has concerning shale gas. In the WEO2010 it predicts that consumption will grow at an annual rate of 1.4 %, meaning that in 2035 it will have increased by 44 %, with unconventional gas providing 35 % of the total. Consumption in China will grow at a rate of 6 % and will absorb 1/5 of the new demand (IEA 2009, 2010a, b, c).

So it is appropriate to analyse unconventional gas and, above all, shale gas, but focusing on the US as it is the country leading the way. The US reached the conventional gas peak in 1973, after that extractions plateaued for a number of years and now they are plummeting. Prices rocketed, which made the exploitation of shale gas possible thanks to the development of hydraulic fracking technology. As is usual in these cases, companies exaggerate the importance of this to obtain new investments and governments use the good news to score political points. But these phenomena end when the exaggerated expectations are not met. Although it has compensated for the decline of conventional gas and meets the increased demand, imports are at the same level.

The development of shale gas faces many problems, despite the fact that the best deposits are being exploited. The exploitable area of a deposit is reduced to less than 10 % of the surface and can be much less if the authorities establish exclusion zones to protect aquifers (Berman and Pittinger 2011). During the first year the extraction rate plummets by 63–85 %. The environmental impacts are huge: "An unavoidable impact of shale gas (. . .) is a high land occupation due to drilling pads, parking and manoeuvring areas for trucks, equipment, gas processing and transporting facilities as well as access roads and transport facilities. Major possible impacts are air emissions of pollutants, groundwater contamination due to uncontrolled waste water discharge (. . .) Fugitive Methane emissions from hydraulic fracturing processes can have a huge impact on greenhouse gas balance" (WI-LBST 2011: 10). For this reason Quebec, France, Poland, Bulgaria, South Africa, North Rhine-Westphalia, several Australian provinces, etc., have forbidden this technique (Nikiforuk 2012).

Some reports are starting to be published that contradict the optimistic predictions. According to the US Potential Gas Committee, the exploitable reserves of shale gas are equal to the consumption of just 7 years. A study by the Massachusetts Institute of Technology (MIT) states that the supply of shale gas in the US will stagnate by the middle of this decade. The US Energy Information Administration states that shale gas will only modestly raise gas supplies between 2011 and 2035 (Coob 2011). In the US the price of gas is low (in 2011 around 4$ per thousand cubic feet) and the profitability of the exploitations requires a price of 8$ (Berman and Pittinger 2011). As a result many companies are closing or reducing their activity (Callahan 2012). The conclusion of a study on shale gas by the Wuppertal Institute and Ludwig-Bolkow-Systemtechnik for the European Parliament is: "The resources of unconventional gas in Europe are too small to have any substantial influence on these trends" (decline of European gas production and increase of demand) (WI-LBST 2011: 78).

Returning to the analysis of conventional gas, up until very recently a large majority of the gas was transported via gas pipelines and the contracts were long-term, due to the enormous costs. But in recent years the infrastructures for sea transport have been heavily developed, to liberalise the sector and broaden the spot market. But the process has not gone far. Around 80 % of the gas purchased by Japan and South Korea has been done so with long-term contracts. Most re-gassing plants in the US are not being used, due to the increase of the domestic supply. Europe is increasingly connected by Russian and Algerian gas pipelines. Gas pipelines are being built: between Canada and the US, Iran and India, former Soviet republics and Asia, etc. (Skrebowski 2010; IEA 2010a).

In the EU gas provides 24 % of its energy and it imports 40 %, which breaks down to 25 % from Russia, 14.5 % from North Africa (Algeria, Libya and Egypt) and 5.5 % from various countries. Internal demand has been growing at an annual rate of 2.5 %. The extrapolation of this trend and the gradual depletion of Norwegian gas would lead to a dependency on supplies from outside the EU by 2030 of 70–80 %. But this prediction is not very probable due to peak gas and the EU's policies: an increase of efficiency and the development of renewable energies. But consumption of gas will continue to grow in the future, although less than in the past, meaning that Europe will continue to be very dependent. As a result the EU, apart from the policies mentioned, seeks to diversify its supplies, reducing its dependency on Russia. But Russia is building two large gas pipelines (the North Stream, through the Baltic Sea to Germany, and the South Stream, through the Black Sea to Bulgaria) which will make Hungary the distribution centre. Their combined capacity is equal to that of the Ukrainian network (80 % of its exports). The EU agrees to the North Stream (imposed by Germany), but not with the South Stream and is instead promoting the Nabuco gas pipeline, which seeks to bring gas from the Near East and Turkmenistan. But the project is at a standstill (Mearns 2010a; Vernon 2010).

It is predicted that in the future Russia will supply 60 % of imports, but there is speculation concerning its capacity to do this. The three gigantic reservoirs that have provided most of the gas are being exhausted. Gazprom (a state company) has been working, in particular, on three new projects (Shtokman, Yamal and Sakhalin). The first one was expected to start production in 2012, but was frozen in 2010. The Yamal project needs strong investment, but it is estimated to have reserves that would make it the world's second largest deposit, over three times as large as the Russian Urengov which is currently number two. The Russian government plans to increase exports by 52 % by 2015. But some analysts think that Gazprom will only be able to increase pumping at an annual rate of 1 % until it reaches a peak by around 2020. And this gas would be absorbed by domestic demand, which in the best case scenario would maintain the current level of exports (Ndefo et al. 2007; Vernon 2010).

3 Peak Coal

Coal reserves are divided into two groups: high (anthracites, hard coal) and low energy value (lignites, brown coal). Their energy intensity varies between 14 and 32.5 Megajoules/kg. The former are slightly less abundant than the latter, as shown by the graph. They are known to be very concentrated, with estimates that six countries (US, Russia, China, Australia, India and South Africa) have 85 % of the world's reserves of anthracite (hard coal). The first four also have most of the reserves of lignite (brown coal). Australia, Indonesia, Russia, US, South Africa and Colombia are the main exporters (Heinberg and Fridley 2010; www.worldcoal.org) (Fig. 10.12).

Since the year 2000 the consumption of coal has been growing at an average annual rate of 3.8 %. But in 2010 it grew by 6.8 %, due to the 8.4 % increase in the Non-OECD countries. This trend is explained by the fact that it is the cheapest fossil fuel and does not require expensive infrastructures for its handling. 58 % of consumption is for electricity production and the rest is divided in three equal parts: residential heating, cement and steel production. In coal-mining countries most of their electricity is produced with coal: South Africa (93 %) Poland (90 %), China (79 %), Australia (76 %), Kazakhstan (70 %), India (69 %) etc. Russia is an exception (17 %), due to the abundance of gas and oil. Over 6 years (2002–2007) China built 500 plants, and India 200. The US has been slowly increasing its consumption. Consumption in China grows at an annual rate of 10 % and in 2010 it consumed 40 % of the coal produced worldwide. Consumption in Europe is stagnant. Table 10.1 shows the ten main consumers (Oster and Davis 2008; Mearns 2010a, b; www.worldcoal.org).

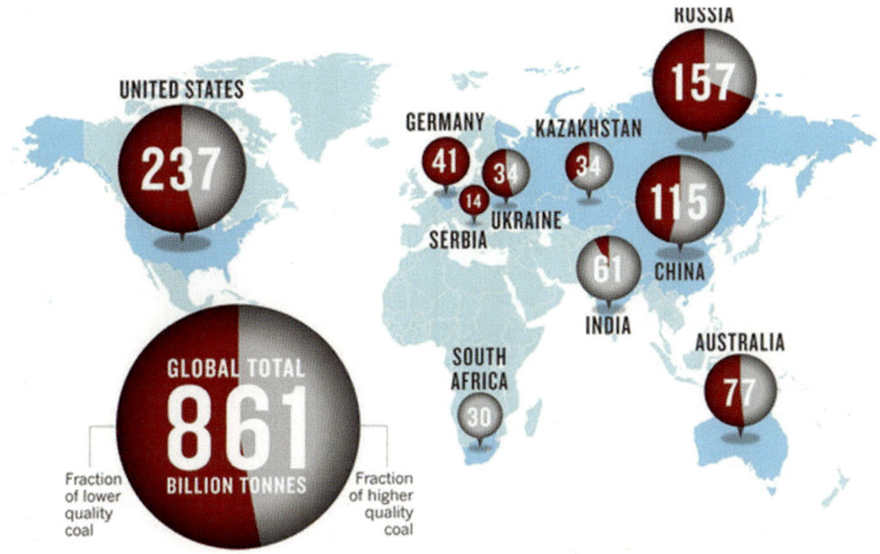

Fig. 10.12 Reserves by countries in billions of tonnes (Source: Heinberg and Friedley 2010-11-24)

Table 10.1 Top ten coal consumers World Coal Association (2010e)

PR China	2516Mtce	South Africa	141Mtce
USA	733Mtce	Germany	105Mtce
India	434Mtce	Korea	103Mtce
Russia	177Mtce	Poland	87Mtce
Japan	165Mtce	Chinese Taipei	60Mtce

Source: www.worldcoal.org

China was self-sufficient until 2009, but since 2010 it started to import coal (177Mt) and became the second largest importer. And it is highly improbable that it will be able to maintain its increasing rate of consumption without massive imports. In 3 years China can absorb the entire export volume of the Asia-Pacific region and, although exporters are increasing their capacity, the challenge of supplying China would be impossible. K. Aleklett president of the ASPO, estimates that its extractions are increasing at a falling rate, as is usual when the peak is approached, and that it will reach it this decade. As a result, analysts predict a high increase in coal prices (Aleklett 2010; Heinberg and Fridley 2010).

For many years it has been stated that there was enough coal to last 400 years, but the World Coal Association declares: "At the current production levels, proven coal reserves are estimated to last 118 years". But this estimate is not valid either, because coal also has its peak. In addition, since 1986 all the studies carried out have strongly revised previous estimates downwards, except for Australia and India. Between 2003 and 2008 the estimated reserves of Germany and South Africa have been reduced to one third. At the start of the century reserves in the US were estimated to last 5000 years and they are currently estimated to last 240 years. Other prominent reductions are Botswana (90 %) and Poland (50 %). A report by the EU

Table 10.2 The global peak and that of the main countries

Country	EJ peak[a] (year)	Ultimate coal production (EJ)
China	2011	4,015.6
USA[b]	2015	2,756.7
Australia	2042	1,714.5
Germany/Poland	1987	1,104.4
FSU[c]	1990	1,070.3
India	2011	862.6
UK	1912	753.0
S. Africa	2007	478.6
Mongolia	2105	279.2
Indonesia	2012	135.5
Global ultimate/peak	2011	13,170.5

Source: Patzek and Croft 2010: 3112

[a]Note that sometimes the peaks of produced coal tones and EJ do not coincide

[b]Excluding Alaskan coal

[c]The Former Soviet Union, excluding the Russian Far East coal

states that "coal might not be so abundant, widely available and reliable as an energy source in the future" (Kavalov and Peteves 2007: 36; Heinberg and Fridley 2010). The National Academy of Sciences in the US (NAS 2007) states that "only a fraction of the previously estimated reserves are actually recoverable".

Up until now we have used units of weight, which are the most common. The different types of coal have very different levels of energy per unit of weight, meaning that energy capacity is the most interesting information. Bearing in mind that the best quality coal is extracted first, the proportion of poorer quality coal increases. In 2010, 6185 Mt of hard coal and 1042 Mt of brown coal were extracted, and "in 2010 Non-OECD brown coal production rose to record levels" (www.worldcoal.org). This means that the peak energy of coal is closer than that of its weight. Many analysts predict that this energy peak will be reached by around 2020. Zittel and Schindler (2007) predict that the peak will be reached by around 2025, but from 2015 the rate at which extractions increase will start to fall. Which is what Heinberg and Fridley predict (2010). M. Höök predicts a peak by 2020, followed by a 30-year plateau. But Patzek and Croft (2010) consider that the global peak (and that of China and India) was reached in 2011, which is very close to the prediction by Aleklett (2010), followed by Indonesia (2012) and the US (2015), as shown in Table 10.2. http://www.worldcoal.org/.

4 Fossil Fuels and Climate Change

After having carried out a general evaluation of fossil fuel reserves and their predictable evolution, it is pertinent to see whether this analysis has a bearing on climate change predictions. We have gone from a concentration of 280 ppm of equivalent CO_2 before the Industrial Revolution to 380 ppm today and its accumulation rate is accelerating. The concentration of methane has gone from 715 ppm in the year 1732 to 1774 ppm in 2005. These and the rest of the greenhouse gases (GHG) have generated a rise in global temperature of 0.75 °C and in Europe of 0.90 °C in 100 years. The difference is due to the fact that climate change occurs faster at the poles than at the equator. During the previous decade the global temperature rose by 0.2 °C. The IPPC estimates that average rises could reach 6° by the end of the century (IPCC, Working Group I 2007: 3–13). The EU wants to limit the increase of emissions to 450 ppm, to limit warming to 2 °C.

There has been extensive criticism of the IPCC, because it has been found that many conclusions are based on unreliable sources and, in particular, because it contemplates (based on estimates by Rogner) exaggerated amounts of fossil fuels and, as a result, an exaggerated rise in temperature. In 17 of the 40 scenarios it contains, the consumption of fossil fuels in 2100 would be higher than today. The IPCC estimates at 11–15 billion barrels of oil equivalent (BOE) the reserves of fossil fuels. On the other hand, the German mineral resources agency puts it at 2.7bn BOE, BP at 3.2bn BOE and the World Energy Council at 3.5bn BOE. Some of the many critical authors are: J. Laherrere (2001), C. Campbell and A. Sivertsson (2003), P.A. Kharecha and J. E. Hansen (2008), Energy watch Group (2007), Rutledge (2007) and M. Höök et al. (2010). These authors reach the conclusion that the predictions by the IPCC "are exaggerated and so are the resulting emissions" (Höök et al. 2010: 24). Kharecha and Hansen estimate that the highest concentration of CO_2 reachable in the twenty-first century would be 580 ppm. Rutledge estimates, assuming reserves of 3.2bn BOE, that a maximum level of 460 ppm would be reached in 2070. This would cause a rise in temperature of 0.8 °C in 2100. In addition, the emissions of other gases, deforestation and industrial agriculture would raise its one degree further: 1.8 °C in total. This would mean 2.6 °C after adding the increase that has already taken place.

The fact is that the importance of managing forests and the use of land is rapidly growing. The change of use of land is becoming an increasingly important factor. A number of studies corroborate Rutledge's opinion on this. A report by the Australian National University reaches the conclusion that natural forests absorb three times as much CO_2 as what was thought until now. It estimates that deforestation has been responsible for 35 % of the historical accumulation of CO_2 and for 18 % of annual emissions. The Institute for European Environmental Policy (IEEP) states that the combustion of diesel oil emits less CO_2 than agro-diesel (produced with palm oil), due to the deforestation that its cultivation causes. A report for the EU Commission estimates that land has twice as much CO_2 as the atmosphere and three times more than the vegetation and that the improper use of land (and in particular peat bogs) causes enormous emissions of CO_2. That is the main reason why it recommends preserving agricultural land and promoting ecological agriculture (Alterra et al. 2007).

Chapter 11
Repercussions of the End of the Oil Age

Keywords Oil scarcity • Geostrategic struggle for essential resources • Economic impact of peak oil • Structural effects • Sectoral effects

It can be considered that a system collapses when it cannot reproduce itself or when it loses or has very debilitated basic functions, and it loses its identity. Civilisations are complex systems, the dynamics of which cannot be predicted. It is impressive how civilisations and political regimes have rapidly collapsed, despite the fact that they seemed stable not long before. The Roman Empire disintegrated over the space of a few decades. The Bourbon dynasty in France or the Soviet regime collapsed much more rapidly. It looked like the industrial civilisation (IC) had overcome historical limitations, due to its access to abundant fossil fuels. However, there is already a clear scarcity of resources, in particular of oil and strategic materials, and the bad performance of the economic system leads to increasingly frequent and intense crises. As a result, the IC has reached a crossroads. It will have to choose between transformation or collapse.

In this chapter first of all we will analyse the geostrategic conflicts caused by the distribution of fossil fuels. Then we will review the causes of the current crisis and will analyse the general repercussions of peak oil. Then we will study its structural effects and, last of all, its sectoral impacts.

1 The Geostrategic Struggle for the Control of Essential Resources

Competition is the privileged mechanism of the market economic system, but it often turns into confrontation when control over resources is at stake, as shown by the history of the twentieth century and the start of the twenty-first century, when the pre-emptive wars launched by the Bush administration to control resources reached

R. Bermejo, *Handbook for a Sustainable Economy*,
DOI 10.1007/978-94-017-8981-3_11,

their high point. This led to the consolidation of a group of affected countries, the core of which is formed by China, Russia, Iran and Venezuela. They have created a close-knit network of economic and geostrategic agreements. Beyond the strategic interests of each country, the strong growth of Asia's main economies is generating a gigantic struggle with Western powers to control global resources. Before reaching conclusions, I analyse the main players in the geostrategic game.

1.1 US

The abundant oil reserves in the US were a decisive element in its hegemonic status since WWII. Before the economic crisis of 2008–2009 it consumed over a quarter of the world's oil (over 21 Mb/day, importing 60 %), despite the fact that it only has 5 % of the world's population. Since oil production peaked in the US in 1971 it has been designing a policy destined to guaranteeing its supply by any means possible. In 1980 president Carter made this policy explicit when faced with the threat of destabilisation in the Persian Gulf due to the Iranian revolution: "an attempt by any outside force to gain control of the Persian Gulf region will be regarded as an assault on the vital interests of the United States of America, and such an assault will be repelled by any means necessary, including military force". Its energy security strategy has been aimed at controlling the main oil producing countries and the largest supply routes. Reagan started this policy with the creation of the "Central Command", a military structure to guarantee access to the Persian Gulf. It then redeployed its armies to control the main sea routes, removing them from the scenarios of the "Cold War" (the North Atlantic, the Mediterranean and the North-east Pacific). Today and apart from the Gulf, it dominates the South-East Pacific (the Chinese routes) and the Gulf of Guinea (given the growing importance of oil from Angola and Nigeria). Bush adopted the policy of pre-emptive wars which culminated with the invasions of Afghanistan and Iraq. In addition, it tried to isolate Russia to prevent its re-emergence (Klare 2007, 2008).

At the beginning it was very successful with its policy to isolate Russia. Most Eastern European countries (Poland, the Czech Republic, Hungary, Estonia, Lithuania, Latvia, Bulgaria, Romania, Slovakia and Slovenia) joined NATO and many of them have American military bases. It managed to bring Azerbaijan, Georgia, Moldova and Ukraine into its orbit. These changes lead to modifications of the routes of oil and gas pipeline projects that suited its strategy. In 2005 it managed to complete the oil pipeline that allows transporting the oil from Azerbaijan through the Turkish port of Ceyhan, passing through Georgia. But since 2005 it has suffered numerous setbacks in this area (which I will explain later), a loss of influence in South America, with the rise to power of populist governments and the subsequent nationalisation of fossil fuels. It was unable to prevent oil prices from escalating. To the contrary, it made things worse, because resistance tothe invasion of Iraq kept the rate of oil extraction below pre-war levels. Its manoeuvres to destabilise the regime in Iran are not yielding results either. In 2008 an agreement was reached to build the Iran-Pakistan-India gas pipeline (which has also been joined by China), despite strong pressure by the US on Pakistan to make it back out from the project. In addition, its interventionist policy is infeasible from an energy and budgetary point of view. The rapid intervention strategy requires armies equipped with large means of transport, which multiplies the consumption of oil. During the 1991 Gulf War, consumption per soldier was 15 l of oil per day (l/day). In Iraq the ratio has risen to 40 l/day and the rapid intervention wars would need around 240 l/day. These data led a report by the Pentagon to declare that the US's war strategy is infeasible (Klare 2007, 14/06/07; Brown 2009: 17). As a result of all this, it is increasingly evident that its leadership is weakening. A number of different bodies acknowledge this. For example, the National Petroleum Council (NPC 2007) states that "many of the expected changes could raise the risks for the US's energy security, in a world where its influence will probably decline as economic power shifts to other nations".

1.2 Russia

It has a key role in the future of fossil fuel energy due to its enormous oil and natural gas reserves. It was responsible for the 40 % increase in the global offer of oil in 2000–2008, it is the second largest exporter of oil and the largest oil exporter. The governments of the Yeltsin era pillaged the Russian State's enormous wealth of natural resources, of which the most important asset were fossil fuels, violently repressing all opposition, and all this with the consent of the West. Putin has reversed much of the path travelled by Yeltsin, returning most of the oil and gas to state control through the companies Gazprom and Rosneft. Gazprom owns 90 % of the gas and has a monopoly on exports. Rosneft controls most oil reserves (Weird 2007).

In addition, it is intensifying cooperation agreements with ex-Soviet republics to keep them within (or to return them to) its orbit. Its main weapon is natural gas, although it has a structural weakness in this field. Its gas exportation networks to Europe are in Ukraine (80 %) and Belarus and it needs to sell gas to Europe, because the domestic market consumes over 2/3 of the total and at below-cost prices. Although it wants the ex-Soviet western republics to pay market prices, this has not been achieved. In addition, they often do not pay their debts, they divert more gas than has been contracted to resell it and when Russia cuts off the supply, they consume their quota by taking it from the flow going towards the EU, causing confrontations with it. The election of a pro-Russian government in 2010 has facilitated relations between both countries. But its strategy consists of building new gas pipelines towards Europe avoiding traditional enemy countries. In addition, it diversifies its client portfolio with multiple gas and oil sale contracts to Asian countries (China, India, Japan, etc.) (Levine 2009).

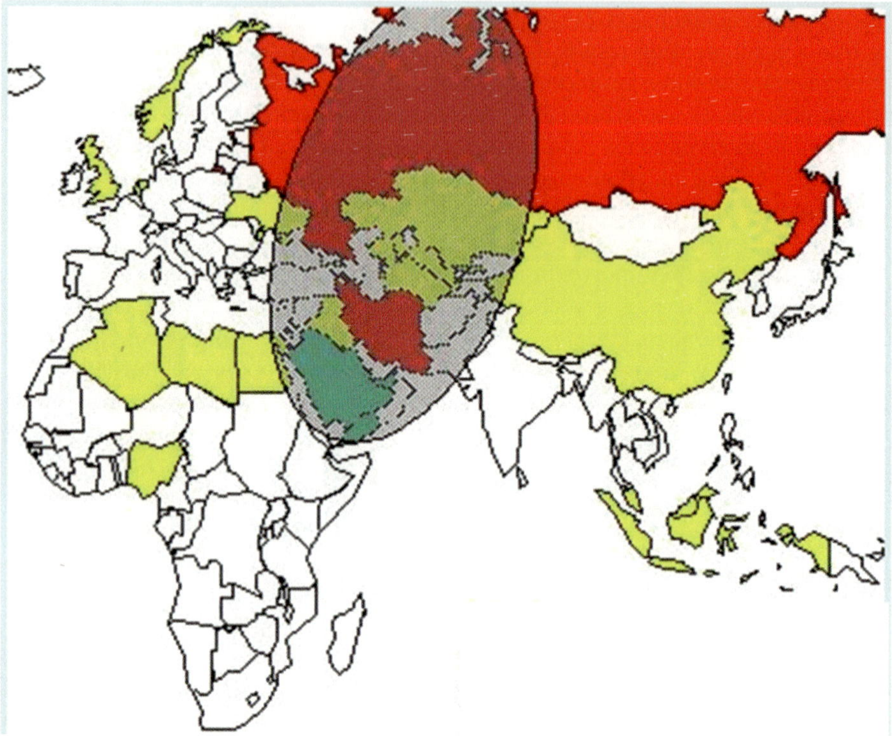

Fig. 11.1 Strategic ellipse that contains around 70–80 % of gas and oil reserves (Source: BTC 2010)

1.3 European Union

The EU imports 54 % of the energy it consumes and this situation is quickly worsening, because its domestic contribution is decreasing. Consumption of oil by the EU-25 in 2008 was around 15 Mb/day and it imported around 80 % (considering Norwegian oil as a domestic contribution). It is estimated that oil extraction in Norway, Great Britain and Denmark is decreasing by over 8 % per year. Russia is supplying 32 % of the oil. The situation of natural gas in Europe is very similar to that of oil. It imports 61 % including to Norway. So its energy vulnerability is very high. In its favour is the fact that at 5,000 km. from the centre of Europe there is an area that contains 70/80 % of the world's oil and gas reserves, as is shown in Fig. 11.1, which makes it easier to make strategic alliances that can guarantee the supply. Member States agree on diversifying its sources of imports, reducing the dependency on Russia, but each country prioritises its own interests. Because the situations are very different: Finland, Slovakia, Bulgaria, Romania and Hungary are 100 % dependent on Russian gas and Austria, Poland, Greece and the Czech Republic, 80 %. Germany will soon go from the current 40 to 60 % (Reguly 2007; European Commission 2008g).

In recent years oil and gas pipeline projects between Russia and many EU States have been approved. Germany and Russia are building an oil pipeline through the Baltic Sea (Northern Stream), avoiding countries that are in confrontation with Russia. Russia has also made an agreement with Bulgaria to build another pipeline (Southern Stream) through the Black Sea to this country. This branches into two pipelines: one goes through central Europe (with Hungary as the distribution centre) and the other through the Mediterranean, to Greece and Italy. In addition, it has an agreement with Bulgaria and Greece to export oil through the Mediterranean Sea. It will be sent by sea to the Bulgarian coast and from there a pipeline will take it to a Greek port, from where it will be distributed by sea (Toshkov 2008).

1.4 China

In China, coal produces 79 % of its electricity. But the need for oil and natural gas is rapidly growing. Until 1993 it was self-sufficient in oil, but in 2003 it became the second largest importing country, with its purchases growing at an average annual rate of around 10 % and its domestic consumption 6 %. Its top priority is access to the world's oil and natural gas, but it knows that it has arrived late to the sharing out and that in this field politics take precedence over the market. Apart from promoting various oil and gas pipeline projects with the ex-Soviet republics, it is reinforcing its presence in Africa (30 % of its imports) and in Latin America. It has reached agreements with Russia, Kazakhstan (an oil pipeline transports 200,000 b/day between Kazakhstan and China), Iran (in 2007 they signed an enormous oil and natural gas sale contract) and many African countries. Investment in Africa has gone from $490 m in 2003 to $9.33bn in 2009 (in 2009 it agreed to invest around $40bn in Russia, Brazil and Venezuela in exchange for supply contracts) (Levine 2009).

1.5 The Foreseeable Future

It is highly probable that the progressive scarcity of fossil fuels (and in particular oil) will acutely exacerbate the historical confrontation over their control. O'Reilly, president of Chevron-Texaco, accurately describes this situation: "we are seeing the start of a dispute over the supplies from the Persian Gulf between the East and West (. . .) and the shift of the centre of gravity towards Asia and, in particular, towards China and India" (www.boston.com/news). The Pentagon has been carrying out regular analyses of the impact of the energy crisis and seeks to be the instrument once again to guarantee the use of the scarce resource. The one from 2010 states: "In the next 25 years, the US armed forces will be continually involved in a dynamic combination of combat, security, involvement, aid and reconstruction (. . .) There will always be an opponent that will try to undermine political stability and free access to the global resources that are crucial for the global economy"

(JOE 2010). It is particularly concerned about the possibility that the “arc of instability” (from northern Africa to south-east Asia) could become the “arc of chaos”, if a consumer nation intervenes militarily and more specifically in the Strait of Hormuz (17 Mb/day transit through it), the Strait of Malacca (15 Mb/day) and the Suez Canal (4.5 Mb/day) (JOE 2010: 4, 27, 28). The German army considers that conflicts over they control of resources will multiply and that they will focus on what it calls the “strategic ellipse”, mentioned earlier. It predicts the proliferation of sabotage to the transport infrastructures of these resources and even possibly wars over their control. The most hard-hit sector by the crisis will be transport, and therefore it predicts a “mobility crisis” and the need to ration the distribution of fuel. Most of the oil will be distributed internationally based on agreements between governments. International institutions will be weakened because governments will prioritise solutions to their own problems. It estimates that the crisis will last 15–30 years. It urges the government to understand the seriousness of the threat and to act swiftly, starting with a vulnerability study. And it proposes that in the long term the army should only use renewable energy (BTC 2010).

2 Analysis of the Current Crisis and of the General Impact of Peak Oil

2.1 Causes of the Crisis and of the Unequal Impact on Different Countries

To analyse the causes of the crisis means contradicting orthodox economists because: they are incapable of foreseeing crises (and therefore of rigorously explaining them); and they will not admit that there is a scarcity of natural resources. Figure 11.2 shows that they never foresee the crises, despite them being recurrent.

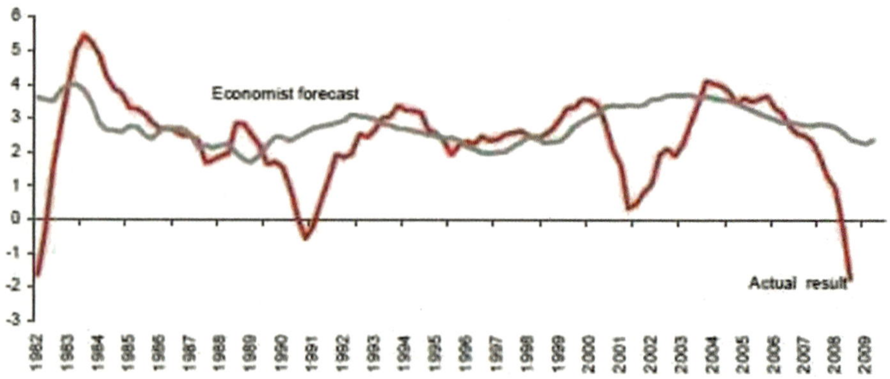

Fig. 11.2 Economists’ predictions and the truth about the US’s GDP development (Source: Cohen 2009a, b)

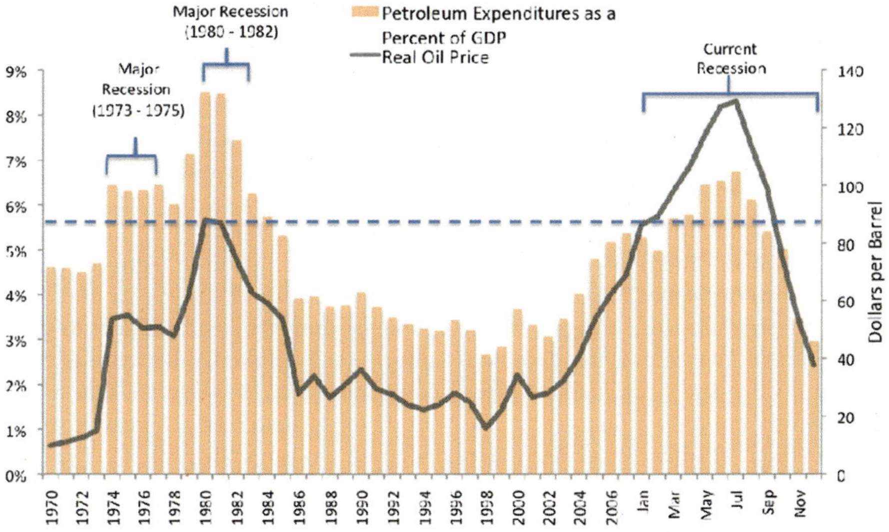

Fig. 11.3 Percentage of expenditure on energy in relation to global GDP (Source: Murphy and Hall 2011: 58)

They argue that there is not a scarcity of resources, because when it appears the price of a resource rises and demand falls, incentivising investments in new explorations, which results in an increase in supply and lower prices. Although this does happen, it is not able to end scarcity. New deposits are increasingly small and harder to access, and the high prices of raw materials make investments more costly, which results in fewer new deposits being exploited (Cohen 2009a).

Many studies show that high oil prices normally precede economic crises, from which we can infer that there is a clear correlation between prices and recession. J. J. Hamilton (2009) shows that out of 11 recessions that took place in the US after WWII, 10 were preceded by sharp rises in oil prices. D.J. Murphy and C. Hall (2011) estimate that when energy spending reaches 5.5 % of GDP there are recessions, because this overspending results in a reduction of other consumptions and of investment. This is shown in Fig. 11.3. But in the 1973–1975 recession at 5.5 ratio was not reached, which indicates that other variables are also involved. In this case prices quadrupled between October and December 1973 and the driving forces behind the 1950–1973 expansion period had been weakened (Tverberg 2011).

In the crisis that started in 2008 the escalation of oil prices played a decisive role. However, the centres of power classify it as a financial crisis. Without denying this factor, the escalation of oil prices and those of many other raw materials (some of them saw price rises greater than those of oil) were a reflection of what J. Grantham (2011), Chief Investment Strategist of GMO Capital, calls a "paradigm shift". The combination of both escalations was a decisive factor in the events that led to the current economic crisis. These escalations reached levels far higher than the thresholds mentioned above and caused high inflation and a strong reduction of spending and investment capacity in the OECD countries due to oil payments. When the barrel reached $135, it cost the US (oil consumption extrapolated to 1 year) one trillion dollars (15 % of tax revenue) (Korowicz 2010: 11). In late 2007 inflation started to rise, reaching in OECD countries figures of 5–6 % in 2008. Central banks responded with the usual reaction: by raising interest rates. This made credit more expensive, which was joined by a fall in revenue due to the higher price of oil and other raw materials. Both factors caused the collapse of the real estate bubble that took place in the OECD. This in turn caused the collapse of the financial system, affected by numerous problems. Some of them originated by the enormous speculation brought by the liberalisation of the sector and other structural problems of the neoliberal economic system (the need for debt to grow to encourage investment, the enormous and growing difference between the real historical value and stock market value of companies, etc.). So the crisis that started in 2008 was not only a financial crisis, "but was a symptom of a bigger crisis, an energy crisis" (Rubin 2009). Many authors agree with this analysis: G. Tverberg (2009b), J. Hamilton (2009), Murphy (2009), L. Kilian (2010), Kopits (2011), etc.

But these and other authors usually go further, by defining the price thresholds that start to stall economic growth. They consider that OECD countries are affected when the price of a barrel exceeds $90 during long periods. Which would indicate that they cannot emerge from the crisis while prices are over $100/b (Kopits 2011). And some authors even calculate the reduction of GDP that is taking place. G. Davies predicts that, if the price of a barrel continues to be much higher than $90, the GDP of OECD countries will lose around 1 % in 2011 (Mearns 2011b). The EU's oil bill went from $280 m in 2010 to $402 m in 2011. This explains, in part (the Euro crisis is also having an influence), why the predictions for the consolidation of economic recovery have not been met in 2011. Unlike what happens in the OECD, analysts consider it probable that China and India can cope with prices of $100–$110/b. The reason that would justify this disparity of thresholds is that the increase of marginal productivity of a barrel of oil is greater in India and China, given that their consumption per capita is far lower than that of the OECD, because the contribution to GDP of a barrel is greater than in the OECD (Skrebowski 2011; Financial Times, Nov. 27).

2.2 *Analysis of the Economic Impact of Peak Oil*

Three methods are used to analyse the economic effects of peak oil: linear decline, oscillating decline and systemic collapse. The first method assumes certain rates of decline of oil extraction and it is considered that the impacts are proportional to prices. The impacts on GDP, the balance of payments, etc., are analysed. It is also supposed that, despite the economic impacts, the financial system remains stable, states maintain the usual services, there is no social instability, etc. So there is no interrelation between the factors (Korowicz 2010: 30, 31).

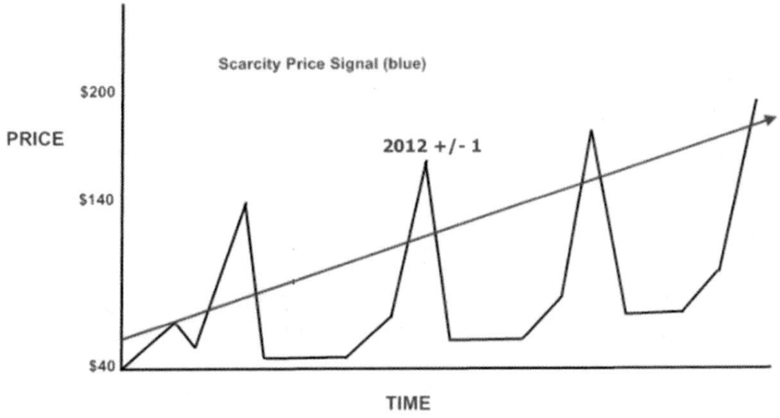

Fig. 11.4 Oscillating decline (Source: Cohen 2009a, b)

The oscillating decline method supposes that there will be successive escalations of oil prices, which will cause recurring crises and slumps in barrel prices. This would relaunch the economy, leading to a new escalation of prices that would reach a higher level. This would cause another new cycle, which would repeat itself until the energy model is changed. In each cycle there will be an economic decline (Cohen 2009a, b; Korowicz 2010: 31). Figure 11.4 reflects this theory and also that the first escalation caused the first crisis cycle.

This theory does not reflect the current reality either. Instead of recoveries that would relaunch consumption and the price of oil to new heights, we are seeing economic stagnation in the main countries of the OECD. This is due to the fact that these countries have a lower impact threshold regarding oil prices than the rest, and which has already been surpassed. In the EU we must take into account the impact of the crisis on the monetary system. All this reduces crude oil consumption in the OECD: 3–4 % per year. But Non-OECD countries increase it by 4 % per year (IEA 2011).

When peak oil becomes manifest, it will worsen the decline of crude oil exports, which has already been happening during the supply stagnation period. The price of a barrel will surpass, also, the price thresholds of Non-OECD countries, which will lead to a long global crisis, because the amount of available energy will be reduced. Many reports predict a collapse: "the decline or collapse of the oil supply will hit all economic sectors, leading to rapid changes in transport" (IPTOES 2010a: 26). The report by the German army (BTC 2010: 58) predicts that "in the medium term, the global economic system (. . .) could collapse" and that the impact of peak oil will last 15–20 years. This interval is also predicted by the Hirsch report (Hirsch et al. 2005). This scenario of a collapse seems inevitable, because governments are not taking steps to quickly break away from oil. The magnitude of the collapse will largely depend on the reduction rate of the oilsupply. R. Hirsch et al. (2010) establish two scenarios, with annual reductions of 2 and 4 %, and consider that the second would be catastrophic. Global GDP would be reduced by 20–30 % in little over a decade and would generate a social crisis capable of destabilising many countries. But the mitigation mechanisms that generate high prices make the second scenario more probable.

This vision of the crisis still only focuses on the effect of oil and is applied mechanically (there is no response of the market to the scarcity of oil). Many other factors are missing, such as the financial system or the scarcity of strategic materials. For this reason a systemic approach to the causes of the collapse is required (Korowicz 2010: 31). The economy needs markets, financing, monetary stability, infrastructures (transport, telecommunications, water treatment, electricity grids, etc.), industrial and food production, health and education services, R&D, respected institutions, socio-political stability, but also a growing flow of energy and materials. The system adapts to changes spontaneously, because there are no global government institutions (GI) capable of controlling its operation (Korowicz 2010: 7). Without aiming to carry out an exhaustive analysis, a number of positive and negative factors are made evident. Among the first there are:

- Expensive oil will continue to cause an acceleration of technological changes to exploit deposits that were not profitable with previous technologies and prices. This factor and the reduction of consumption as a result of the crisis will slow down the falling flow rate of oil.
- During this decade the main renewable energies will reach cost parity, which will accelerate the increase of its market share and its application to the production of hydrogen.
- The movement of societies regarding energy emergency is rapidly being strengthened, encouraging their economies to make them more resilient to peak oil (Bermejo 2011: 304–308).

But there are very important negative factors that will delay the changes:

- Peak oil will reinforce the international status of oil exporting countries and the trend towards the nationalisation of oil. But its high rate of increase in consumption will reduce exports even more (BTC 2010: 26).
- Confrontations between States over access to deposits in disputed territories will worsen. Geostrategic tensions between powers over long-term supply contracts will also worsen. This is what has been predicted by many analysts and reports by government bodies (BTC 2010: 73).
- China's role in stabilising the global economy will not be possible in the future, because it will be increasingly affected when the price of oil exceeds its threshold, worsening the global crisis.
- It is not probable, not even in the medium term, that powers will agree on a substantial reform of the financial system. Smaller measures will undoubtedly be adopted, such as weakening the role of tax havens, controlling somewhat financial speculation, etc. But they will not prevent the financial system from playing a prominent depressive role in economic activity.

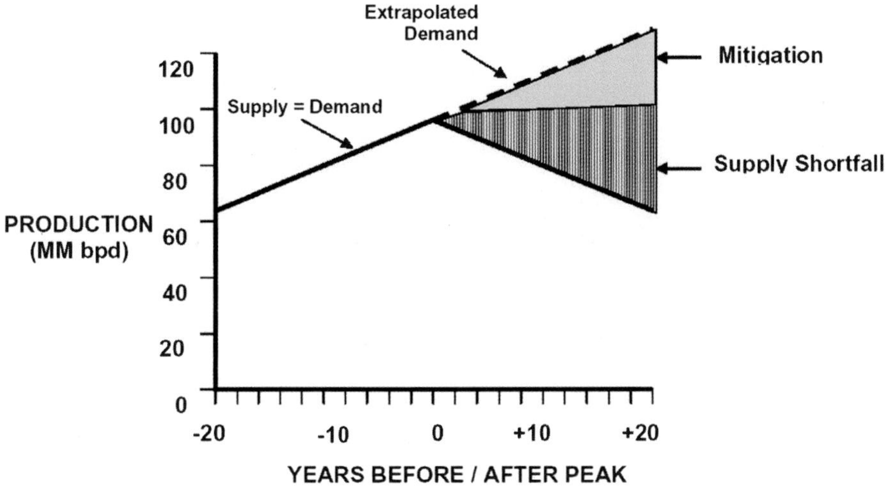

Fig. 11.5 The worst crisis scenario (Source: Hirsch et al. 2005: 64)

- It is not probable that the growing environmental impacts generated by the extraction of resources from the Earth's crust will curb (significantly) its exploitation. Meaning that in the medium to long term the essential services provided to us by the biosphere will be decisively deteriorated and will become an increasingly important crisis factor.

So the collapse will mean the break-up of the financial and monetary system, of the globalised productive system, of critical infrastructures (transport, electricity grids, health, etc.) (Korowicz 2010: 13). But we must not forget that countries have very different levels of vulnerability and for this reason they will suffer very different impacts. Figure 11.5 shows two types of reduction of consumption due to falling supply, supposing that the substitution process of fuel takes place in 20 years. As the difference between supply and demand expands, the drops in consumption are increasingly serious. The Triangle (representing mitigation) in Fig. 11.5 indicates the magnitude of the savings in oil that can be achieved without traumatic measures. But it represents less than half of the required reduction of consumption. The lower triangle shows that the above measures will not be enough, meaning that the consumption of fundamental products and services will have to be reduced.

These two types of consumption are shown as a continuous scale in Fig. 11.6. The elasticities of consumption are organised in relation to prices from high to low (between 2.0 and 0.8). An example of high elasticity to prices is the reduction of the use of cars in towns or cities that have an efficient collective transport system. But as prices continue to rise due to growing scarcity, increasingly essential activities are being reduced (electricity, industrial activity, collective transport, heating in buildings, health services, safety services, etc.). The drastic reduction of these

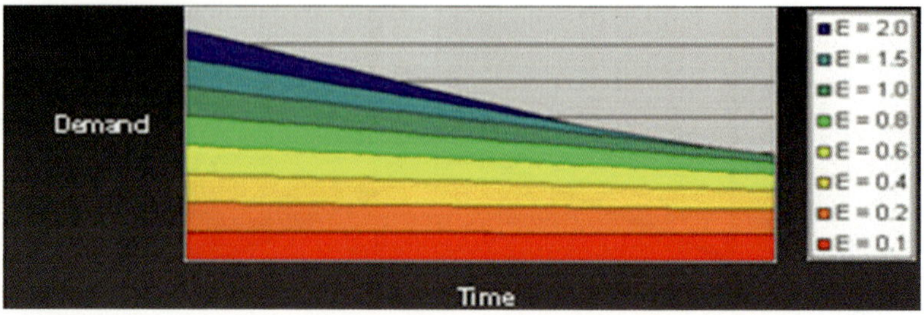

Fig. 11.6 Different elasticities of consumption segments to prices (Source: Vail 2008a, b, August 20)

activities could lead to the de-legitimisation of government institutions and chaotic situations.

Most societies will suffer a very serious impact, but its intensity will depend on the characteristics of each one. The impact will be softer in proportion to the predominance of the following factors: administrations with healthy finances; low relative dependency on oil; high energy efficiency; large oil and/or gas resources; strong development of solar technologies; high capacity for technological change; a not excessively exposed economy; highly diverse economic fabric; solid institutions; high social cohesion. The extractive countries of the OECD (Norway, Canada, Australia, etc.) will be among those less affected (BTC 2010).

As has already happened in the past, States will become more interventionist, due to the need to encourage rapid transformations to overcome the crisis. The BTC expects a reduction of the free market mechanism (petrol rationing, political alliances among states to guarantee the supply of fossil fuels, protectionism, etc.) and more self-centred economies due to the rising costs of trade. The latter will be drastically reduced, particularly long-distance trade, due to the rise of energy costs and falling economic activity. J. Rubin and B. Tal (2008), economists of the bank CIBC, state that "globalisation is reversible". Expensive oil strengthens regional trade. The BCT recommends the creation of redundant and highly resilient infrastructures and decentralisation. But necessity will force the creation of more diverse, self-sufficient and de-centralised economies. This phenomenon will mainly take place in the energy, industrial and agriculture sectors. It will increase support for hydrogen of renewable origins. The activities of repair, re-manufacturing, recycling, etc. will be strengthened. In the financial sector measures that will mean a certain nationalisation of credit will be applied and community financing systems will be strengthened. Priority will be given to investments in collective transport and railway transport for goods. J. Rubin (2009) says that he would not be surprised, "if the new, smaller world that emerged was more reliable and pleasant than the one we are about to leave behind".

In the sphere of international relations the BTC (2010) predicts that they will be weakened and that there will be more serious confrontations over access to increasingly scarce energy, particularly in the strategic ellipse and in seabed deposits. It is possible that this will be the first consequence of peak oil, but afterwards it seems probable that they will be strengthened to facilitate the collective transition.

3 Structural Effects

In this section I will briefly explain the structural problems of the international trade, finance and monetary systems and how peak oil can transform them.

3.1 Trade

Historically strong and lengthy increase in the price of fuel has had a depressive effect on trade, particularly long distance international trade, and redirects it towards regional trade, as shown in Fig. 11.7. During the period from 1960 to 1972 the export indicator in relation to global GDP grew by 50 % and over 60 % during the period from 1987 to 2002. During both periods oil was very cheap and export duties were reduced. On the contrary, during the 1974–1986 period we see that, despite the fact that global GDP grew at an average annual rate of 3.5 %, the above indicator remained the same, due to the strong escalation of the price of oil. There was also a reduction in long distance trade in favour of shorter distance trade. During the period from 1966 to 1973 the average trade distance increased by 30 %, but during the period from 1974 to 1986 long-distance trade dropped by 30 %. Between 1973 and 1980, period during which the nominal increase of the price of a barrel of oil was 400 %, the proportion of long-distance imports (from Asia and Europe to the US) fell by 32 %, while regional trade (Latin America) increased by 30 % (Rubin and Tal 2005, 2008).

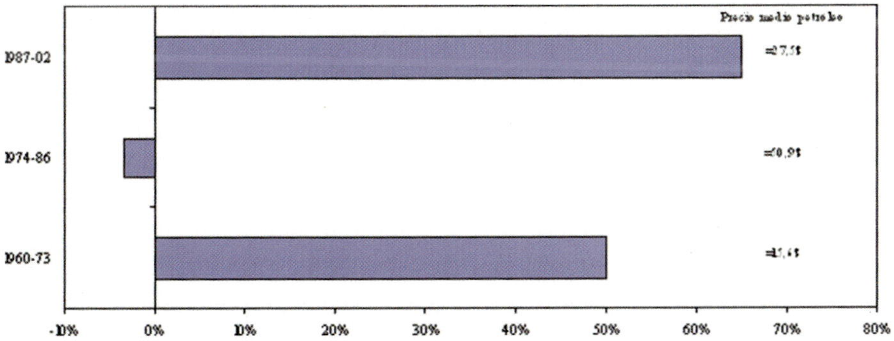

Fig. 11.7 Ratio of worldwide exports in relation to global GDP (Source: Rubin and Tal 2008)

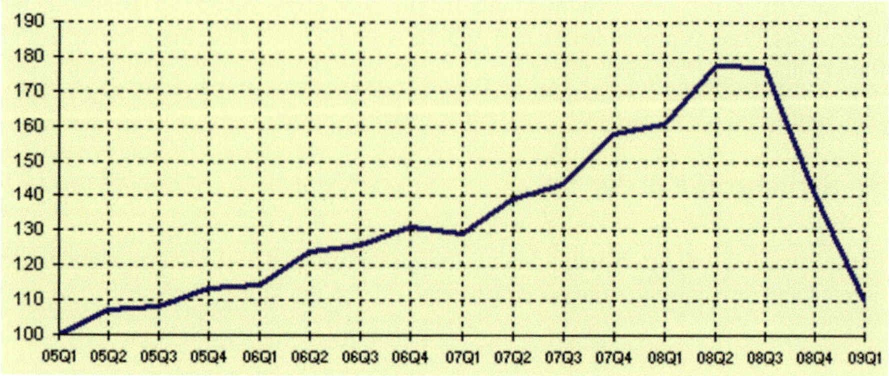

Fig. 11.8 Evolution of global trade 2005–2009 (Source: www.wto.org)

On the other hand, the monetary blindfold once again prevents us from seeing reality. Until at least 2007, global trade apparently continued to experience strong growth. According to the WTO, in 2006 global GDP grew by 3.7 % (the second highest since the year 2000) and global trade did so by 8 %, with the trade of goods reaching over 9 %, subtracting inflation. However, in physical terms the trade of fossil fuels and metals stagnated, although in monetary terms it was the trade segment that grew the most, due to rising prices. The crisis caused the collapse of trade measured in commercial value from the third quarter of 2008, as shown in Fig. 11.8 (www.wto.org; WTO 2007: 4–10).

The analysis of trade equating the costs of transport to duties make the effect of oil prices on it more evident. In the year 2000, when the barrel was at $20, transport costs were equivalent to a 3 % duty for the US. In 2008 they were equivalent to a duty of 9 %, with $130/b. At $150/b the equivalent duty is 11 % and at $200/b the duty gets to levels of the early 1960s, when the liberalisation process had hardly begun. Each time the distance increases by 10 %, transport costs rise by 4.5 %. These ratios explain the reduction of long-distance goods trade with a high weight/price ratio, such as the case of steel and iron products, appliances, footwear, industrial machinery, etc., which is what happened in 2008. These are typical products of Chinese exports. In 2004 they represented 52 % of Chinese exports to the US and in 2008 they dropped to 42 %. These events caused a strong rise in domestic production and imports from neighbouring countries, such as Mexico and the Caribbean (Rubin and Tal 2008).

Cases of trade regionalisation due to the action of the market and of governments are starting to be widespread. The Chinese government is aware of its economy's vulnerability to escalating oil prices, because it depends heavily on exports (100 million Chinese people work in the export industry). And to reduce it, it seeks to: increase own consumption; reinforce regional trade; and to do it mainly by railway (it is promoting the creation of a railway network spanning a large part of Asia), because it avoids sea routes, controlled by the American fleet. The extension of the EU intensifies regionalisation. This is also happening with Russia and the ex-Soviet republics of Asia, in South America and the region encompassing the Persian Gulf and northern Africa, and in sub-Saharan Africa (Bergin and Glick 2006; Dubyne 2007).

3.2 *Financial*

Private banks generate the vast majority of money by granting loans, appearing in digital form: in the accounts of the customers who request these loans. They fulfil their payments through transfers to other accounts, credit cards, checks, etc., which are elements of the money supply. Central banks still print money, but this is only a very small fraction (around 3 %) of the money supply. Each loan that is granted produces a cascade of new loans, because most of this money is not usually spent immediately, but in instalments and part of it goes to other accounts within the bank. So the creation of money and economic growth are determined by a constant increase in debt. But central banks require that they have part of their funds available, normally less than 10 %. In a growing economy the generalised increase of wages allows most debtors to fulfil their financial obligations. It is a system based on confidence in perpetual growth. But when the economy ceases to grow the system collapses. In this context banks drastically restrict credit, and cease to offer loans and obtain funds in the inter-bank market (as they have done traditionally to maintain their liquidity), because there is a big lack of confidence in the solidity of the rest of the system.

It has been stated that the weakness of the financial system was in the international sphere, as it lacks institutions capable of regulating a market characterised by speculative excesses. And on the other hand, at a national scale there are institutions that regulate the markets and avoid excesses. In addition, the certification companies would inform the market on the solvency of companies and the value of their financial assets. But this has turned out to be untrue. In the neoliberal frenzy some powers deregulated their financial systems and applied fiscal policies that encouraged speculative processes. Traditionally, two types of banks have coexisted: commercial and investment banks. The former have a high number of customers who deposit their income (that could be wages, pensions or profits, mainly from SMEs) and with these funds they lend money. It is a business that generates a moderate profit margin and is quite safe, because the bank is aware of the financial situation of its customers. Investment banks organise large sales of shares, public debt securities and manage the mergers and purchases of companies. It is a much riskier business, but it can provide greater profit margins. The separation of both entities seemed reasonable, because in the case of the mixed bank, if the investment division makes a managerial mistake, the commercial division would be forced to offer loans to the entities involved. The US, the leader in the deregulation process of the system, repealed in 1999 the law (approved during the Great Depression) that separated both banks and considerably reduced the ratio of capital that the investment bank had to keep in reserves, thus increasing the available funds for its business, but also the risks. And the Bush Administration decreed a tax reduction for the highest incomes, thus rewarding speculators. This process meant that commercial banks gradually reduced their traditional business practice and increasingly acted as investment banks. This process meant that credit became more expensive and scarce for SMEs, micro-companies and the general population. In addition, the rating agencies cannot be objective, because they live off the payments of customers (Paris 2009a; Townsend 2013).

The process described created a breeding ground for an enormous development of speculation, which ended in collapse of the banking system. The speculative nature of the financial system and its high volatility are determined by the large amounts of funds that seek to maximise their profitability in the international markets. Which has negative repercussions. Banks knew that the speculative policy was risky, but most of them were drawn in to apply it, because those who started with it obtained hefty profits and then ran the risk of being absorbed by them. This panorama is described by the Bank for International Settlements (BIS 2007: 7) as "irrationally exuberant" and explains it as a natural trend: "There seems to be a natural tendency in markets for past successes to lead to more risk-taking, more leverage, more funding, higher prices, more collateral and, in turn, more risk-taking". And this warned of the high risk of collapse of the financial system. A prediction, shared by many authors, that has become true, although the magnitude of the collapse has been so big that few predicted it, due to the lack of transparency of the financial system. The crisis led to the collapse of many banks and governments injected enormous public funds to re-float them and thus bring back confidence in the banking system, which is an essential requirement for its operation. But this policy generates an enormous feeling of discrimination in the rest of the business world and in societies, because it means the privatisation of profits and the socialisation of losses (Stiglitz 2009; Paris 2009a, b; Townsend 2013).

3.3 International Monetary System (IMS)

One of the pillars of the supremacy of the US is that the dollar is the hegemonic currency in international trade. The monetary system implemented after WWII made the dollar the global currency, but it was convertible into gold, although in a restricted way (only governments could demand gold in exchange for dollars). At first no one was interested in making this exchange, given the strength of the dollar, motivated by the vigour of the American economy and its military might. But the subsequent weakening of its economy led to an increasing outflow of gold, which is why president Nixon made the dollar inconvertible into gold in 1971. Which meant that the US acquired the privilege of paying its debts by printing dollars and without any risks. Being the reserve currency, its volume must grow in order to facilitate the increase in transactions. This is achieved by having a negative balance of payments, so what is a problem for the rest of the world, for the US is a need of its monetary role. Though it is evident that such power has its limits, as a high and sustained foreign deficit leads to flooding the global economic system with dollars and with it, its devaluation, made worse as it is not backed up by gold. So in 1971 it started to lose value quicker than in the past, with the result that within the OPEC some favoured selling oil in more stable currencies. In addition, when in 1973–1974 its price quadrupled in Germany, France and Japan sought to buy oil with their own currencies, faced with the danger of exhausting their reserves of dollars. Both movements were aborted by the US. In 1975 it reached an agreement with Saudi Arabia whereby the US pledged support to the Saudi regime in exchange for it selling its oil in dollars and maintaining a low price. The OPEC was forced to approve the measure. Since then the US (and in particular the Pentagon) have defended the role of the dollar by all means. The manoeuvre meant that oil importing countries had to build up their reserves of dollars and exporting countries were forced to maintain the strength of the dollar, in particular by investing in these countries, thus draining dollars from the global market. This flow allows it, in addition, to cover the budgetary deficit because part of the investment is in public debt securities, and to maintain an artificially low interest rate. This virtuous cycle avoided the collapse of its economy. During the 30 years after 1971 global reserves of dollars increased by 2,500 % (Jones 2005: 105; Engdahl 2006; Petrov 2006).

However, the predominance of the dollar and the US's policy of depreciating the dollar by maintaining a low interest rate, as a way of reducing its chronic balance of payments deficit, go against the interests of the rest of the world: oil and gas exporters see that such a policy reduces their profits; States that export to the US are forced to align their currencies with the dollar to avoid damaging their sales; and the volatility of the dollar harms the rest of the world. Low interest rates worsen inflationist tensions in economies undergoing a strong expansion, such as those in the Persian Gulf or China (Johnson 2008).

This situation leads countries to develop defence mechanisms. Fossil fuel exporting countries tend to dispense with dollars by: selling fuel in other currencies; replacing investments in dollars with others denominated in different currencies; and a reduction by their central banks of the amount of dollars in their reserve funds. Some of these trends are shared by most OECD and emerging countries. The number of countries that have decided to sell oil and gas in currency other than the dollar is growing. This is usually easy for Persian Gulf countries, because most of their trade is with the EU (30 %) and Japan. In 2008 Iran announced that it had ceased trading in dollars. Venezuela, Ecuador, UAE and Indonesia are following this policy with different intensities. But although they all want to replace the dollar, there are discrepancies as to how to do it: the moment to start and the pace of development. The creation of hydrocarbons markets based on other currencies is a further factor that weakens the dollar. The Gulf Cooperation Council (GCC) created a hydrocarbons stock exchange (Dubai Mercantile Exchange) that carries out transactions in a common currency. Russia will sell gas in euros to the countries that will receive gas from the new gas pipelines (White 2009).

Last of all, the global situation in which the US imposed the current IMS over 50 years ago has changed drastically. At the time it was the most important oil extracting country, one of the main energy and food exporters and held 80 % of the world's gold. Now it is the country with the highest debt, the one that imports the most energy and it has a smaller reserve of gold. In 2009 it had a deficit of $1.4 trillion (11.2 % of GDP) and the aggregate debt was $5.4 trillion (Ferguson 2010). So there is a widespread request for a change of the IMS. J. Stiglitz (Nobel Prize in Economics) states that the current system is "relatively inflationist, unstable and unfair" and asks for its replacement (UN News Service, 10/09/09). The Bank for International Settlements (BIS 2007: 6) states that "the dollar continues to be clearly vulnerable to a sudden loss of confidence by the private sector". And oil is a key factor of the change. The international IMS based on the dollar without any gold backing is harmful for the world. It is only explained by the fact that the US still retains some leadership that is enough to maintain this "extraordinary privilege", according to a report by the Pentagon. Which means that "any action that goes against the hegemony of the dollar represents, de facto, a declaration of war" (The Economist, 10/11/05). But the demand for a change of system cannot cease to grow. It has been demanded by the UN's Global Financial Crisis Inquiry Commission, the president of the IMF, China, Russia, France, Japan, India, Brazil and the Gulf Cooperation Council. But these countries go further: the media has informed us that, at least, during the autumn of 2009 they met secretly to agree on another IMS (JOE 2010: 19).

4 Sectoral Effects

In this section I analyse only the impacts of escalating oil prices on industry and agriculture because, as the energy and transport sectors were the most hard hit and will suffer big transformations in the future, I dedicate several chapters to them.

4.1 Industry

During the first oil price escalation from 2005 to 2008 several trends in the industrial sector were observed: a displacement of energy-intensive industries from OECD countries to oil and gas exporting countries; a reduction of the length of product chains; an extraordinary development of the industry that produces renewable energy capture systems; a rebirth of production in some industrial sectors in OECD countries as a result of the higher transport costs of goods with a low ratio of commercial value and weight; a decline of road vehicle and aircraft production companies; the strengthening of railway, naval and bus manufacturing companies. Some of these trends continued during the subsequent crisis.

During the period from 1985 to 1999, when oil was extremely cheap, global product chains were developed, thanks to low transport costs. A typical example is that the most work-intensive stages are carried out in Asia and the rest in Europe and the US. The escalation of oil prices greatly reduced these displacements and aborted projects. In 2008 Tesla Motor (a pioneering company that makes electric cars) cancelled a project in which the batteries were to be built in Thailand. In 2008 Chinese exports to the EU and the US were reduced due to transport costs. The same could be said about components for cars, presses, furniture, etc. In 2008 Ikea built its first furniture factory in the US (Rubin and Tal 2008; Rother 2008).

Investment in the sub-sectors of heavy industry and basic chemistry has been intensified in the Persian Gulf, due to low energy prices. In this region the capacity for ethylene production (the main raw material of the petrochemical industry) increased by 7 million tons in 10 years and is expected to grow to 32 million at the start of this decade. Transnational companies from the metal industry are also searching for cheap energy in this area. ALCAN is building an aluminium plant in Saudi Arabia that will produce 2 million tons by 2016. A Malaysian consortium has reached an agreement with the Saudi government to invest $30 billion in plants to produce aluminium, steel, copper, etc. Dubai is building another aluminium plant in Dubai that will produce up to 1.4 Mt/year. But there are signs of change. The city of Masdar in Abu Dhabi, apart from using only solar energy, hopes to become a global centre of technological development in this field, with a specialised University and a Technology Centre. The strong opposition that the Saudi industrial model has generated, led the government in 2009 to decide to create six large cities that will foster the knowledge economy in 15 years (www.odac-info.org; Ambah 2008).

There is an extraordinary growth in the production of systems to capture renewable energy. The wind and photovoltaic power installed worldwide has been growing at an average annual rate of 30–60 %, but the crisis is having an impact on wind energy. In Europe, installed renewable power has grown more than conventional power in 2008, 2009 and 2011 (EPIA 2012).

As we will see in the next chapter, the escalation of oil prices had a strong impact on the use of road vehicles and on air transport, and plunged the automotive sector and in particular the large American companies into a strong crisis. The main cause of the higher impact in the US was its inability to move away from classic models, which consume large amounts of petrol. General Motors was the company most reticent to change its policy. Companies are looking for provisional solutions (hybrid and electric cars, etc.), but the most important ones declare that their goal is a hydrogen fuel cell vehicle. So the scarcity of oil and the lack of a mature alternative will determine a reduction of the size of the sector in OECD countries and the development of more efficient vehicles at a global scale. Air transport was also hit hard by rising oil prices and the subsequent crisis, meaning that it will be forced to carry out a thorough restructuring process.

On the other hand, the railway sector is showing clear signs of strength despite the fact that the crisis had an impact on the traffic of goods in OECD countries and in Russia. SCI/Verkher (2008) predicts an annual increase of 4.5 % in tram sales over the next 5 years. Around 300 cities have published development or improvement plans for tramlines. Another study by the same company reaches the conclusion that railway markets will grow during the period from 2009 to 2019 at a global scale at the following annual rates: 1.5 % (metropolitan commuters); 2.6 % (rest of passengers); 3.6 % (goods) (SCI/Verkher 2009).

4.2 Agriculture

Industrial agriculture uses large amounts of fossil energy (in particular oil) and many analysts state that its scarcity will greatly reduce the global production of food. Lester Brown (2009: 8) states that "The tripling of the world grain harvest over the last half-century is closely tied to oil". But it will be much more difficult to increase global production "with expensive oil and a declining supply". But not only is the production energy-intensive but also the distribution. Oil is used to move the agricultural machinery and to pump water, in transport, in the manufacture of consumables and in the industrial manipulation and refrigeration of food. Gas is used to produce fertilisers. In the US food production accounts for a fifth of its total energy consumption (21 %), while the rest of the main consumption items are: 14 % transport; 16 % processing; 7 % packaging; 7 % restaurants and 32 % refrigeration and domestic preparation. The total energy consumed in production is distributed as follows: 28 % fertiliser production, 7 % irrigation, 34 % in fuel used by agricultural machinery and the rest in the production of pesticides, grain drying and harvesting (Ho 2005; Lawrence 2005).

Transport is the factor that is increasing energy consumption the most, as the distances travelled by food are increasingly longer. This has led to the popularisation in English-speaking countries of the food-miles indicator, which describes the distance travelled from the farm to the plate. In OECD countries the average distances travelled by fruit and vegetables varies between 1,500 and 2,500 miles (2,500 and 4,000 km). In GB the average distance travelled increased by 50 % from 1978 to 1999. The food-miles ratio increased by 15 % during the 10 years prior to 2002. Food represents 25 % of the goods transported by lorries. But to be able to estimate the energy consumption associated with transport, we must take into account the other factors of the product chain. For example, lamb meat raised in natural pastures in New Zealand and transported by sea to GB (18,000 km) emits four times less CO_2 than local lamb partially fed with animal feed (Braclay 2012; http://en.wikipedia.org/wiki).

The theory that the scarcity of oil will reduce the production of food, because industrial agriculture is more productive, has no scientific basis. Many authors have shown that traditional agriculture, based on the simultaneous cultivation of various crops, is by far more productive per surface unit, and that a vegetable garden produces several orders of magnitude more per surface unit than industrial agriculture. It has also been proven by research carried out at the Rodale Institute in Pennsylvania by a team led by David Pimentel, between 1981 and 2002. It was carried out on 6.1 ha of land and they used normal cereal and pulses rotation systems in three agricultural models used (agro-industrial monoculture, integrated ecological agriculture with and without livestock). The productivities were similar, except in cases of drought, where ecological agriculture performed far better, as its rich soil retains more moisture. Although global food production is rising, the hungry population, which fell from 825 million in the 1990s, rose above 1 billion in 2009, due to the exportation of crops (Renton 2008; Brown 2009: 10–12).

Ecological agriculture consumes two to seven times less energy than industrial agriculture. Until recently it was focused in OECD countries, but it is growing much more rapidly in emerging countries. In China production has quadrupled in 4 years. In Brazil it grows by 40 % a year. In 2010 there were 37 Mh in the world dedicated to ecological agriculture. A large part of this surface is dedicated to pastures, meaning that Australia and New Zealand lead the world regarding surface area. However, OECD countries have 80 % of the agricultural surface area and in Europe it grows by 9 % a year (www.soilassociation.org; Willer and Klicher 2009).

Chapter 12
Towards Sustainable Transport at the End of the Fossil Fuel Era

Keywords Dominant transport model • Transport economics • Oil and transport • EU transport strategy • Sustainable transport

In the past mobility grew strongly, as well as transport speed and the average length of journeys. However, mobility has been slowed down during recent years due to rising oil prices and the economic crisis, which primarily affect most OECD countries. The transport sector suffered the biggest impact due to its high dependency on oil: In the world 95 % of the energy consumed by transport is oil (in the EU consumption reaches 96 %). In particular, in 2007 and during the first half of 2008, high oil prices caused changes to the traditional mobility model in the countries most vulnerable to the phenomenon. Mobility, speed and journey length were reduced. Although most of these changes vanished later, oil scarcity is causing a slow but persistent rise in oil prices, which will cause structural changes. This dependency on oil is caused mainly by the dominant role of road transport in the sector, and to a lesser extent by air transport, which has been the mode with the highest rate of growth.

This chapter contains the following issues: an analysis of the current model of transport; a study of the economic impact of the construction of new infrastructures; an assessment of the impact of rising oil prices on the transport system; a critical review of the EU strategy on transport; and finally some basic elements of a transport strategy aimed at achieving the sustainability of the sector.

R. Bermejo, *Handbook for a Sustainable Economy*,
DOI 10.1007/978-94-017-8981-3_12,

1 The Dominant Transport Model

1.1 The Consolidation Process

Growing mobility is fuelled by several factors: higher income, the population's preference of using the car as the usual mode of mobility, and economic globalisation. This last factor is producing a huge increase in international trade and longer and more frequent passenger journeys. In the EU the expansion of the "single market" is another factor which is pushing in the same direction. Governments contribute to this development by giving priority to investment in transport infrastructures in their budgets.

As Fig. 12.1 shows, during the 1995–2006 period domestic freight transport in the EU-26 grew (measured in tonnes-kilometre) at an average annual rate of 2.8 %. In the case of passenger transport (measured in passengers-kilometre) the increase was 1.7 %. Thus, freight transport growth has been greater than that of passengers. As the population growth has been lower, there has been an increase in mobility during the period mentioned. But the growth of different forms of transport has been very different. Average road freight transport grew annually by 3.5 %, while rail freight transport had the lowest rate of growth, at only 1.1 %. According to the latest data from Eurostat, ship freight transport accounts for one third of total intra-EU transport, and for three quarters of extra-EU freight transport. However the growth rate of sea transport has been less than that of road transport. Intra-EU passenger transport during the same period has been dominated by road transport (it accounts for over 83 % of the total). But air travel has seen the highest growth rate. The latest data shows growth of 5.3 % during the 1993–2008 period. On the other hand, railway passenger transport has decreased. Overall, rail demand was reduced by 1 % during the 1990–2006 period (Focus Groups 2009: 6, 7).

Also the evolution of energy efficiency by mode varies greatly. Since 2009 road transport activity has grown by 61 %, while energy consumption has risen by 29 %. So energy efficiency (energy consumption per unit of traffic) has grown by 20 %. Most of this improvement has been due to the technological development of passenger vehicles, because in road freight transport minor improvements have

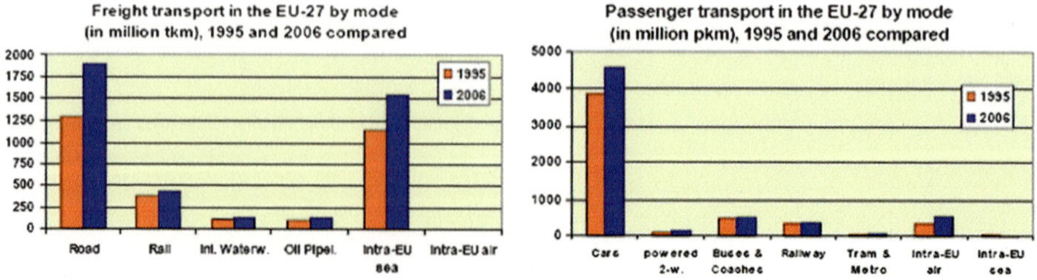

Fig. 12.1 Intra-EU freight and passenger transport by mode between 1995 and 2006 (Source: Focus Groups 2009)

Table 12.1 Energy consumption in 1,000 tonnes of equivalent oil per unit of traffic

Rail	0.020	0.021	0.019	−3 %
Inland navigation	0.05	0.032	0.034	−32 %
Road	0.145	0.119	0.116	−20 %
Aviation	1.058	0.899	0.899	−15 %
Maritime	0.004	0.004	0.004	−6 %

Source: Focus Groups 2009: 8

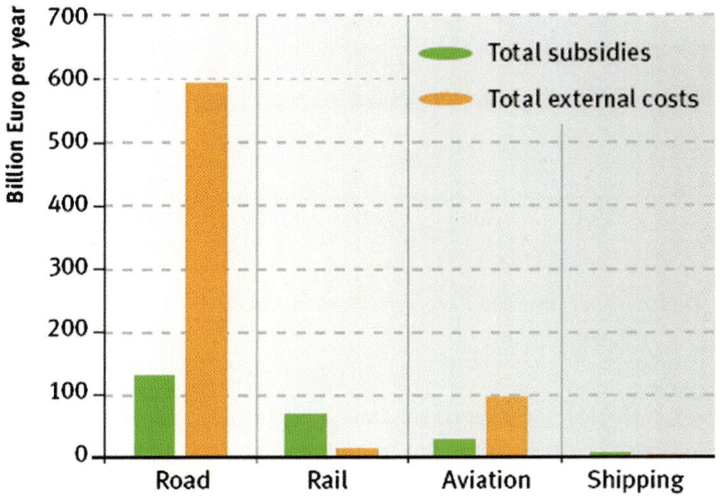

Fig. 12.2 Subsidies and external costs in the European Union (Source: UIC/CER 2008: 26)

been produced. Transport by sea is the most efficient mode, and has improved over the last decade, not as a result of technological development in particular, but by achieving a higher load capacity. In aviation there has also been a remarkable growth in efficiency. Air traffic has reached a staggering growth rate of 110 %, and energy consumption has grown by 78 %. As a result, fuel consumption per unit of traffic has improved by 15 %. It is curious that inland navigation improved its efficiency by 32 % and maritime navigation only by 6 %. The lowest efficiency improvement was obtained by rail transport: 3 % (Focus Groups 2009: 8) (Table 12.1).

Road transport uses 74 % of total oil transport consumption. The rest of the energy consumptions by modes are: Aviation (15 %); maritime (7.8 %); railway (2.2 %); and inland navigation (1.1 %) (Focus Groups 2009: 7).

The dominant role of road transport is the result of many historical policies: a fiscal policy; priority given to investment in roads; low implementation of policies aimed to demand proper management; etc. Figure 12.2 shows that road transport receives the biggest subsidies, and despite emitting large amounts of pollutants their costs are not normally internalised. The subsidies are especially significant in road freight transport. The European Commission states that "motorway maintenance would cost six times less if cars were the only vehicles to use the motorways. This benefit is not offset by any corresponding differential between the charges paid by

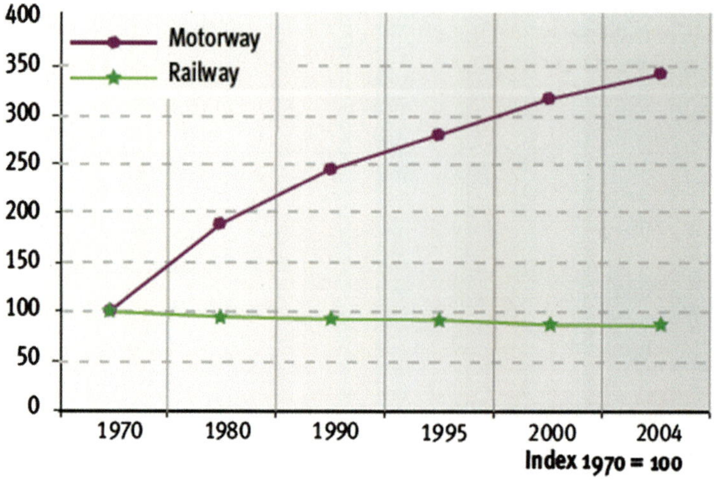

Fig. 12.3 Length of rail lines and motorways (km) in EU-15 (1970 = 100) (Source: UIC-CER 2008: 25)

heavy goods vehicles and by private cars" (2001a: 23). We will see later that the Swiss government is financing a rail line mainly for freight transport through tolls on heavy trucks.

These policies have condemned rail transport to a secondary role. In the EU-15 the length of rail lines has been falling steadily between 1970 and 2004. At the same time the length of motorways built has been multiplied by a factor of 4. The European Commission reports "that in the last 30 years an average of 600 km of lines have been closed each year in Europe, while at the same time the motorway network was increasing by 12,000 km a year". But "there are branches and lines which today would have been extremely useful for coping with saturation on parts of the rail networks" (2001a, b, c, d, e: 32) (Fig. 12.3).

In spite of these policies it is evident that there is a strong need for the use of railways as an efficient, secure and massive freight transport, and as a mass passenger transport in metropolitan areas, as Fig. 12.4 makes clear. It shows that rail can transport 45 times more people than cars per hour and infrastructure width, six times more than conventional buses, and nearly two times more than a bus with a separate lane (UIC/CER 2008: 16).

1.2 Wide Range of Applications of the Dominant Transport Model Between Countries

Although most countries in the world are developing the traditional model of transport, there are significant differences between them. Here we briefly describe

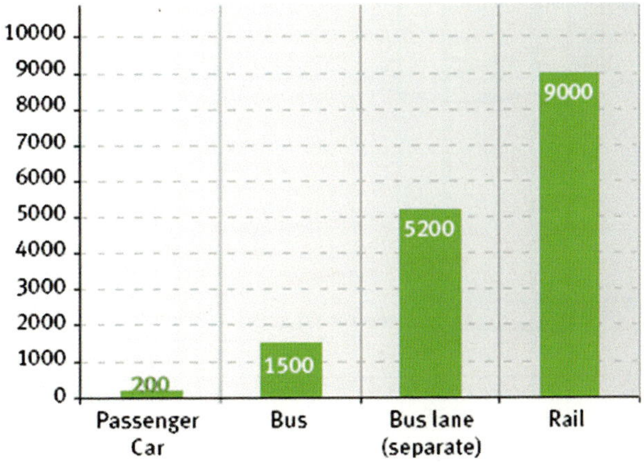

Fig. 12.4 Capacity of urban transport modes per metre of infrastructure width (person per hour) (Source: UIC-CER 2008: 16)

two extreme types of policies. On the one hand, we have the Spanish policy, and at the other extreme the Swiss one. In Spain all governments have maintained during the last decades a policy which gives priority to a massive build-up of transport infrastructures. During the first decades they promoted the construction of many motorways, and during the last two decades priority were given to the construction of a network of high-speed rail lines, with the aim of connecting Madrid with all the provincial capitals. This priority does not mean eliminating investment in other infrastructures. On the contrary, the construction of motorways, airports, ports, etc. has continued intensively.

This strategy is defended based on several arguments, but two are by far the most significant: one is economic and the other related to unfair treatment. The economic argument has two dimensions: the construction of infrastructures produces huge economic development, and frequently the regions defend the construction in their lands of big infrastructures on behalf of the state (like high speed railways-HSR), because supposes the mayor investment in history (this issue is analysed in the next section). The argument of unfair treatment is put forward by regions that demand the same type of infrastructures, which they do not yet have. This is a typical political argument. The economic argument leads governments to celebrate the fact that Spain has built more kilometres of motorways or HSR lines than any other country in Europe. In 2008, Spain reached the top position as regards kilometres of high speed railway lines, and in 2010 it reached the second place worldwide, behind China and surpassing Japan. Also, Spain is the European country which has the most kilometres of motorways and the most airports. The Strategic Plan of Infrastructures and Transport (in Spanish, PEIT 2005–2020) is the political expression of the arguments mentioned. It establishes the following objectives to be reached by 2020: to build 9,000 km of HSR lines and 6,000 km of motorways. Accomplishing the first objective means that Spain will have in 2020 50 % of the European HSR network. In order to reach such objectives Spain has been investing close to 2 % of its GDP, while France, which occupies the second position in the

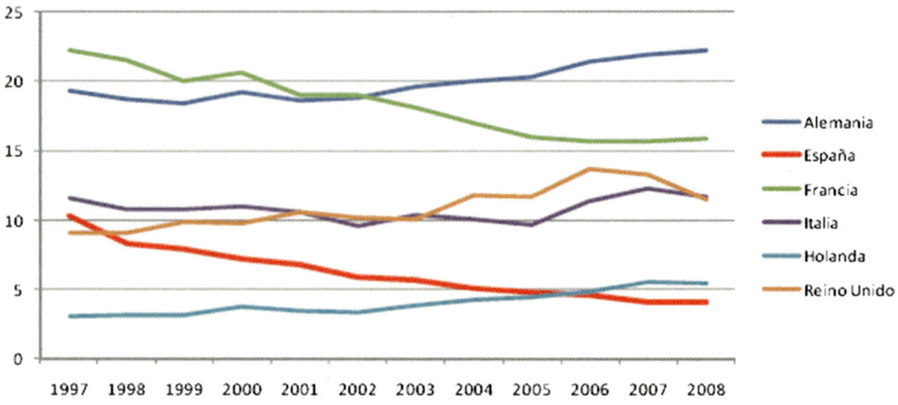

Fig. 12.5 Evolution of rail modal quota of freight transport in several EU-15 countries (Source: Ministry of Public Works and Transport 2010: 16)

HSR ranking, has been spending less than 1 % of its GDP. The rest of the countries spend less than France or do not spend on HSR at all (Ministry of Public Works and Transport 2005) (Fig. 12.5).

The above figure shows the evolution of the rail modal quota of freight transport of five European countries, and it shows that Germany is by far the leading country, and that its quota has been growing since 2000. The Netherlands' quota has been increasing slowly but steadily since 1997. The average quotas of Italy and the UK have increased during the 1997–2008 period. France was the leading country in the first part of the period, but its rail quota has been falling until 2005, and from this year it has maintained it. Spain's quota has been falling during this period. It is remarkable that France and Spain, the two countries which have invested the most in HSR lines, have lost a great part of their previous freight quota. It has to be remarked that those heavy losses of quota happened in a context of significant growth of rail freight between 1999 and 2007. In the EU-15 the increase was 22 % (EEA 2011: 41).

Switzerland has developed the most dense rail network in Europe, followed by Germany, the Czech Republic, Hungary, Belgium and Austria. It has the greatest ratio of railway use (49 journeys per person and year), followed by Austria (25 journeys). Other remarkable policies are demand management, the development of inter-modality, and a permanent policy of modernising and extending public transport networks. It has the greatest car-sharing system of the world in relation to population. Perhaps Switzerland's most outstanding policy is to shift goods form road to rail. In 1994 it approved by referendum that long-distance freight transport must be done by rail. The EU criticised the decision, but reached an agreement with Switzerland: at least 70 % of freight transport should be by rail. Since then this country has been building an adequate infrastructure to meet the objective, financing it mainly by taxing heavy lorries. The European transport policy for 2010 (White Paper) hailed this policy by declaring that "taxation on heavy lorries is a textbook example" of the policy it promoted. Between 2000 and 2010, the volumes in net tonnes of freight transport by rail have increased by 17 % (European Commission 2001a, b, c, d, e: 15; EEA 2011: 48).

2 Transport Economics

Governments usually justify new transport infrastructures with their supposed great contribution to economic growth. However, they never analyse ex-post the validity of this argument, despite having the data to do so. There are mature infrastructures and as a result the construction of big infrastructures is not justified on a narrow economic basis. Their opportunity costs are usually situated in the lowest posts in comparison with other kinds of investments. But frequently, expenditure on transport is not profitable in economic terms, and often does not take into account the social and environmental factors. This applies especially to the EU plan of building a high speed rail (HSR) network through the continent. But here we develop a general approach to the economics of investment in transport infrastructures. We treat the HSR case in the section which deals with the EU transport policy.

Numerous reports (some official) and authors reach the following conclusions: developed economies have mature transport systems; for this reason, investment in new infrastructures is not a good economic decision, because it improves little the global efficiency of systems; the best option is to invest in the overall efficiency of the system to solve inefficiencies like congestion, inter-modality, etc.; decisions of investment must not be based on generalisations about the relationship between transport and economy, but on a deep context-specific appraisal; and assessment methodologies are not reliable, because they do not take into account (or they do not measure properly) several factors. For the analysis of these conclusions we take into account two reports commissioned by the UK government and another by the French Accountability body. They are *Transport and Economy* (Standing Advisory Committee on Trunk Road Appraisal (SACTRA)); *The Eddington Transport Study* (R. Eddington); *Le Reseau Ferroviarie* (Cour de Comptes).

Developed countries have been building complex transport networks for over a century and, although these systems have to be adapted to new challenges, the construction of big infrastructures is not a profitable decision:

"It can no longer be expected that the impacts of domestic transport improvements will be transformational in economies such as the UK". This conclusion leads to the next: "In mature economies such as the UK, with established transport networks, the benefits from improved transport are likely to be the greatest when focusing on congestion or bottlenecks. But not all transport projects will deliver growth benefits" (Eddington 2006: 11, 17).

The SACTRA Report draws conclusions which are similar to those of the previous report: "Our studies underline the conclusion that generalisations about the effects of transport on the economy are subject to a strong dependence on specific local circumstances and conditions". This conclusion leads to the following:

"Another is to reinforce the importance of subjecting proposed interventions to context-specific appraisal" (2000: 8, 20).

The Eddington report also defends a case-by-case appraisal and taking into account all impacts. Besides, it measures the benefits of such an approach: "Detailed assessment of the impact of transport projects forms the bedrock of project appraisal, covering economic, environmental, and social impacts (. . .) Assessment of overall benefits on a project-by-project basis could be increased by up to 50 % in some cases" (2006: 11, 14).

Cost-Benefit Analysis (CBA) is the typical tool used for the assessment of economic impacts produced by investments in transport infrastructures. The European Commission issued a brief document on the subject in 1997 (and has since updated it twice, in 2002 and 2008). On that same year it also issued the *Guide to Cost-Benefit Analysis of investment Projects,* which seeks to be a guide for all types of large investments. The aim of the Guide is "to be helpful to managing authorities, public administrators and their advisers in the Member States, when they examine ideas or pre-feasibility studies at an early stage of the project cycle" (European Commission 2008j: 14). It defines six steps: Presentation and discussion of the socio-economic context and objectives; clear identification of the project; study of the feasibility of the project and of alternative options; financial analysis; economic analysis (in this step social and environmental externalities are considered and given a monetary value, as well as indirect effects, apart from economic data); and risk assessment. This methodology ends up by calculating a performance indicator. So CBA is a logical methodology, but the Guide makes it clear that the result cannot be considered to be scientific: "CBA is applied social science and this is not an exact discipline. It is largely based on approximations, working hypotheses and shortcuts of lack of data or because of constraints on resource of evaluators" (European Commission 2008a, b, c, d, e, f, g, h, i, j: 15, 16).

However, CBA is frequently applied to back previously adopted political decisions, especially in the case of projects for transport infrastructures. Normally, the number of passengers which will use the infrastructure is exaggerated, and the expenditure predicted is far greater than previously estimated. Externality costs are always undervalued, and do not take into account impacts due to climate change: "Current methodologies do not fully encompass the environmental and social impacts". These are the main conclusions of many ex-post analyses. The Cour de Comptes Report cites data from and studies carried out by Aalborg University, which has analysed 258 transport infrastructure projects in 20 countries, reaching the following conclusions: the average increase of all projects from the estimated cost was 28 %, and in the case of rail projects the percentage was 45 % higher. The greatest known cost deviation took place in the new airport terminal in Madrid: 600 %. Besides, the Cour de Comptes report analyses French HSR projects obtaining an average cost deviation of 25 %, and a 50 % average reduction in the number of passengers previously estimated. This report concludes that "priority should be given to expenditure on equipment renewal, rather than to investment in new infrastructures" (Cour des Comptes 2006: 138). The Eddington report states that "there are many examples around the world of projects founded on speculative demand forecasts, which did not deliver their purported economic benefits". The previously cited SACTRA conclusions lead to the last one: "Both of these conclusions apply as much to restraint policies as capacity enhancements" (SACTRA 2000: 8, 20).

3 Changes in the Transport Model Due to Escalating Oil Prices Produced During the 2002–2008 Period

Rising oil prices produced behavioural changes during 2002–2008 period, especially in 2007 and the first half of 2008. During the second half of the year oil prices collapsed and the economic crisis was evident. In the first short period of time many cases of reduced speed, mobility and length of journeys were reported. However, the intensity of these conducts varied widely between modes and countries. It was particularly evident in poor countries which could not afford oil imports, but also the United States suffered more than other developed countries due to having the lowest taxes on petrol among OECD countries. In the EU the impact was milder because the euro was highly appreciated in relation to the dollar. In May 2008 the exchange rate was 1.6$ for 1€. As Fig. 12.6 shows, the freight transport downturn is remarkable. It began in 2007 and was acute from the second half of 2008, caused

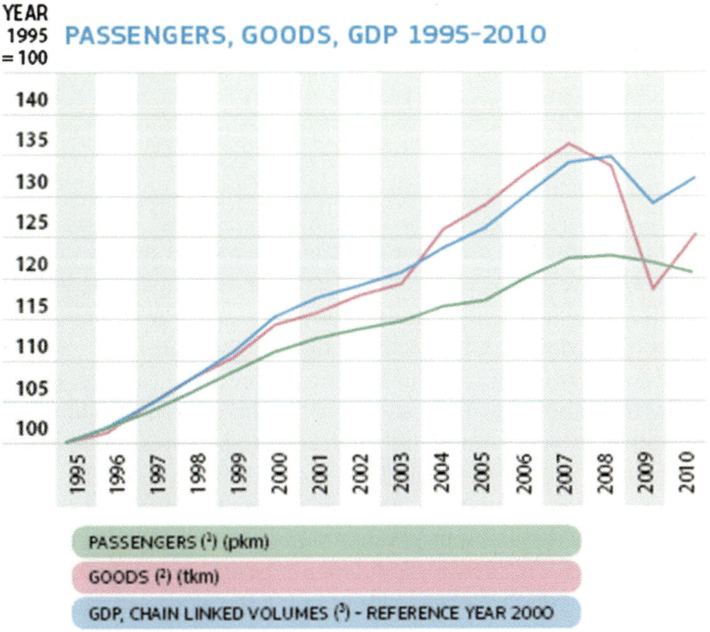

Fig. 12.6 Passengers, goods, GDP 1995–2010 EU-27 (Source: European Commission 2012: 21)

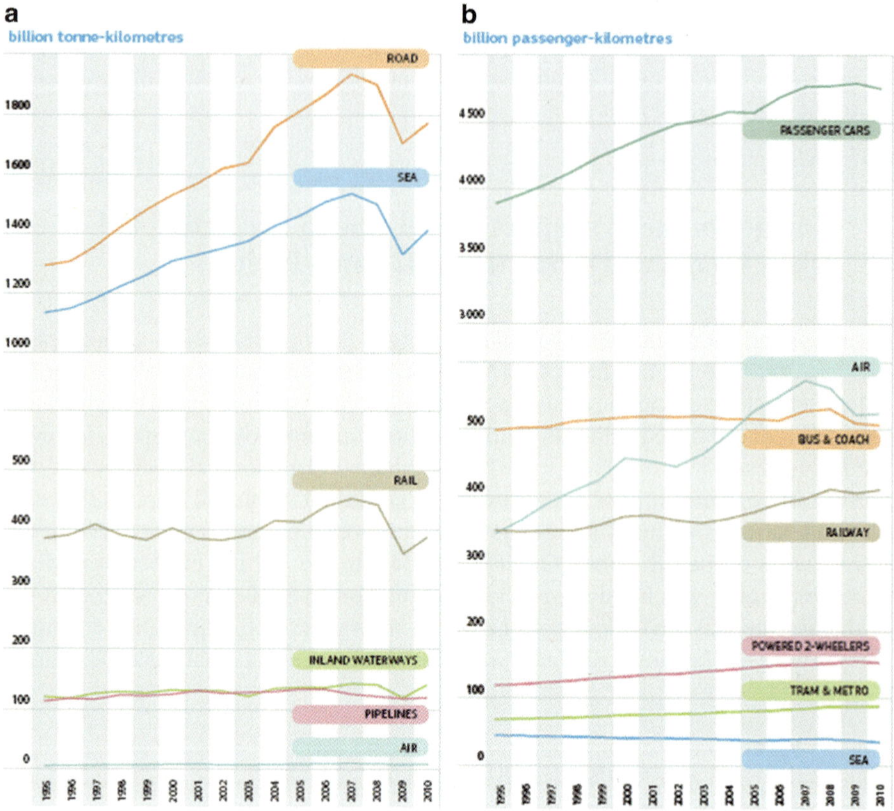

Fig. 12.7 (**a**) EU-27 performance by mode of freight transport-1995–2010 (European Commission 2012: 35). (**b**) EU-27 performance by mode of passenger transport 1995–2010 (European Commission: 2012: 45)

by the economic crisis. The effect on passenger flows was milder than that on the freight; its growth rate was heavily reduced until the second half of 2008. After that it began to fall slowly but steadily and did not recover the growth trend in 2010, as freight transport did, despite GDP recovery during that year.

In the following two Fig. (12.7a, b) we can see the evolution of freight and passengers by mode between 1995 and 2010. In the case of freight the slowdown of the main modes is very similar, although road transport suffered a reduction of traffic slightly more severe than that of sea and rail. But it has to be remarked that the increase of rail transport before 2007 was by far less than that of sea and road transport. This means that rail transport lost some of its market quota in that period.

As regards passenger traffic we should highlight the intensity of journeys by car, decreasing only a little in the second half of 2009, the worst economic year, and its growth in the period before 2006 was the second biggest, behind aviation. But this mode suffered the greatest downturn from 2007 onwards, showing once more its

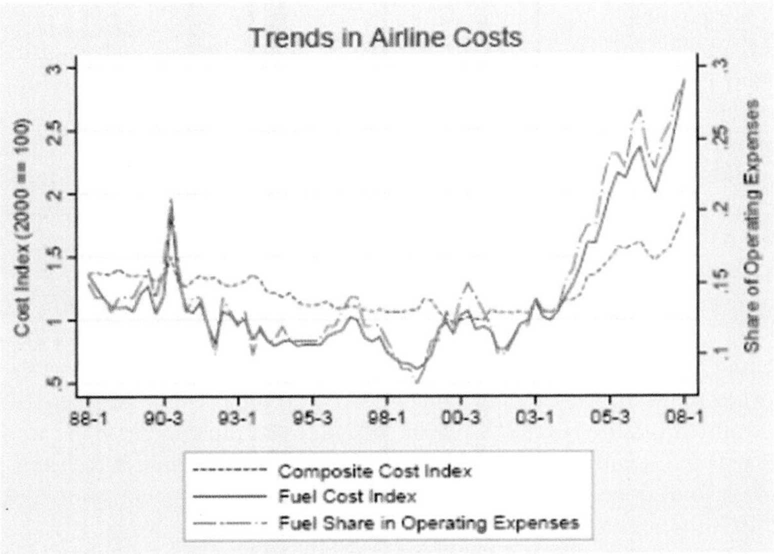

Fig. 12.8 Trends in airline costs during 2002–2008 period (Source: Hummels 2009: 60)

great vulnerability to expensive oil. Railway traffic increased on average throughout the whole period and above all when oil prices started to rise, and only suffered a slight decrease during the second half of 2008 and the first half of 2009.

So companies of all modes suffered the consequences of high oil costs all over the world, but the impacts on each mode varied widely, though most of the responses were similar. Among the most common and quick initiatives adopted for consumption reduction were: the substitution of fast vehicles for slow ones; a reduction in the weight of vehicles; the elimination of non-profitable routes; improvements in the guidance of vehicles; change of modes; etc. These measures were the immediate response to expensive oil, but after that companies implemented some structural measures: changing old inefficient vehicles for new and more efficient vehicles; staff reduction; eliminating non-profitable lines; company takeovers; transport logistics companies diverted freight to more efficient modes; etc. (Bermejo 2008: 146–150).

Air transport was especially impacted by expensive oil, because fuel costs are higher in relation to total costs than in all other modes. In Fig. 12.8 we see that fuel costs were diminishing until 2000, but this tendency changed abruptly from 2004 onwards. In 2002 fuel costs amounted to 9.9 % of the total management costs. In the first quarter of 2008 these costs reached 35 % of the total and even 40 % for the low cost companies, due to the low salary costs. Aerial enterprises all soon implemented short-term measures, such as: eliminating the delivery free goods; a reduction of the structural weight of aircraft; reduction of water and fuel to minimal standards; speed reduction; etc. Easy Jet, BMI, Air New Zealand, Air Canada, Brussels Airlines, Southwest Airlines; Jet Blue, etc. adopted a reduction of 20 km/h in long journeys.

It was not possible to implement further reductions, because the IATA prohibits it in international corridors. However, in regional flights there is no such prohibition, and many companies (above all in emerging countries) use propeller aircraft. They reduce fuel consumption by up to 70 %. In 2007, 400 units of propeller aircraft and 250 units of jet aircraft were sold. Other implemented measures were: eliminating non-profitable routes, price increases, staff reduction, etc. In the first quarter of 2008, 25 firms went bankrupt in the world. In 2007, in the United States more than 400 airports reduced flights, and about 30 lost all regular flights (Milmo and Adetunji 2008; Hummels 2009: 15).

However, air freight transport was especially hit by escalating fuel costs. Between 1951 and 2004 air freight grew (in ton-miles) annually 11.7 % around the world. But taking a period in which oil prices were increasing, freight transport showed a downward trend. In the EU, the value of airborne imports fell between 2000 and 2007, from 25.1 % in 2000 to 18.1 % in 2007, due to a rapid increase in cargo costs, while costs fell steadily until 2000. In contrast, ship freight grew in the same period from 41.5 % of import value to 50.3 % in 2007. A similar phenomenon took place in the 1970s (Hummels 2009: 3–6).

Road transport also was hit by rapidly increasing fuel costs in 2007 and the first half of 2008. In some OECD countries road freight was widely reduced, with less severe impact on passenger traffic by car. In the USA this transport fell by 1.5 % in the first half of 2008, truck numbers dropped 3 % (over 45,000 trucks, most of them operated by small truckers) and miles/vehicle by 4.2 %. In addition, truck speed was generally reduced to save fuel. Scattered data shows that in the first semester of 2008, the number of passengers in metropolitan public transport increased by 15 % in relation to the previous year, with these companies having many problems to cope with the abrupt surge. In 2008, registrations of new cars fell by 7 % in the EU area and by 11.3 % in the USA (Krauss 2008; Utchitelle 2008; OECD 2009a).

Between 2007 and 2008 world railway traffic was strengthened remarkably, as the UIC statistics show. Traffic in passengers per kilometre increased by 3.9 % in Europe, 2.4 % in Japan and much more in India (10.8 %) and China (7 %). In 2008, USA rail enterprises had ordered 3,000 new vehicles. The increasing railway traffic in OECD countries was not a hurdle for taking measures to reduce energy consumption. For example, First Group and TransPennine Express asked drivers to go more slowly (UIC 2009; Milmo and Adetunji 2008) (Fig. 12.9).

In rail freight, while non-OECD countries maintained a tendency of strong growth, OECD countries suffered a slowdown. In the EU this traffic increased by 6.1 % during the first semester of 2007; in the whole period considered, traffic was reduced by 4.9 %. In the USA the slowdown was less severe (1.7 %). On the contrary, strong increases were produced in India (8.4 %), Russia (5 %) and China (3.5 %). The differences between India and China were due to the fact that the Indian economy is more self-sufficient than that of China (UIC 2009; COM(2010) 474 final) (Fig. 12.10).

Traffic was also reduced in inland and maritime transport. In the EU maritime freight traffic suffered a similar downward trend to other modes. Maritime passenger transport has been suffering a mild downward trend in recent decades, but in the

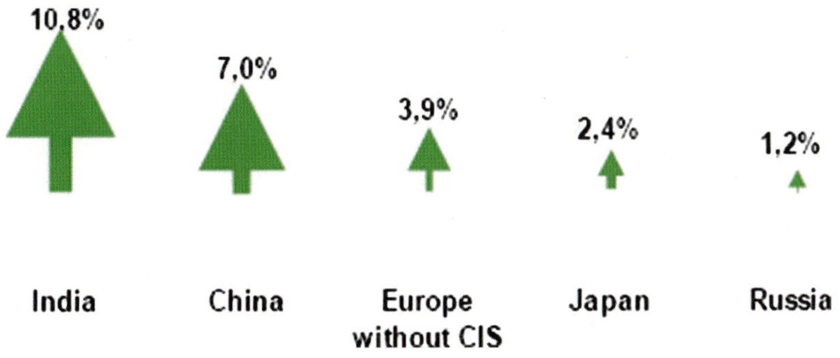

Fig. 12.9 Passenger traffic 2007–2008 (evolution in %) (Source: UIC 2009)

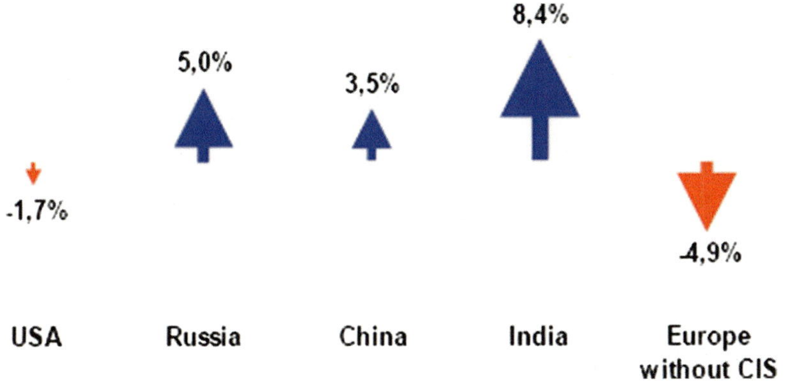

Fig. 12.10 Freight traffic 2007–2008 (evolution in %) (Source: UIC 2009)

period considered this trend was stronger. Maritime companies took measures to reduce fuel costs. In Northern Europe, ferry companies exchanged new vessels for slower old ones, because they consumed less energy. Freight cargo companies reduced speed in long-distance shipping, like Hapag-Lloyd (from 23 to 20 knots) and Maersk (from 24 to 20 knots). This company reported a reduction in fuel consumption of 40 % in the Europe-Asia line. Some companies began to utilise giant skysails (located at a height of 100–300 m and controlled by computer) to propel partially cargo ships, reducing fuel consumption by 10–15 % (Milmo and Adetunji 2008). Since then the technology has been greatly improved.

4 Critical Analysis of the EU Transport Strategy

Since the 1990s the EU has been developing a transport strategy aiming to liberalise all modes of transport and to construct adequate infrastructures needed for the implementation of a single European market. Although most of the modal networks were mature, the national rail systems were seen as the major hurdle to reach the EU's objective, because they were built mainly with a national perspective, so the railway market was fragmented. And to construct an integrated rail network for passengers (mainly high speed lines) and freight (with less emphasis) is the main endeavour of the UE transport policy. Peripheral to this aim is sustainability, but with a distorted approach. Although there have been some policy changes until today, the objective has become stronger than ever.

In December 1994, the European Council (held in Essen) approved 14 infrastructure transport priority projects. In 1996 the EU adopted a Decision (No 1692/96/EC) on guidelines for the development of a trans-European transport network, known since them as TEN-T, and comprising the EU-15 territory. The Decision was amended several times during the following year. In 2004, the TEN-T received the biggest amendment, extending the TEN-T to many new member States. In 2007 it was amended to include other enlargements. As a result, the Commission compiled a new list of 30 priority projects to be launched before 2010. The infrastructures are intended to be completed by 2020, and the European Commission approved the budget to aid State governments. Most of these projects were HSR lines: "Community funding has concentrated on major high-speed rail projects" (COM(2010) 44 final). But at the end of the last decade, the Commission began to take into account rail freight, and in 2010 approved 15 rail multi-modal corridors for freight and passengers which included the 30 priority projects. In parallel, the EU has taken many liberalising decisions concerning the sector.

From here on we analyse the EU transport policies from several points of view. The analysis is based mainly on three documents: the *European transport policy for 2010: time to decide* (COM(2001) 370); the *Roadmap to a Single European Transport Area – Towards a competitive and resource efficient transport system* (COM(2011) 359 final) White Paper; and the *Working document accompanying the White Paper* (SEC(2011) 391 final). The issues to be analysed are:

1. Unrestrained mobility
2. Creation of an European Single Transport Market
3. Economic impact of the huge investment needed, especially for the HSR lines
4. Government capabilities of investment
5. Priority given to HSR
6. Sustainability: oil scarcity, climate change and other impacts due to infrastructure construction

The approach to mobility of 2001 WP runs counter to that proposed by 2011 WP. The former proposes implementing 60 measures to produce "a marked break in the link between transport growth and economic growth" (2001a: 12). The 2011 WP states that "curbing mobility is not an option", because "mobility is vital for the internal market and for the quality of life of citizens as they enjoy their freedom to travel" (2011a: 3, 5). We can comprehend the future challenges of this policy by taking into account the EU forecast about traffic increasing. Freight transport would increase by 40 % in 2030 and by over 80 % by 2050 in relation to 2005. Passenger traffic would grow 34 % by 2030 and 51 % by 2050. The Commission forecasts high congestion levels in interurban networks "without countervailing measures" (2011b: 13). Thus, restraining an ever-growing mobility is inevitable.

The EU is making strong and persistent efforts to liberalise transport markets and other sectors. The declared main argument to defend this aim is that it promotes efficiency, which strengthens economic growth and sustainability. But liberalisation promotes increasing mobility. Besides, this policy creates huge oligopolies which always manipulate markets. So this policy can only function properly if strong anti-trust measures are put forward.

The analysis of the economic impact of the European transport strategy has several dimensions: profitability, budgetary problems, political constraints, etc. The Working document accompanying the White Paper (from now on, the accompanying document) declares that "the EU is endowed with a dense transport network" (European Commission 2011b: 95). It has the densest transport network in the world (5 million km of paved roads, 66,000 km of which are motorways, more than 216,000 km of rail lines, including around 6,000 km of HSR lines, etc.). The rail network is a little longer than that of the USA (203,600 km), in spite of being by far a smaller territory (European Commission 2011b: 95). So the EU transport network is not only a mature one, but the most mature of the world.

However, Europe *needs* to dedicate massive funds to build many new infrastructures: "A 'core network' of corridors, carrying large and consolidated volumes of freight and passengers traffic with high efficiency and low emissions" (European Commission 2011a:13). The Roadmap estimates the investment for 2010–2030 at over €1.5 trillion. The completion of the TEN-T network requires about €500 billion until 2020 (European Commission 2011a: 14). And taking into account the conclusions of Sect. 2, its economic impact will be a minor one, and negative if taking into account the opportunity costs. The USA and Japan do not follow the European strategy, because they believe that it is unprofitable. Some North European countries have decided to improve the system's efficiency, because they believe that the construction of new lines is too expensive. The USA does not have a national rail network for passengers, but it has developed a dense network of airports. Neither utilises a single electric market: there are 100 electric markets. And all USA governmental administrations have not considered this situation to be an obstacle for the national economy. However, we will see in Chap. 14 that the EU is working to develop a single electric network.

Between 1970 and 2000 the rail freight share fell from 21.1 to 8.0 %. The 2001 WP remarks the disparity of policies of the European Union and the USA: In the USA the railway system "accounts for 40 % of total freight compared with only 8 % in the European Union". Since then the share has been narrowly maintained, but because of several problems (mainly a lack of international network integration and reliability) the EU freight railway system "faces difficulties to maintain its present position, let alone improve it" (COM(2010) 474 final; European Commission 2001a, b, c, d, e: 26).

The main cause of this situation is that investment in railways has been focused on HSR lines. There is no other similar endeavour in the world. The EU dedicated to it more than 83 % of its budget for 2007–2013 (COM(2010) 474 final). However, the Europeans States has to make by far the greatest part of investment (70 % in the 2007–2013 period). Up until now we have analysed the impact on economics of infrastructure building, but in the case of the particularly costly HSR lines the question to study is if they reach a minimal socio-economical profitability. We have seen that the Cour de Comptes Report does not consider many of these lines to be profitable. The European Commission defines minimal standards of socio-economical profitability: "Only under exceptional circumstances (a combination of low construction costs, plus high time savings) could a new HSR line be justified with a level of patronage below six million passengers per annum in the opening year; with typical construction costs and time savings, a minimum figure of nine million passengers per annum is likely to be needed" (2008a, b, c,d, e, f, g, h, i, j:84). In Spain no line reached six million passengers in the first year, and they did not accomplish the first requirement. And in other countries, many lines did not reach the normal standard. Besides, the EC standard is debatable. The Commission is not a reliable organism, because it is promoting HSR, meaning that it has a positive bias in favour of HSR. In Sect. 2 of this chapter, we have commented that the current CBA methodology has many shortcomings, and one of the most significant is that the integration of external costs is very poor. The Eddington Report states that "current methodologies do not fully encompass the environmental impacts of projects" (2006: 14).

Another problem (which is associated to the analysis made) is the financing of the TEN-T project, and in particular the HSR lines. Although the Roadmap states that "Member States need to ensure that sufficient national funding is available in budgetary planning" (European Commission 2011a: 14), it is wishful thinking. The accompanying document declares that State expenditure on transport has been falling during the last decade, despite the fact that during most of this period economic growth was remarkable. But due to the crisis, member States are strongly cutting budgets to reduce debts, and consequently they are reducing their investment in infrastructures. The outcome of this situation is that many TEN-T projects are "facing planning complications and budgetary constraints", and that "EU financing is proving to be too weak" (European Commission 2011b: 9). The Commission proposes that this obstacle can be overcome by "the internalisation of external costs and infrastructure use charges" (2011a: 14). But the 2001 WP proposed these measures and others, such as doubling the EU contribution to the projects, in the section that has the apropos title of "The headache of funding" (2001a, b, c, d, e: 58, 59). Thus, there has not been any significant improvement in this field, and probably there will be none in the future, due to the unanimity required to change fiscal policy.

The 2011 WP states that "the transport system is not sustainable", which is obvious. Its vision of sustainability contains basically energy efficiency, reduction of oil dependency and CO_2 emissions. In recent years, declarations on oil scarcity and its grave consequences for the economy have become more and more frequent. But the last WP contains the most advanced treatment of oil in EU literature. It states that "oil will become scarcer in future decades", and that "if we do not address this oil dependence, people's ability to travel – and our economic security – could be severely impacted". Despite the improvement, there is no sign of urgency. But oil prices continue to rise, although at a less intense pace than in previous years, due to the economic crisis. And we know that technological change is rather slow. To address the problem, the WP proposes measures to reduce the road and air modal quota in favour of rail, to increase vehicle efficiency, and to introduce alternative fuels. In essence, the tools are investment in rail infrastructure and in new technologies, as happens in other fields: "Growing out of oil" will be supported "by a cluster of new technologies", because "innovation is essential for this strategy" (2011: 3, 10, 12). The importance given to the oil risk reinforces the Commission's emphasis on reducing CO_2 emissions until it became de facto the only parameter to measure sustainability. And broadly speaking, any action or technology which reduces the amount of the above CO_2 is labelled 'green'. Although the Roadmap analyses other environmental impacts, it only establishes the objective of CO_2 reduction (60 % by 2050 with respect to 1990) (European Commission 2011a: 3).

The promotion of HSR in order to reduce the modal quota of aviation is defended, among other arguments, based on the reduction of CO_2 emissions due to its potential for capturing many passengers from aviation. But this argument is weak and downplays the CO_2 emissions of rail transport, due to the consideration that nuclear energy does not emit CO_2. This premise does not take into account the emissions produced by embedded energy in the installations and, above all, the energy consumption of the life cycle of fuel extraction and processing. Some reports from several European universities conclude that CO_2 emission can reach up to the same amount as a natural gas station. The Commission foresees that air travel will increase by 120 % between 2005 and 2050, but 90 % of the above percentage would be consist of journeys beyond 1,000 km, because in journeys below this range HSR would obtain a greater share than aviation: "High Speed Rail could undertake some 176 million more passenger kilometres by 2050 relative to 2005, outpacing the increase in aviation (some 67 passenger kilometres for journeys below 1,000 km)" (European Commission 2011b: 27). This forecast is debatable because it does not take into account that aviation will remain far cheaper than HSR, due to having the lowest costs, in relation to other modes, for infrastructure building and management, even without considering the low-cost airlines phenomenon. Below the range of 500–600 km time-reduction is the biggest factor in favour of trains. Thus there is no consistent argument to forecast that HSR will take most of the passengers in the 500–1,000 km range.

Besides, aviation is improving energy efficiency more than other modes: "Fuel burned per seat in today's new aircraft is 70 % less than that of early jets" (European Commission 2011b: 30). But this data appears in a report from 1999. Since that date there have been great improvements in efficiency. Boeing and Airbus report fuel consumption below 3 l per passenger/100 km in their largest aircraft. The Commission also reports that recent studies forecast efficiency improvements due to innovation "of the order of 35–45 % by 2025 and 60 % by 2050" (European Commission 2011b: 31). And the 2011 WP declares that "the EU aviation should become a frontrunner in the use of low-carbon fuels" (European Commission 2011a: 7).

The EU, by focusing on CO_2 emissions, downplays other environmental impacts, such as habitat fragmentation or the loss of huge areas of natural land to the construction of overland infrastructures. The European Environmental Bureau (EEB 2008a) denounces the land and habitat fragmentation caused by road and rail infrastructures, and informs that a study by BirdLife, T&E, WWF and EEB has found that 1,000 Nature sites are potentially threatened by the TEN-T project. But the threat is greater due to the fact that the study does not assess the impact on corridors between habitats. Aviation needs a tiny area of land in comparison to the surface needed by overland modes, and it does not fragment the territory.

Thus, the 2011 WP does not change the EU's basic strategy, despite the analysis of the situation and the changes proposed by the Green Paper (GP), previous to the WP. The GP acknowledges that "considerable delays in the completion of many projects" and Member States' "investment decisions are essentially driven by national objectives", and it questions "the methodological soundness of the selection" of projects. Consequently, it calls for "a fundamental review of TEN-T policy rather than just reviewing and possibly updating outline plans and priority projects" (COM(2009)44). As a concluding remark, EU policies are driven by the political interest of the main countries and by the strongest free market orthodoxy in the world, but changes will come through the painful effects of unsolved problems, mainly non-sustainability.

5 Some Proposals for Advancing Towards Sustainable Transport in the EU

Before laying down some proposals it must be emphasized that the transport sector is one of the most unsustainable sectors, because it is the sector with the biggest rate of growth, it has a strong dependency on oil and it has huge environmental impacts. For these reasons it is difficult to envisage a sustainable alternative. But, being the most vulnerable sector to rising oil prices determines the urgency and inevitability of undertaking profound changes in a relatively short period of time. The main changes that have to be promoted are: a reduction of mobility; a change of the current modal split to sharply reduce oil dependency; and a change of the technologies and fuels employed. In order to reach these objectives several tools must be implemented: fiscal policy, new priorities for investment, a policy of demand management, raising awareness among people about the challenges lying ahead, etc.

Reducing mobility is one of the most difficult changes to obtain, because it means a radical change in the dominant culture and in the economic structure. But we have seen that high oil prices reduced mobility in the past, and when the current crisis begins to recede, oil prices will rise rapidly again. Besides, there are many good examples of policies aimed to create proximity: promotion of compact cities, decentralisation of services (each main part of cities must be equipped with them), integration in cities of productive activities, and so on. Changes in the economic structure by decentralising the economic system are highly challenging but, as we have seen in the previous chapter, high oil prices deter long distance trade and strengthen regional trade. The international fossil fuel trade has been decreasing in OECD countries since 2005, and the IEA forecasts that this trend is irreversible. A strong backing for renewable energies would accelerate the process.

Changes in fiscal policy and in infrastructure investment are key drivers for reducing the share of the most unsustainable modes, and will also reduce mobility. We have seen that road transport is the mode which receives the most subsidies, especially heavy freight, but all modes are subsidised. By eliminating perverse subsidies and charging proportionally to the environmental impacts caused, the most efficient and less polluting modes and vehicles would be reinforced. Also, by heavily charging for the great environmental impacts of infrastructure construction this activity would be restrained. The 2011 WP declares that "many branches of transport are treated favourably in terms of taxation, in comparison to the rest of the economy: tax treatment of company cars, VAT and energy tax exemptions on international sea and air transport" (2011a: 15). But similar statements have been issued routinely over the past two decades without results.

We have seen that the main transport priority is to strongly and rapidly increase the share of freight of the most efficient modes, in detriment of road freight: "Road freight emissions are still increasing" (European Commission 2011b: 19). In order to reach this objective, most investment in transport must be directed to rail and short-distance shipping, and the indicated fiscal policy has to be implemented. The 2001 WP informs that "companies attach as much importance to goods as to passengers" (European Commission 2001a, b, c, d, e: 13) But, as road freight costs increase due to rising oil prices, many corporate associations are asking primarily for a reliable freight rail service. In order to reach this objective it is necessary to remove the many existing hurdles (poor cross-border links, different technical parameters used by each country, the need to upgrade many old lines, etc.). Although priority must be given to freight, also passenger transport at speeds below 200 km/h. But with congested lines it would be necessary to split them into two dedicated to freight or passengers. The 2011 WP establishes the objective of transferring 30 % of road freight to rail and sea transport by 2030 (2011a: 9). But, viewing the EU's priority of investing in HSR, the objective will not be met.

In the case of mid and long-range passenger traffic, we have seen that the EU is dedicating most of the investment on building a European HSR network in order to transfer most of the air traffic to this network in the range of 500–1,000 km. The arguments put forward have been changing over time, but some of them are: a reduction in CO_2 emissions, a reduction of rail congestion, energy efficiency, etc. The congestion issue can be solved without HSR lines, apart from being only a problem mainly in some central European areas. Some countries are employing HSR. J. Whitelegg (2009) informs that in some lines double-decked trains are used, like the Zurich-Paris line or the Zurich-Basle line. On this line 60 trains run per day. Besides, aviation will remain the dominant mode for passengers from 500 km onwards, because it offers the cheapest service, has greatest rate of efficiency improvements, and its environmental impacts are not as big as with other modes. But the subsidy on kerosene has to be done away with.

Apart from the need to redirect investment in transport, its amount must be reduced, because the main chapter of investment in transport (investment in HSR) would disappear. The remaining investment capacity has to be redirected to improve (among other issues) vehicle efficiency, to the development of renewable energies, and to the improvement of the production and implementation of renewable hydrogen and fuel cells. Hydrogen is the unique universal fuel capable of replacing oil, and fuel cells are the best technology to use hydrogen, as we will see in Chap. 14.

Chapter 13
Solar Economy Elements

Keywords Renewable energies • Energy efficiency • Main renewable technologies • Smart grids • Renewable energy storage

1 Introduction

Initially, the promotion of efficient and renewable energies appears to be no more than a technological revolution (replacing current energy technologies that use fossil fuels with others that have the ability to efficiently capture and use solar energy in its multiple forms). However, given that it is a diffuse source of energy, renewable energy is suited for a decentralized energy model, especially in the electric and heat sectors. As a consequence, renewable energy sources have a great potential for changing the actual centralised energetic paradigm. On the other hand, it supposes a great challenge in many fields. For example, fossil fuels are easily handled resources of great energy density (i.e. a lot of energy per weight unit), not to mention that they are the raw material of a vast number of products omnipresent in our daily lives, for which oil is their main source. Thus a sustainable energy alternative must be capable of providing the same services as the current model.

Renewable energies offer many benefits: sustainability, abundance, low environmental impact, adaptability to a decentralized economy, a guaranteed energy supply, a generator of much more employment than the conventional per-investment-unit factor; etc. Nevertheless, the multiple obstacles hindering their expansion must and can be overcome. For example: insufficient technological maturity requires large R + D investments to replace non-renewable energies, i.e. to create the necessary transformation capacity. (This investment could possibly be greatly reduced if short-range advantage is taken of the existing capacity for energy savings and efficiency). Another example of an obstacle is the shortage of some of the essential materials needed to build capture systems in the current state of technological development, etc. In addition to the obstacles inherent in new energy system development, others are those derived from the interests of the existing power

R. Bermejo, *Handbook for a Sustainable Economy*,
DOI 10.1007/978-94-017-8981-3_13,

system. On the one hand, proponents of the existing system want to hold on to the current centralized electricity production model, while at the same time expanding current state-managed grids, firstly to an all-European level and then to surrounding countries. This grid would have the benefit of feeding off conventional sources and massive renewable production installations located in privileged spots. Nevertheless, such a system is highly vulnerable and unnatural since renewable energies are diffuse. Furthermore, that existing power system drives growth in agro-fuels and nuclear energy (according to the EU to maintain technological leadership), whilst demanding large subsidies for conventional energies, thus draining away the enormous funds necessary for rapid expansion of a solar economy.

Moreover, some governments and academic sectors contend that renewable energies will never be an alternative to conventional energy systems. They claim such limitations as being sparse and expensive with minimal collection of net energy.

In this chapter we will see, in response to the above claims, that wind-produced and photovoltaic electricity is already less expensive in privileged spots, and that quality is rapidly expanding. However, when comparing costs, different criteria must be considered. Thus "grid parity" is the most negative measure for renewables, since it means that renewable electricity cost today is equal to that of conventional sources. On the other hand, "dynamic grid parity" is a less strict and more real criterion, since its definition is "the moment when the present value of long term net earnings (considering revenues, savings, cost and depreciation) of the electricity supply of a FV installation is equal to the long-term value of receiving traditionally produced and supplied power over the grid", in a specific market segment and country. Although this is the EPIA definition, it is valid for any renewable energy. Renewable energies have two great advantages over conventional ones, i.e. they do not require combustibles (which are getting more expensive) and maintenance costs are much lower (EPIA 2011, September). And all this without taking into account the enormous subsidies for conventional energies. Chapter 8 showed the IEA estimated energy subsidies to be over half a billion dollars per annum (most of which was spent on conventional energies). Furthermore, renewables reduce fossil fuel dependence (improving balance of payments and energy safety) and CO2 emissions, which in the EU means a savings on the purchase of emission rights.

The low net energy argument (difference between obtained and inverted energy) arose two decades ago, which makes no sense now although it might have done then. The majority of analyses give a much higher net than inverted energy for both wind and photovoltaic power. The latest estimates on photovoltaic give a net energy of 15 on polycrystalline silicon panels and 40 on those of thin film in areas with good radiation. And net energy grows quickly (Bardi et al. 2011). We have seen that the mean net energy for crude oil is 9 (10:1) and dropping.

Renewable energies are distributed globally and numerous studies show their capacity is much greater than our needs. According to the conclusions of a Conference on renewable energies organized by the *U.S. Department of Energy* (DoE), and attended by 200 scientists from America, Asia and Europe, the radiation over 0.16 % of the earth's surface could supply 20 Terawatts (Tw) of photovoltaic

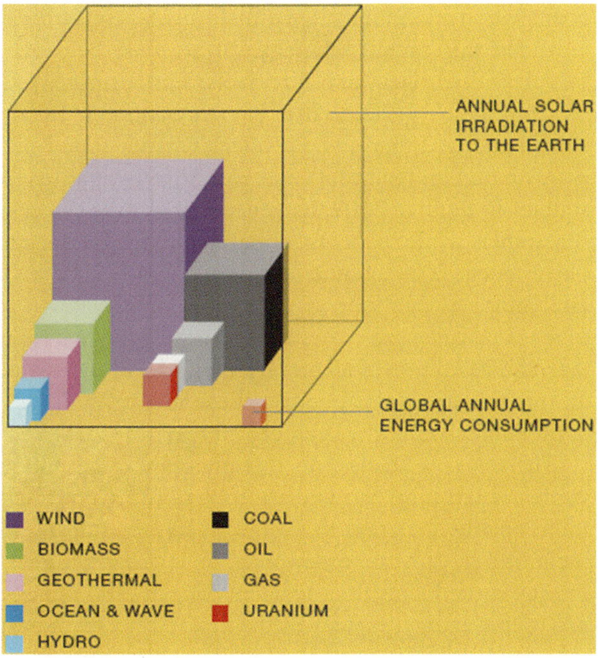

Fig. 13.1 Physical capacity of renewable energies and fossil fuels (Source: EPIA 2010c)

electricity (with a yield of 10 %), when total world consumption is 13 Tw (Lewis et al. 2005: 9). Figure 13.1 shows all the renewable sources contemplated can provide more energy than that consumed now, furthermore, solar and wind have greater capacity than all the conventional sources, even though nuclear and coal capacity is exaggerated.

This chapter defines master lines of action for a sustainable energy system, assessing its current development and foreseeable evolution. However, it only centres on electricity generations, and is limited to the three main technological systems (wind, photovoltaic and solar-heat electricity) for reasons of space, besides other general elements: efficiency, grids and storage. The following chapter will analyse renewable fuels.

2 Expansion Strategies of Renewable Energies

In recent years bodies like the European Renewable Energy Council, Global Energy Renewable Council, *German Advisory Council on Global Change* and others have proposed renewable energy development strategies. These facts plus the need to stop climate change and very likely the (undeclared) concern for the shortage of oil, led to the Brussels European Council (March 2007) approving an "Action Plan", with the following renewable energy targets for 2020: 20 % increase in energy efficiency, 20 % contribution of renewable energies to total consumption; and negatively a 10 %

contribution minimum of agro-fuels to total consumption. The 2009/28/EC directive amended this target, whereby 10 % must be reached by electricity (electrical vehicles), hydrogen and agro-fuels. Furthermore, two important events occurred on 26th January 2009. IRENA (International Renewable Energy Agency) was constituted in Bonn and President Obama stated the master lines of his energy policy based on renewable energies. Representatives of 75 States approved IRENA statutes in Bonn in 2009. Its general targets are "to help generate massive growth of renewables while reducing global rivalry for access to fossil energies" (The Federal Government 2009). Some specific targets defined by the Statutes are: "to act as a facilitator and catalyser, as well as providing support for all those questions related to renewable energies" (www.irena.org). President Obama's energy plan established that, in 2005, 25 % of electricity would be of renewable origin (10 % in 2012). And all this was to obtain "energy independence" and for "the US not to be hostage to dwindling resources" (Tverberg 2009a).

There was a shadow in this panorama: the enormous campaign in favour of nuclear energy by the nuclear lobby. The Commission supported this more and more openly declaring the "objectively and transparently" analyzed need, besides including it with the renewables in low carbon energies (COM(2010) 639final). However, the Fukushima accident (March 2011) caused a generalized reverse in the use of this energy. Japan was the first to reject it. Nevertheless Germany's rejection was much more important, since they passed a bill in 2011 to eliminate nuclear energy (23 % of quota in 2010) by 2022 (8 of the 17 plants immediately) and because they decided to create a long-term renewable energy system. The first step was to increase renewable electricity percentage from 30 % to 38 % by 2020. The German decision meant a decisive blow to the idea (driven by the energy industry) that we cannot live by using only renewables (Rickerson and Jungjohann 2011).

Profitable energies have been maturing in a few European countries, and on reaching a certain degree of maturity began spreading, first to other European countries and afterwards outside Europe. Over 100 countries have a renewable energy target, and in many the aim is to reach 15–25 % in primary energy or electricity by 2020. China is greatly developing renewables and raising their targets periodically. The Chinese government strategy for 2010 foresees an accumulated renewable power of 500 Gw (although hydraulics dominate) by 2020. Reaching 15 % of primary energy (Martinot and Junfeng 2010). The Commission projects that, with existing dynamics in 2011, renewables will meet the 20 % target by 2020. However, a strategy is being prepared for 2050. The growth of renewables is such that the renewable energy power installed in the EU and USA greatly exceeds that of conventional energies between 2008 and 2011, except in 2010, due to a massive startup of gas power stations. In the EU it was 53 % of the total in 2008 and 58 % in 2009. According to a study financed by the Commission, this ratio is expected to be almost 80 % between 2015 and 2020 (COM(2010) 639final: 5). Figure 13.2 shows a dramatic change produced in the electricity mix by the new renewable energies for the period 2000–2011. These went from little more than 5 % of the quota in 2000 to surpass 17 % in 2011, which together with large hydraulics satisfied almost 1/3 of the demand. The need to shut down many European power stations (already

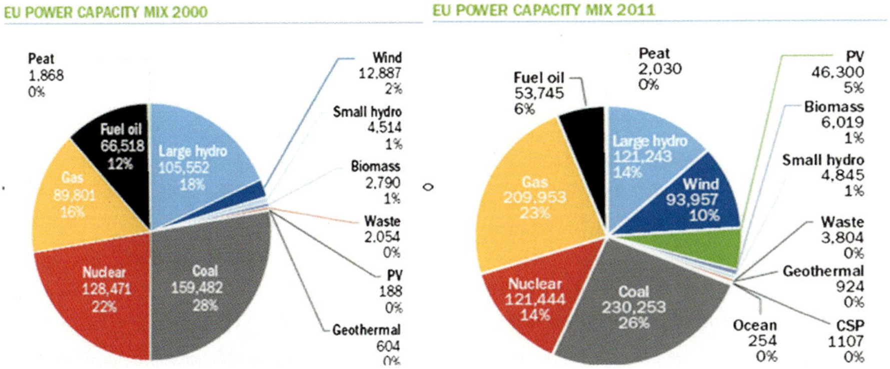

Fig. 13.2 EU power mix 2000–2011 (Source: EWEA 2012: 8)

over 30–40 years old) during this decade presents an extraordinary opportunity to accelerate the expansion of renewables.

Furthermore, in 2010 the European Renewable Energy Council (EREC) presented two strategies with different bases for 2050. The first was based on projections from each renewable sector, and as such is unreliable. Its targets are to reach 100 % renewable in 2050, for both electricity and primary energy (EREC 2010b). The second called ®evolution, promoted with Greenpeace and drawn up by the German Institute of Technical Thermodynamics, aims to reach 97 % renewable electricity and 92 % heat in 2050. The remaining 3 % of electricity will be covered with cogeneration and natural gas power stations. And most of the remaining 8 % in heat will be petrol used in 38 % of vehicles. It establishes two scenarios, i.e. basic and advanced. The second foresaw a quicker development of renewables, an earlier modal transfer of passengers and goods to the railway, and intensive use of electric vehicles, which would reach 14 % of the market in 2030 and 62 % in 2050. This means transport electricity demand in 2050 will reach 1,240 Twh/a, against 850 Twh/a in the basic scenario. Going for efficiency means a reduction in current consumption of around 40 % in 2050. It foresaw that 60–70 % of the energy generated would come from small power stations (the rest would be produced by large marine wind farms and large solar-thermal plants in the solar belt), belonging to many small producers. Thus small electricity companies will play a major role as service providers. The electrical system will have multiple local grids (*smart grids)* and large transport lines (*supergrids*). Wind and photovoltaic energy stand out in the advanced scenario with around 500 Gw each, followed by biomass, geo- and solar-thermal with powers of approximately 100 Gw. The scenarios are based on several premises: in 2050 oil will be €124/barrel; gas shortage not expected; although its price is expected to rise considerably. All the renewables together would reach cost parity in 2030; furthermore land shortage would limit biomass development (EREC and Greenpeace 2010). The main flaw in these strategies is ignoring the oil and gas ceilings. This means the former plays an even more important role in transport in 2050, while ignoring the gas ceiling means renewables will not reach cost parity until 2030, contrary to the majority of forecasts. Lastly, there is no specific treatment of hydrogen, despite it being a critical transport factor and a renewable energy system.

A study of several German companies and bodies for the European Parliament on the technical capacities of the new generation of electricity generation technologies, although not a strategy, serves to contrast with previous studies. It does not analyse land wind energy because it considers it mature. Marine wind, photovoltaic and thermal-solar energies comprise the main technologies. The capacity of the first equals 85 % of current consumption (this consumption is greatly surpassed on including land wind energy). Photovoltaic capacity is greater than current and thermal-solar is 50–67 % of consumption. Biomass capacity depends on technologies used; however, it is more limited than the three mentioned due to land shortage. It does not establish geothermal capacity due to existing uncertainties and it greatly reduces biomass capacity while strengthening the thermal-solar role. In 2020–2030 costs will greatly drop for everyone; nevertheless, it undervalues the reduction capacity of photovoltaic cost, while geothermal continues to be the most expensive (LBST 2011a).

The "Energy Roadmap for 2050" Commission's Communication is its first attempt to reach an energy strategy for 2050, which is determined by the EU commitment to reducing the GHG emissions between 80 and 95 % in relation to those extant in 1990. The Communication states the mere extension of the policies designed to reach the targets for 2020 would only reach a GHG reduction of 40 %. Thus it urges to define a strategy as soon as possible to reach a point on the indicated interval also it would be costly to delay application of the necessary policies. The reason being that, in this decade, 30–40 year old infrastructures need replacing, which should be done pursuant to the targets of a future strategy, since it would be more costly later. However, by attempting to advance in defining the strategy, one comes across numerous uncertainties, i.e. future oil prices; "to what extent will shale gas be viable in Europe?"; the role of Member States on nuclear energy; or the evolution of planetary action on climate change. Yet, despite defining several hypotheses, one reaches several common conclusions:

"**Electricity will tend to have a much more important role than at present** (. . .) it will almost double its quota in the final energy demand"; "primary demand drops between 32 and 41 % in 2050"; "electricity prices will increase until 2030 and then drop"; the renewable electricity quota will range between 55 and 97 % in 2050 according to hypothesis; centralized and decentralized generation systems (nuclear and gas power stations) "will have to collaborate more and more" (COM(2011) 885 final). Despite the open panorama and the strong impulse towards renewable energies, one can conclude the Commission still commits its standard errors, such as: supporting nuclear energy, gas, agro-fuels, etc. The following chapter will deal with alternative fuels to oil.

Furthermore, it states that "for transport a combination of several alternative fuels will be necessary to replace oil (. . .). Biocarburants are likely to be an essential option for aviation, likewise long distance road and rail transport when it cannot be electrified."

3 Efficiency

World economic growth is causing a greater increase in electrical consumption, i.e. 1 % increase in the former becomes a 2 % growth in the second. Although for decades numerous studies have been showing us there is an enormous savings potential and profitable energy efficiency, especially electricity due to the great inefficiency of our energy model. Inefficiency factors are chiefly the low performance of explosion motors, non-renewable electrician generation and that buildings are inefficient. The energy performance of a car is around 20 %. Two thirds of primary energy is dissipated in final energy conversion processes. A study by the US President's Science and Technology Council of Advisors shows that electricity generation distribution and performance is only 32 % and that less efficient transport means dominate. Apart from the car performance mentioned, that of the plane is 25 % and the ship 40 %. EU buildings consume 40 % of energy. Thus improvement of the energy system efficiency involves transformation of these sectors (PNUD et al. 2001: 13; President's Council of Advisers 2006).

The PNUD and other institutions estimate industrialized countries can profitably save between 25 and 35 %, chiefly in the energy conversion process in services (transport, lighting, heating, etc.) (PNUD et al. 2001: 13). The UE is 50 % more efficient than the US and aims to improve efficiency 20 % by 2020. The German Council WBGU (2003) recommends an EU efficiency growth of 1.4 % per annum to reach 1.6 % sooner. However, in the 1990s it dropped from 1.4 % to current 0.5 %. With 1.6 % efficiency it would be multiplied 3-fold in 2050 in relation to 1990. The EREC-Greenpeace (2010b) strategy aims to reduce energy consumption from 73,880 Pj/year in 2007, 2050 to 45,000–46,000 Pj/year in 2050. However, the Commission claims that with the current efficiency improvement rate only 10 % would be achieved in 2020 (COM(2010) 639final).

4 Main Technological Systems

4.1 Wind

The EEA estimates the gross capacity (excluding social, economic and environmental considerations) of wind energy is 20 times (marine wind would provide 6–7 times) the estimated EU consumption in 2020 (EEA 2009). As wind technology develops, the power available will grow, since the largest generators increase the energy produced per surface unit. There will be a drop in costs (14 % per duplication of installed capacity in the period 1984–2011) increasing the profitable wind speed threshold. And use of longer blades will increase to make use of low speed winds. In 2010 generators of 4–7.5 Mw of power were commercialized (those over 5 MW were only installed at sea) and the aim is to reach 15 Mw, but it is not expected to go further. Another factor affecting costs is farm size; however, there is no consensus as

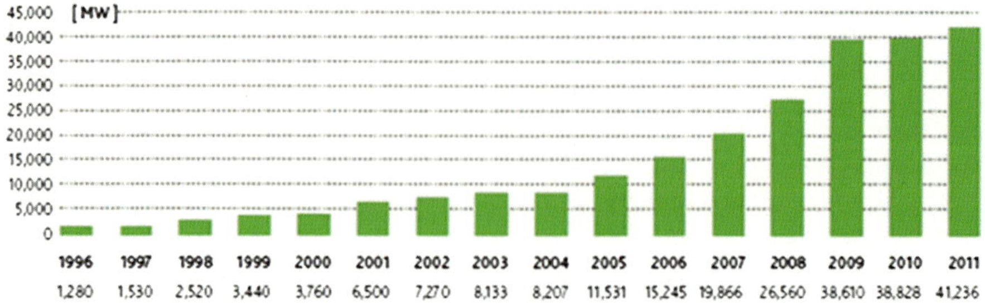

Fig. 13.3 Global annual installed capacity (Source: GWEC 2012)

to whether small or large are more cost-effective. Initially scale economies indicate large; although the bureaucratic costs are greater, because they also require high voltage transmission lines. Companies building small farms allege that they achieve scale economies on promoting several farms at the same time. In the 1980s costs dropped in Europe from €350/Mwh to an average of €60/Mwh in 2009. The Bloomberg study "New Energy Finance" reached the conclusion that in the US on average, wind farms are already close to gas power station costs (€52/Mwh, against €46/Mwh) and that grid parity will be reached by 2016. Taking into account that electricity is more expensive in Europe, the two sources are consistent (Crowell 2011; BNEF 2011).

Throughout the previous decade the annual power installed worldwide grew at an annual rate of 28 %. However, in 2010 and 2011 it only grew slightly, and this last year reached 41,236 Mw. Although the panorama was not uniform: Europe and North America had reduced their annual installed power, yet said reduction was particularly compensated by the strong growth of Asia. The historical leadership of Germany, Denmark and Spain is losing relative weight in the EU due to the drive of other countries (particularly eastern Europe and GB for offshore wind), as well as in Europe globally due to Asian dynamics. Nevertheless, offshore wind development will strengthen Europe's global position (the maturation of offshore wind will increase the number of countries exploiting the same) and initially that of North Sea and Baltic countries. China continued to be the leader in 2011 (17,600 Mw, although 6.9 % less than in 2010, due to limitations imposed by the government resulting from the many accidents occurred) followed by the USA (6,800 Mw, although they are expected to install around 12,000 in 2012), India (3,000 Mw), Germany (2,086 Mw), Brazil (1,500 Mw) and UK (1,293 Mw). In terms of regions, Asia reached leadership of capacity installed in 2009 (thanks to the boost from China) and increased it in 2011 (21,298 Mw). It is followed by Europe (10,281 Mw), North America (8,077 Mw), Latin America (3,203 Mw), Pacific Region (2,858 Mw); and Africa and Middle East (1,093 Mw) (EWEA 2012; GWEC 2012) (Fig. 13.3).

Despite lower annual growth, the cumulative capacity installed is still very important, and approaching a quota of 250,000 Mw in 2011, the vast majority being onshore. China was also the leader in cumulative capacity (63,000 Mw), followed

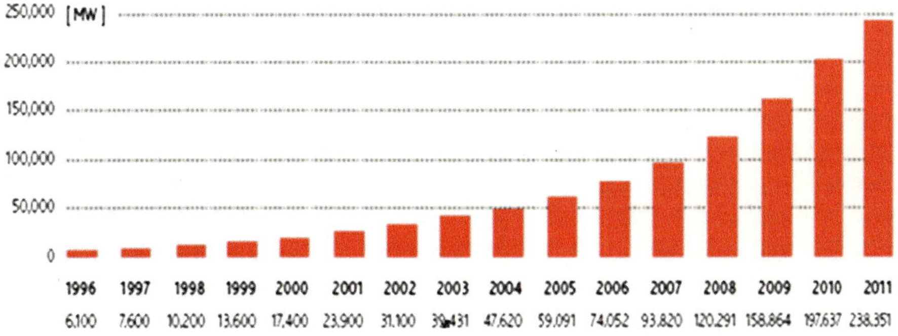

Fig. 13.4 Global cumulative installed wind capacity 1996–2011 (Source: GWEC 2012)

by the USA (47,000 Mw), Germany (39,060 Mw), Spain (21,674 Mw). By regions, Europe is the leader (96,616 Mw), followed by Asia (82,398 Mw), North America (52,184 Mw), etc. In the EU wind produces 6.3 % of electricity. Countries with a higher quota are: Denmark (25.9 %), Spain (15.9 %), Portugal (15.6 %), Ireland (12 %), and Germany (10.6 %) (GWEC 2012; EWEA 2012: 11) (Fig. 13.4).

The *European Wind Energy Association* (EWEA 2010) foresees 180 Gw onshore and 40 Gw offshore by 2020, and 400 Gw by 2030 (250 Gw onshore and 150 Gw offshore). And since wind is stronger and more constant at sea, the 150 Gw offshore would produce as much electricity as onshore. Its electricity market quota would go up from 4.2 % in 2009 to 14.3–16.6 % in 2020 and 26.2–34.3 % by 2030. The graph in Fig. 13.5 shows the EWEA forecast for installed capacity increase. The upper curve shows cumulative onshore capacity, which tends to reduce from 2020 (describing a classic S shape in product development) to 250 Gw. The lower curve describes a similar evolution to that of onshore, but with a 15 year delay until reaching 150 Gw by 2030 and continuing to grow. Nevertheless, it is highly likely the economic crisis will considerably reduce the installed capacity figures, although not so much market quotas, since consumption will be less than expected. However, China's new targets are: 200 Gw, 400 Gw and 1,000 Gw by 2020, 2030 and 2050, respectively (Yuanyuan 2012).

Onshore wind on a fixed shelf is taking off in Europe due to its long experience in wind energy and the fact that the North Sea is very windy, plus its continental shelf is superficial, enabling installation of turbines secured to the seabed. The most abundant offshore wind is expected to compensate the former and then surpass the installation overcost. To reach the target mentioned of installing 40 Gw in 2020, a mean annual growth of 28 % is required. The UK is the leading country, because GB 49 Gw are pre-approved. However, offshore take-off is lagging behind. In 2009 582 Mw were installed and in 2010 883 Mw (and it was thought take-off had occurred), nevertheless there were 866 Mw in 2011. It is being installed in depths under 20 m and at distances from the coast under 20 km, although in the future they are expected to be installed up to 60 m of depth. China installed 200 Mw offshore in 2011 and has a target of 10.50 Gw for 2020 (EWEA 2010; 2012).

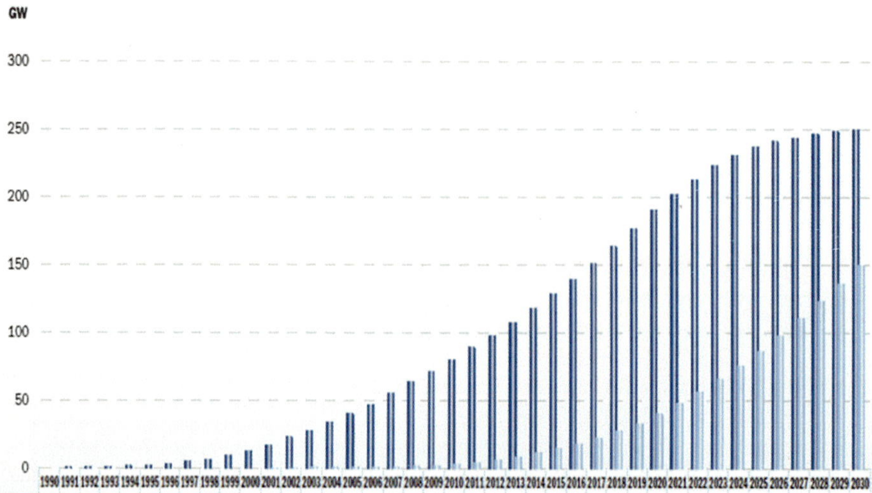

	1990	1991	1992	1993	1994	1995	1996	1997	1998	1999	2000	2001	2002	2003	2004	2005	2006	2007	2008	2009
Onshore	0.4	0.6	0.8	1.2	1.7	2.5	3.4	4.7	6.4	9.6	13	17	23	28	34	40	47	55	63.5	72
Offshore	0	0	0	0	0	0	0	0	0	0	0	0	0	1	1	1	1	1	1.5	1.9
Total	0.4	0.6	0.8	1.2	1.7	2.5	3.5	4.8	6.5	9.7	13	17	23	28	34	41	48	57	65	73

	2010	2011	2012	2013	2014	2015	2016	2017	2018	2019	2020	2021	2022	2023	2024	2025	2026	2027	2028	2029	2030
Onshore	80	89	98	108	118	128	139	151	163	176	190	202	213	223	231	237	241	244	246	248	250
Offshore	3	5	6	9	12	15	18	22	27	33	40	48	56	66	76	87	98	110	123	136	150
Total	83	93	105	117	129	143	157	173	191	209	230	250	269	288	307	323	339	354	370	385	400

Fig. 13.5 Cumulative on- and offshore wind energy capacity in the EU (Source: EWEA 2009b: 45)

However, the continental shelf in a large number of coasts worldwide soon reaches greater depths, so floating turbines should be the main option for most countries with coastline. The EWEA (2009a: 51) states that, "to harness the offshore wind capacity of deeper waters such as the Norwegian coast, the Atlantic Ocean, or the Mediterranean Sea, floating designs are required". As most of the world population is concentrated on the coast, installing floating turbines in the sea at approximately 10 Km from the coast is the ideal option, since the construction of long-lines are not required. Moreover, installation and repair can be done on land, so specialized vessels are not required. A large group of companies are taking steps towards its commercialization. They use different anchoring technologies, new turbine designs, different support structure types (dividing into individual, platforms for several turbines and mixed (turbines and wave catchers)). It goes beyond the scope of this text to review existing initiatives; furthermore, the data provided would soon be surpassed, due to the explosive dynamics of technological development and multiplication of projects. Most projects aim to research the commercial stage by 2020, although there are companies capable of achieving this by 2015. Europe leads this field in dozens of projects (with very different developments), where Scandinavia stands out (Norway (the leading country), Denmark, Netherlands, Sweden, UK, Germany, France, Italy, Portugal, Spain, etc.). Norway installed the first turbine (Hywind, 2.3 Mw), which operated between 2009 and 2011. A new prototype will be installed in 2012. In 2011, Principle Power installed a turbine in Portugal as the first step to installing 250 Mw. Malta has reached an agreement with

Table 13.1 New and cumulative photovoltaic capacity by country

	Country	2011 Newly connected capacity (MW)	2011 Cumulative installed capacity (MW)
1	Italy	9,000	12,500
2	Germany	7,500	24,700
3	China	2,000	2,900
4	USA	1,600	4,200
5	France	1,500	2,500
6	Japan	1,100	4,700
7	Australia	700	1,200
8	United Kingdom	700	750
9	Belgium	550	1,500
10	Spain	400	4,200
11	Greece	350	550
	Slovakia	350	500
13	Canada	300	500
	India	300	450
15	Ukraine	140	140
	Rest of the world	1,160	6,060
	Total	**27,650**	**67,350**

Source: EPIA 2012

the Swedish company Hexiconto to install at least 36 turbines (54 Mw) by 2014. The Japanese government aims to install 1 Gw by 2020, and in 2011 at least seven consortia are developing projects. The Fukushima project has an installation target of one turbine (2 Mw) in 2012 and two more (7 Mw each) in 2013 and 2015. In the USA, offshore wind development has been hindered by bureaucratic problems, although in 2012 there were six advanced projects in the eastern states (Maine (e) International Consulting, 2012; Quilter, 2012; Shahan 2012).

4.2 *Photovoltaic*

Photovoltaic technology has multiple properties: great energy potential; minimal maintenance costs due to absence of mobile parts; long life (over 30 years); versatility, as it can produce electricity from a few milliwatts to hundreds of Mw; they can be installed in buildings forming a part of façades and roofs, saving on building materials; and like all solar technologies production adapts very well to variations in daily demand. The EU27 could produce all the electricity need covering 0.70 % of its territory. According to the AIE, all the electricity consumed worldwide could be obtained using 4 % of the desert surface (EPIA/Greenpeace 2007: 11, 14; Linch 2010).

Unlike wind energy, photovoltaic grew strongly until 2011, when 27.7 Gw was installed against 16.6 Gw in 2010 (70 % growth). As shown in Table 13.1, Europe

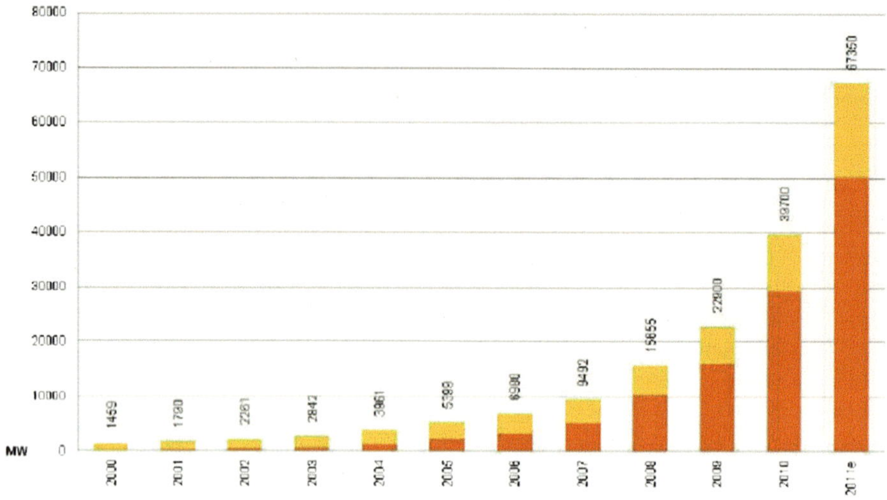

Fig. 13.6 Global cumulative installed photovoltaic capacity 2000–2011 (Source: EPIA 2012: 5)

maintains strong leadership (75 % of new capacity) although dropping (80 % in 2009). However, leadership is due to Germany and Italy (60 % of newly installed capacity), with a cumulative installed capacity of 24,700 and 12,500 Mw. EU electricity market quota went from 1.15 % in 2010 to 2 % in 2011 (EPIA 2012). Although China is in third place (2000 Mw in 2011), their progression is very quick. It is highly likely that the first seven countries on the table and India will lead the installations this decade, even though positions will greatly change. China, USA, Japan, Australia and India are destined to become photovoltaic powers due to their growth dynamics and potential. The EPIA foresees in the future that Europe will continue having notable leadership in installation (although it will lose relative weight up to 50 % in 2014), followed by the USA, Japan (whose aim is to reach 28 Gw in 2020), China, India (both decided in 2009 to install 20 Gw each in 2020) and the rest of the world (EPIA 2010a, 2012; Chew 2010).

The annual growth of installed capacity determines an explosive growth of cumulative power. In 2008–2011 that capacity was multiplied by 10. In 2011, the European cumulative installed capacity was over 50,000 Mw (EPIA 2012) (Fig. 13.6).

Nevertheless, on the photovoltaic panorama different instability factors appeared. The market is highly concentrated in Europe, particularly in Italy and Germany. Moreover both countries have announced they aim to strongly reduce the capacity installed annually. Furthermore, European hegemony in installation contrasts with that of Asia in module production, which is also becoming stronger. The EU and USA accuse China of disloyal competition. The German company QCells was number 1 worldwide until 2009; however, in 2010 it suffered great losses due to Chinese competition, and in 2012 it is no longer among the top ten companies.

The US government initiative to apply customs duty to Chinese panels led to a trade war between both countries. In any event the USA and Europe not only run the risk of becoming installers only unless steps are taken to prevent this. Another problem is that governments find it difficult to foster balanced development of photovoltaic energy, as prices are falling so quickly, that the subsidy which seemed reasonable at the beginning of the years, is no longer so profitable a few months later. This causes a flood of installations and dramatic cutbacks in the premium, which may put a brake on its development. Lastly, there is excessive production capacity that keeps growing, leading to rapid price reduction and the disappearance of many companies (Lacey 2013; Mint 2013).

Crystalline silicon is the dominant technology with a market share of 95 % in 2005. The rest of the market had fine laminate panels of various kinds differentiated by material used: amorphous silicon (Si-TF), cadmium telluride (CdT), Indian selenide, copper and gallium (CIGS), copper and Indian sulphide (CIS), etc. These panels lost market share over most of the last decade due to lower production costs which did not compensate a much lower efficiency. However, the increase in price of crystalline silicon in 2007 and 2008, due to shortages, encouraged investment in the panels leading to their resurgence. The results were: a fast rise in performance combined with new production techniques and mass production caused a cost reduction. They achieved 18 % market share in 2009; however, the strong price reduction in silicon panels reduced their market share to 11 % in 2011. However, it is not expected to disappear for several reasons: First, Solar continues to be the absolute leader; a lot is being invested in developing thin-film; and due to their low weight they are better than crystalline in buildings. Despite the hegemony of TCd panels that prevailed up to 2011, the panels CIGS and CIS are considered to have greater increase potential in performance and cost reduction. And 2011 was a breakout year for CIGS, with a production of 1.2 Gw, with some companies surpassing 17 % efficiency (Montgomery 2012a, b; Shiao 2012).

Photovoltaic panel prices are dropping fast. They are estimated to drop 20 % per production duplication. The average price dropped 2€/w in 2009 to 1.3€/w in 2010, to 0.7–0.8€/in 2011 and at the beginning of 2012 to 0.65€/w. Thus it is becoming the cheapest energy in countries where electricity is relatively expensive, with high insolation, and the price drop is quickly increasing in these areas. Thus "as the photovoltaic industry enters its post-feed-in tariff phase" (Mint 2012 (26, 04)).

Finally, the photovoltaic concentration systems (PCS) (comprising high performance cells on which sunlight is concentrated using mirrors) is considered to have decisive commercialization advances. This is because it is benefiting from the price drop in crystalline cells, so efficiency soars, plus they use much less space per power unit than photovoltaic panels without concentration systems. Efficiency increased from 27 % in 2010 to 30 % in 2011. After years or installing very few megawatts, the sector seemed to take-off in 2010 (16.6 Mw installed and 28 Mw cumulative 16 Mw in Spain), and although there are no total figures for 2011 the take-off continues (even though it maintains a negligible level in comparison to the photovoltaic), because they were building 660 Mw mid-2011. The vast majority were in the USA (where solar heating power stations were reconverted to CPV plants), but also in Spain, Portugal, Mexico and Australia. There are also large scale project provisions for India and South Africa. These systems are expected to reach cost parity before 2015 in arid areas of high radiation (EPIA 2010e; Cameron 2011).

4.3 Solar Thermal Power

Solar-thermal electric power stations (also called Concentrated Solar Power (CSP)) have a long history of R + D and commercialization, albeit intermittent. They can produce electricity in power stations from 15 Kw to hundreds of megawatts. To be profitable these power stations require high radiation, found in the so-called sunbelt, whose width ranges from southern Europe to South Africa, with band radiations of 2000–2800 kwh/m^2. They also require a large surface, i.e. 1 sq. km is required for parabolic cylinder power stations of 50 Mw. However, less surface per energy unit produced is required than for reservoirs (flooded area) and coal power stations (taking into account the mining surface necessary). Only 0.04 % of the Sahara surface is necessary to meet the EU-25 electricity demand and 2 % for the world (IEA et al. 2009: 14; 69).

The graph in Fig. 13.7 shows existing basic technologies, where two stand out: parabolic cylinder (PC) and the tower or central receiver. Furthermore, there are Fresnel (curved) mirrors and those of the parabolic disc. PC power stations comprise mirrors, as indicated, which concentrate the radiation in pipes, which normally heat synthetic oils (350–550 °C) that convert water into steam using an interchanger, which moves a turbo-generator. It is the most mature technology and has a 15 % performance. They have a heat accumulation system incorporated (except those of the parabolic disc), whose capacity increases with the technological maturation process and are already able to function up to 12.5 h with stored heat. Fresnel technology uses curved mirrors and the receiver piping is not connected to the mirror system, thus avoiding turning problems of the PCs. Nevertheless, its performance is only 8–10 %, and its degree of maturity low. The tower power stations consist of large mirrors (heliostats) turning with the sun and projecting their rays onto a central tower, where a fluid is heated to 800 °C. Performance is 20–35 %, depending on the design. Like with the PCs it stores heat and new models do so for 15 h' functioning. The tower power stations will be hegemonic in the future because they achieve higher performances as they obtain higher temperatures with lower transmission losses in relation to the PCs, although their technology is more immature. The parabolic discs concentrate heat on the focus, where there is a motor to convert it into electricity. Their performance is around 25–30 %, although technology is less mature (Kearney and ESTELA 2010: 2: IEA 2010c: 31 and a performance of 15 %).

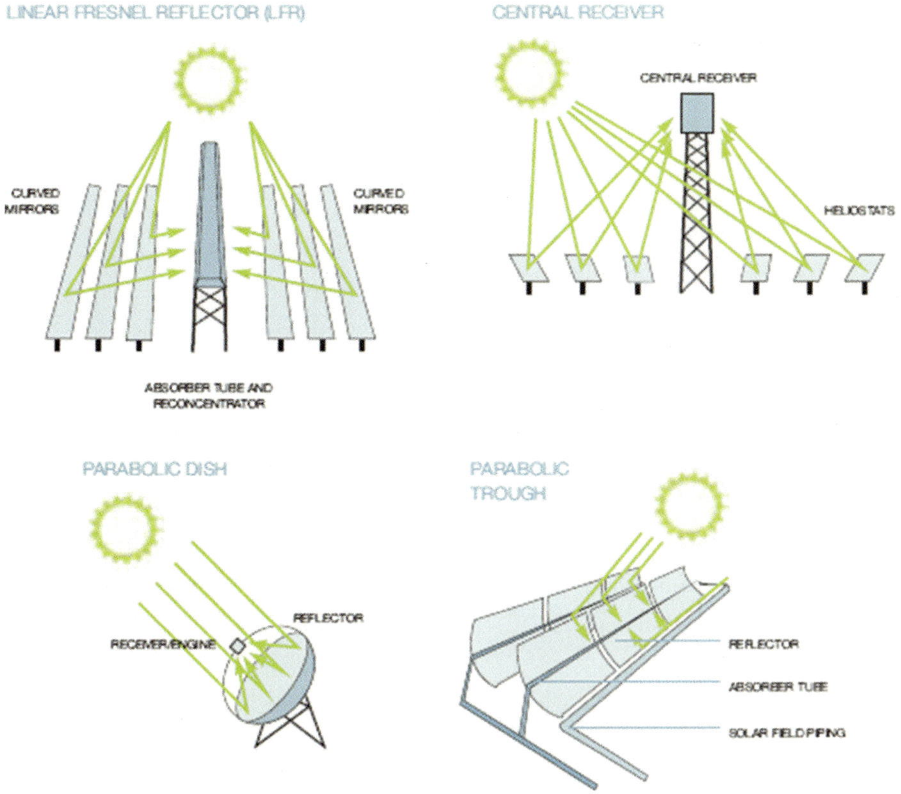

Fig. 13.7 Basic functioning of the four technologies (Source: ESTIA et al. 2005)

After installing 256 Mw (CP, tower, parabolic-cylinder and parabolic disc) in California around 1990, it only began accumulating new capacity from 2007, with a tower plant of 10 Mw in Spain and another PC of 60 Mw in Nevada state. In 2008 a PC plant of 50 Mw was installed in Spain and the first Fresnel power station of 15 Mw. In 2009 a capacity of 176.4 Mw (almost all in Spain) was installed. At the beginning of 2012 around 2 Gw was installed worldwide, two more would be installed in Europe in the short-term and another two in the USA in the next 2 years (DiPaola 2013). The PCs accumulate 88 % of capacity installed and 98 % of the capacity of power stations under construction. Both countries will consider leading capacity development of power stations under construction. Both countries will continue leading CSP development at least throughout this decade. However, these are extending along the sunbelt. Tunisia shortly foresees installing 40 power stations during 2010–2016 and Morocco five power stations with a total of 2,000 Mw for 2020. Saudi Arabia aspires to be an exporter based on the CSP electricity. Abu Dhabi will finish a 100 Mw power station in 2011 and wants to reach 1,500 Mw in 2020. Jordan is building A 100 Mw PC power station. Egypt and Algeria are also installing PC power stations to supply steam for the gas ones. Israel, South

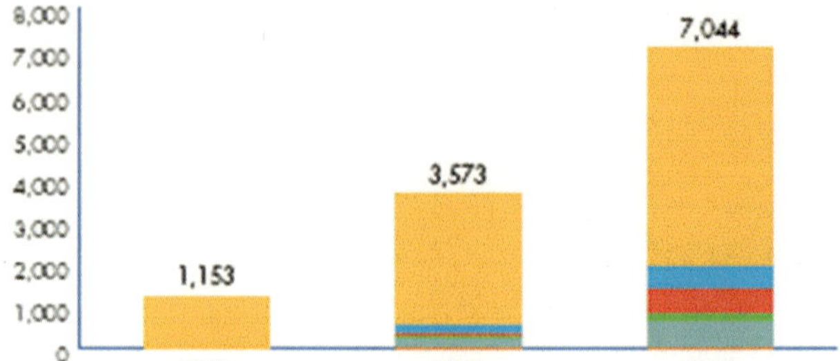

Installed Capacity	2011	2015	2020
Cyprus	0	50	75
France	0	203	540
Greece	0	30	250
Italy	5	62	600
Portugal	0	180	500
Spain	1,153	3,048	5,079

Fig. 13.8 Installed capacity in Europe (Source: ESTELA 2012: 33)

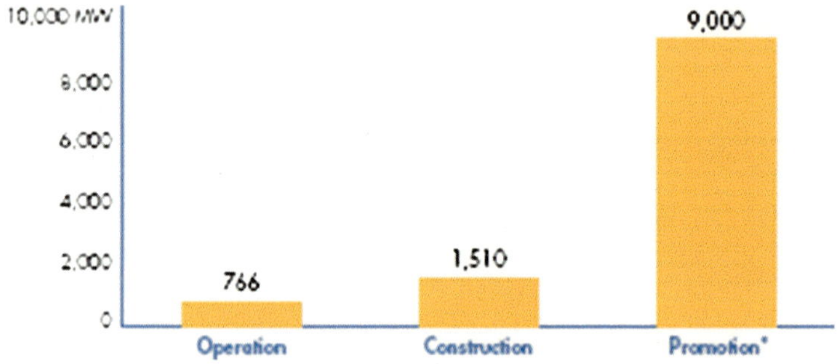

Fig. 13.9 Installed, in construction and in promotion capacity (Source: ESTELA 2012: 33)

Africa, Australia, etc., are also promoting electrical PC power stations (http://social.csptoday.com; www.solarmillenium.de; Kearney and ESTELA 2010: 7).

In 2011 the STE installed capacity in Europe was 1,158 Mw. Six countries have established their own STE targets in their "National Renewable Action Plan" for 2020, altogether over 7 GW. The graph in Fig. 13.8 shows that Spain is (and will be in the future) the leading country in Europe. Current promotion applies to projects either with signed contracts or corresponding to short term national programmes.

In the rest of the world the situation of CSP is as shown in the graph in Fig. 13.9: in operation (766 Mw), in construction (1.510 Mw) and 9,000 Mw are being promoted. The high level of promotions means that most installations in the future will be outside Europe. The estimation of the leading regions by 2050 is: North America (340 Gw), India (150 Gw), South Africa (150 Gw), Middle East and North Africa (MENA, 150 Mw), South America (1e0 GW), China (80 Gw), Australia (60 Gw) and Europe (30 Gw) (ESTELA 2012: 33–34).

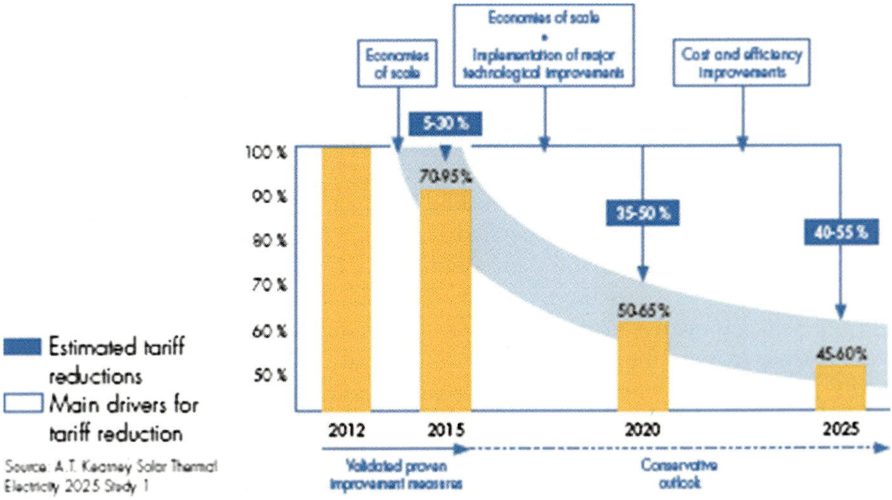

Fig. 13.10 STE learning curve (Source: ESTELA 2012: 35)

Both figures show that take-off is occurring in 2012. That means that CSP is at the very early stage of its learning curve, and there is a wide range of opportunities for quick cost reduction, as the graph in Fig. 13.10 shows.

5 Grids and Energy Storage

Historically it has been believed that an efficient electrical system should be based on large generation plants and long transport lines. Yet the extraordinary development of renewable energies and their diffuse nature have opened the gates to a decentralized electrical model of distributed generation (DG). DG involves a set of small producers (usually renewable electricity) and users connected to a local grid, which is likewise connected to the general grid. Current DG dynamics tend to dilute the limits between generators and consumers, and other sectors. Buildings can generate their own electricity (via photovoltaic panels and/or hydrogen batteries) and evacuate excess to the grid. By using electrical vehicles (battery and hydrogen cells), the transport sector acquires electricity storage and generation capacity. These grids not only manage supply but also demand. All this leads to a level of management complexity unknown in the past. Yet these complex systems can be efficiently managed via the use of information technologies and communications (ITCs), leading to the creation of intelligent grids. A European Parliament study defines DG as that "which is made up of **relatively small scale generation systems** (low and medium voltage: 110 kv and less). **Primary energy is often renewable** (. . .) **and frequently available locally**" (the emphasis is original). Yet, "despite the trend towards a more sustainable energy system, current European legislative structure tends to foster large scale renewable energies and frequently support for developing and integrating small and micro generation is lacking" (European Parliament 2010a: 11, 77).

The EU, electrical utilities and the renewable energy sector are fostering the creation of a European grid for expansion towards countries within their environment. The aim is to create a liberal electric market via the increase in capacity of transfrontier transport, thereby creating a Super Grid which enables transport and distribution of electricity generated in areas with very high capacity like the offshore windfarms of the North and Baltic Seas, likewise the solar plantations of the Mediterranean and Middle East areas. This would be a high voltage continuous current grid, compared to the traditional alternating current, to reduce transport losses. This is what the Commission proposes in their Communication "Energy 2020. A strategy for "a safe competitive sustainable energy" is defined as a *smart grid*, as opposed to the generalized use of the term to define intelligent decentralized grids. Furthermore, it claims "the new interconnections with our neighbours should have the same priority as the intra-EU projects. Such connections are essential not only for our neighbours but also ensure EU supply safety and stability" (COM(2010) 639 final: 10). Before analysing this stance, I shall explain the main initiatives.

Parallel to the deployment of offshore windfarms in the North Sea (140 Gw is expected to be installed), studies and proposals for generating electricity evacuation grids are multiplying. In 2010 all the coastal countries of the North Sea reached an agreement to create evacuation lines and the European Commission boosted this project with 500 million euro. However, at a Summit of Mediterranean States in 2008 the *Mediterranean Solar Plan* (MSP) was launched. Its aim is to install 20 Gw in the region by 2020 using the three technologies explained here. A prerequisite of the Plan is the construction of two evacuation lines to Europe, one between Greece and Turkey, and the other between Italy and Tunisia. The MSP with the support of two consortiums (made up of large energy companies and European banks) drove the Medring and Desertec projects with EU support. Medring aims to create an electrical grid which covers all Mediterranean countries, which will also have numerous powerful connections between north and south Mediterranean. The Desertec project aims to expand the previous grid to the Persian Gulf. The planned investment is 400,000 million euros. Both initiatives aim to draft studies to assess their profitability (Tverberg 2009a, b, c, 06/12; IEA et al. 2009: 69; Blau 2010).

Prior to examining the projects I shall analyse the arguments given in favour thereof: free electrical market – a competition incentive, producing efficiency and low prices, supply safety; need to increase trans-frontier transmission capacity to face the intermittency of renewables; need to import renewable electricity to guarantee European supply; Member States failing to achieve their targets can import renewable electricity.

The defence of free trade is paradoxical (although foreseeable), because sector liberalization has led to the European electrical system being essentially controlled by three companies (as acknowledged by the European Commission). The intermittent unpredictable nature of renewables is exaggerated, converting the current situation (strong dominion of onshore wind generation) into a structural characteristic of renewable electricity. Nevertheless, this problem is reduced, because it has been demonstrated that interference is greatly reduced when different regions use it and wind energy companies inform the operator about the energy evacuated to the grid with great precision 24–36 h in advance. Furthermore, this situation will improve as the renewable *mix* becomes more diverse. The CSP systems can store energy for production runs of 15 h. The fact that offshore wind has greater potential than onshore is not taken into account or that it is distributed along the rest of the coasts. Installing offshore wind at approximately 10 km from the coast to meet the demand of the majority of the population concentrated on the coast as happens and is proposed by Portugal, constitutes a DG. Furthermore, the reliability of conventional power stations is exaggerated. In the US they are idle 10.6–12.3 % of the time, and stoppage without warning is 2.8–4.2 % of the time depending on technologies. Finally, States and regions prefer to develop their own capacities for supply safety and because it allows them to create their own industry. For example, the northwestern states of the US prefer to develop offshore wind instead of importing wind electricity from the central plateaus. The same occurs in the Atlantic Arc regions with offshore energies. The occasional misalignments between supply and demand can be settled (without resorting to the electricity market) via computerized storage and management systems of microgrids (Hansen and Lovins 2010).

Despite that, a 100 % renewable and decentralized system needs an important energy accumulation capacity. There are numerous ways of doing this (Redox batteries, compressed air, hydraulic pump, hydrogen, etc.), however, when storage of large energy quantities is required for long periods (over 48 h), alternatives are greatly reduced. Pumping water from a reservoir to a larger one is the most efficient and used system (80 % performance), followed by compressed air (70 %), the performance of H2 is low (30–40 %). Nevertheless, many countries have few hydraulic resources. The capacity for compressed air energy is 60 times less per volume unit than hydrogen, therefore that is the best option (Wurster et al. 2009: 21; LBST 2011b: 11). With this electricity system there is no need to increase trans-frontier transport capacity.

The supply safety argument has no basis. There are two types of safety, i.e. facing the demand peaks and the risk of a supply cut. The foregoing system is unbeatable in both cases. The former, the DG, can increase capacity in little time and adapt to its needs, since it uses small scale modular technologies. Plus the intelligent grids can also manage the demand. However, most power cuts (and they are frequent) are due to grid damage resulting from climate problems (wind, snow, drought, sun storms, etc.). Thus the more transport is reduced the safer it will be. Reliability of the current systems is over a thousand times less than that of the DG, and this enables saving of energy lost on the large transport lines (5–8 %) (Makover and Pernick 2002; Swisher 2002).

The Desertec project is in fact an extension of Medring, so I will analyse the latter first. Moreover there is already a voluminous study co-financed by the EU and its conclusions are devastating: The EU "is well provided for regarding intense solar radiation, likewise excellent wind conditions in some countries. Member States are able to offer renewable electricity at competitive prices on the European markets today, tomorrow and always; even taking into account the high cost of providing continuous current electricity and transmission losses in 1,000–2,000 km". Thus, "any long term decision to establish transport corridors to export solar and wind electricity should be based on commercial considerations. Supply safety or the desire to diversify supplies should not be the primary impulse of said decision". These conclusions are reinforced by the situation of North African countries. The majority of these countries use little or no renewable energy, moreover electricity prices are so low in some countries that "they deter the establishment of a solar electricity industry". Lastly, the report positions itself against subsidising private initiatives, because "their generation costs cannot compete with public company cost generation" (MED-EMIP 2010). To this, one would add the extreme vulnerability of transport lines to terrorist attacks. In addition to this are the high construction costs of transport lines (some analysts consider this cost may reach 25 % of the Desertec project); and it is estimated that energy loss will be 10 % even if direct current is used at very high voltage (IEA et al. 2009: 69 and ss).

Finally, one has to understand that there are serious points of divergence among the companies involved regarding the new EU electrical system, i.e. between renewable energy production organization and the large traditional electrical utilities, and the managing bodies of national grids, even though they are agreed on market liberalization and dreams of expansion. The latter two groups essentially aim to maintain the conventional grid model, even though the increase in complexity of electricity management leads them to use ITCs. In other words the centralization would be maintained basically. This is the position defended by the *European Regulators Group for Electricity & Gas* (ERGEG) (an EU consultancy body which groups gas and electricity grid regulators): "the difference between current grids and the intelligent grid of the future is mainly grid capacity to efficiently and effectively handle greater complexity than the current" (ERGEG 2009:11). However, professional associations (EWEA, EPIA and ESTELA) and Commissions grouping the renewable sectors state that in the future "generalized generation based on nuclear and fossil fuel power stations will be progressively replaced by a combination of large scale renewables together with a growing number of local and distributed generation", replacing the traditional function of nuclear and fossil generation to supply basic electricity. Furthermore, intelligent technologies (sensors, intelligent meters, programmable systems of energy use and telecommunication networks) will be at the service of "a trans-European grid" that connects *smart grids*, which will achieve flexible consumption and production; and reduce consumption. Storage will play an important role in this system. However, regulators could lose their current role as defenders of large traditional electrical utilities to become similar to "air controllers" (CAN Europe et al. 2009).

Chapter 14
Renewable Hydrogen Economy

Keywords Alternative fuels • Agro-fuels • Renewable electricity for transport • Hydrogen renewable • Fuel cells • Hydrogen role in transport

Renewable energies usually claim to be the alternative to oil. Renewables provide us with electricity, heat and fuels from biomass. Thus, these latter appear first as an energy alternative to oil. In fact the EU and USA are strongly backing the application of so-called bio-fuels (I prefer to call them agro-fuels, because the root *bio* has positive connotations, which are not justified in this case). They are also promoting lightweight road vehicles using battery-powered electricity. In this chapter we will see that the electric car, whose battery power limits it to short journeys, is just that, and that the massive production of agro-fuels has enormous environmental and social impacts. Furthermore it is limited by land availability and thus provides a poor energy balance. So having discarded the foregoing alternatives, only hydrogen associated to fuel cells is left as the sole alternative to oil.

1 Agro-fuel

With increases in oil prices we are seeing a proliferation of state policies strongly supporting the replacement of fossil fuels with agro-fuels (AF). And a former UE objective was that they would reach 10 % of total fuel consumption by 2020. Governments use the following premises as the basis to justify their policies: existence of biomass capacity which can be used without prejudice to food production; existence of appropriate technologies via large investment in R + D; a highly positive energy balance, and as such a reduction in GHG; positive environmental effects; and

R. Bermejo, *Handbook for a Sustainable Economy*,
DOI 10.1007/978-94-017-8981-3_14,

beneficial economic impacts (GNP and employment growth) (Bringuezu et al., 2007). Nevertheless, numerous university reports and those of the EU Joint Research Centres (JRC) coincide in their rejection of these premises. This criticisms supposed that energy directive 20720/20 modified the previous aim with another where the same target will be reached jointly by electricity, hydrogen and renewable AFs.

Prior to assessing such premises, it is necessary to specify the types of AF, their uses and what they currently mean. However, the dimension of this work requires us to be selective, thus we center on first generation fuels; and among the so-called second generation, those that are cellulose based. The fuels obtained are: ethanol (petrol substitute) and agro-diesel (alternative to diesel). Ethanol is obtained via fermentation of plants rich in starch or sugar (cereals, cane sugar, sugar beet, potato, etc.). Normal vehicles only admit petrol and bio-ethanol mixes, not usually exceeding the 80/20 ratio. However, the majority of car manufacturers have put flex cars on the market equipped with sensors which adapt engines to any AF proportion. Agro-diesel is obtained from soy, rapeseed, sunflower oil, palm oil, etc. It can provide 5 % to the mix in conventional vehicles; however, for fixed route vehicles like buses, this percentage may reach 30 %. Prepared engines use only agro-diesel (Patzek et al. 2005).

AF production grew little more than 0.5 Mboe in 2006 to nearly 2 Mboe in 2010. Bearing in mind that world oil consumption is approximately 85 Mb/day and that 70 % is used in transport, it seems that AF represent almost 3 % of all transport liquid. An important international market is being created with Brazil, Indonesia and Malaysia as main exporters; although many countries are following this policy. The US and Brazil account for 90 % of ethanol production (Brown 2010).

Although calculating the energy balance is complex (because many factors that are difficult to assess intervene), highly accurate analytical results are not necessary to have an approximate idea of this balance. There is a consensus regarding the energy balance of sugar cane (to produce ethanol) that it is the best (around eight units are obtained per unit invested), followed by palm oil (over three units obtained per unit invested). Thus the use of tropical crops would be justified from the energy viewpoint and would enable ethanol and agro-diesel to be obtained. Cereals are acknowledged to have the least energetic balance while governments cite balances in the order of 1/1.5, many analysts (including some with wide research experience like Pimentel and Patzek (2005) and Hartmut Michel, Nobel Chemistry Prize), claim the balance is negative (more energy is invested than that obtained) or slightly positive. Acknowledging that the balances were moderately positive, it would be much more efficient to subsidise energy saving. Furthermore, it is the least efficient technology of energy benefit from biomass. The UK Biomass Strategy establishes five forms of biomass energy use ordered pursuant to "a clear hierarchy". The most efficient is to use it to obtain heat (industrial and domestic) and the least efficient is AF production (Defra 2007: 7). Weindorf and Altman, corroborate this conclusion (as the graph in Fig. 14.1 shows) showing the energy produced per surface unit of different biomass uses, in addition they compare it to photovoltaic and wind production.

In an attempt to reinforce the policy adopted by a number of governments, the energy balance of agro-fuels will be greatly increased once those of the second

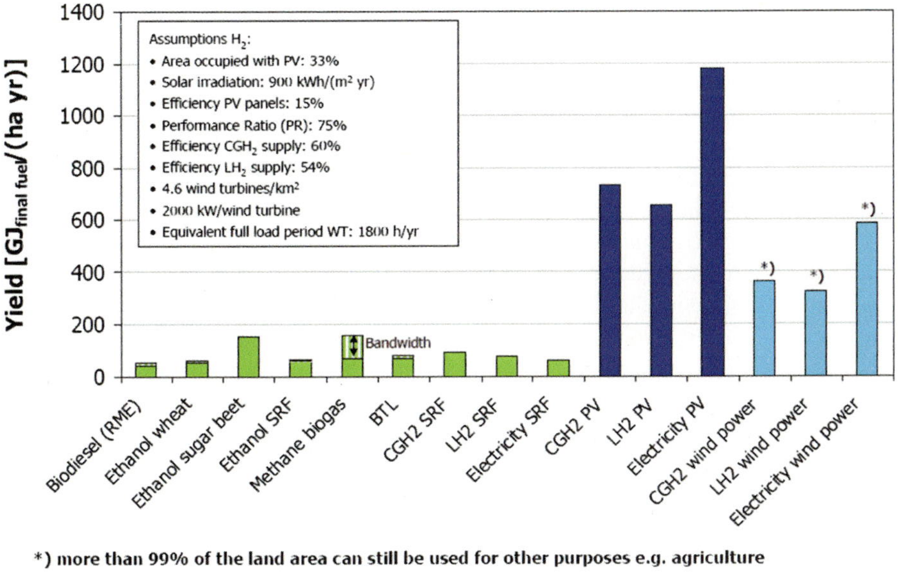

Fig. 14.1 Yield (ha year) of biofuels compared to hydrogen produced from wind power or PV (Source: Schmidt et al. 2009)

generation are commercialised: these will be cellulose fuels (which will become the third generation). In the first case the entire plant can be used, not just its fruit as is current practice. Large R + D funds are destined to contribute to this end. Cellulose is thought to be easily separated from lignine and decomposed to facilitate fermentation, thus achieving large energy production quantities. However, cellulose is the main component of cell walls of tall plants, protecting them from weather conditions, animals, and chemical attacks of fungi and bacteria, by providing them with a highly resistant structure. This implies that a lot of energy must be inverted to obtain it, i.e. chopping up the plant and then subjecting it to steam action or attacking it with very hot highly concentrated acids. Even more difficulty lies in enzyme action which requires more time and exhibits low performance (Patzek 2006: 39–40). The DoE Feedstock Roadmap strategy identifies over 60 obstacles in cellulose ethanol development. A Department of Agriculture (USDA) report states that a second-generation agro-refinery costs 3–4 times more than first-generation ones. Reality demonstrates this, i.e. at the beginning of 2010, the EPA (Environmental Protection Agency) reduced and changed the production target of 10 million gallons of cellulose ethanol for that year to 4 million gallons of methanol (Rapier 2010).

The amount of farming land required to produce AF depends on the energy balance. However, prior to quantifying how much land is necessary for AF, it should be pointed out that land is required for multiple uses apart from food production. We need it to produce raw materials for the chemical and pharmaceutical industries, for other energy uses; and to produce materials (agro-plastics, textile fibres, paper, wood, etc.). P. Schmidt et al. (2009) stress that "biofuel potential in the EU is scarcely sufficient to meet the non-mandatory European biofuel target for 2010 of 5.75 %". The fact is that the EU is importing huge quantities of palm oil, because it is cheaper than domestic feedstocks. The graph in Fig. 14.1 shows its low energy performance per surface unit in relation to other biomass energy technologies, and above all renewable energies. This panorama gets worse with the rapid growth phenomenon of land purchase or rental for various agricultural uses (food, AF, etc.) by transnational companies and non-OECD country governments, which already meant 15–20 million hectares in 2009, occurring in a clear deficit of available land context. This situation has been reported by the International Food Policy Research Institute (IFPRI), which is financed by 60 countries (Phillips 2009).

The environmental impacts caused by the product chain of the AFs are very important. It is obvious that many tropical countries are destroying forests and other eco-systems to obtain land. In Indonesia it is estimated that 6 million hectares have already been deforested. The cases of: Indonesia, Malaysia, Argentina, Uruguay, Paraguay, Colombia, etc., are widely documented. In Brazil much of the land being used is the result of tropical forest destruction, although indirectly of late, because they are using farmers' lands and are compensated by getting land from the forest. These crops are even less sustainable than those destined for human consumption. They use transgenic seeds, pesticides banned for human crops, fertilizers even more intensive than on crops for human consumption, etc. The use of water on crops (although it varies very greatly according to the plants) and in agro-refineries is very intense. The Footprint Network organization estimates it consumes 1,400–10,000 l of water to produce a litre of ethanol and 14,000–20,000 l to produce a litre of agro-fuel. Large NOx emissions are produced in these processes (Phillips 2009; www.waterfootprint.org).

Lately, however, studies on the balance of GHG emissions caused by changes in the use of land, showing very negative results, have achieved great relevance. In the case of crops on lands resulting from forest burning, it is estimated that the crops need to absorb CO_2 for 70–300 years to compensate for previous emissions (Leake and Swindford 2007). In the case of the African palm plantation in Indonesia it should be capturing CO_2 for 840 years to compensate for emissions caused by forest destruction (Webster, 2010). Every year this country burns a forest surface equal to Wales in size, which is why it is the third country that emits the most CO_2. An Ispra JRC (Italy) report concludes "its effect (on climate change) cannot be claimed to be positive" and "the costs of using biofuels exceed profits" (Edwards et al. 2008). However, an "Institute for European Environmental Policy" (IEEP) study with the collaboration of the JRCs, based on other European Commission ones, reaches devastating conclusions, despite the conservative hypothesis base. Some of the premises: the AFs will provide 92 % of the 10 % which alternative fuels would reach by 2020; and claim 6.9 new MH (the area of Belgium). These would be achieved by dedicating forest, meadow and non-intensive crops to intensive crop lands for AFs. The EU expects to import 50 % ethanol and 41 % agro-diesel in 2020, although the trend exceeds these percentages. The result being that the indirect land use change (ILUC) associated with AFs will

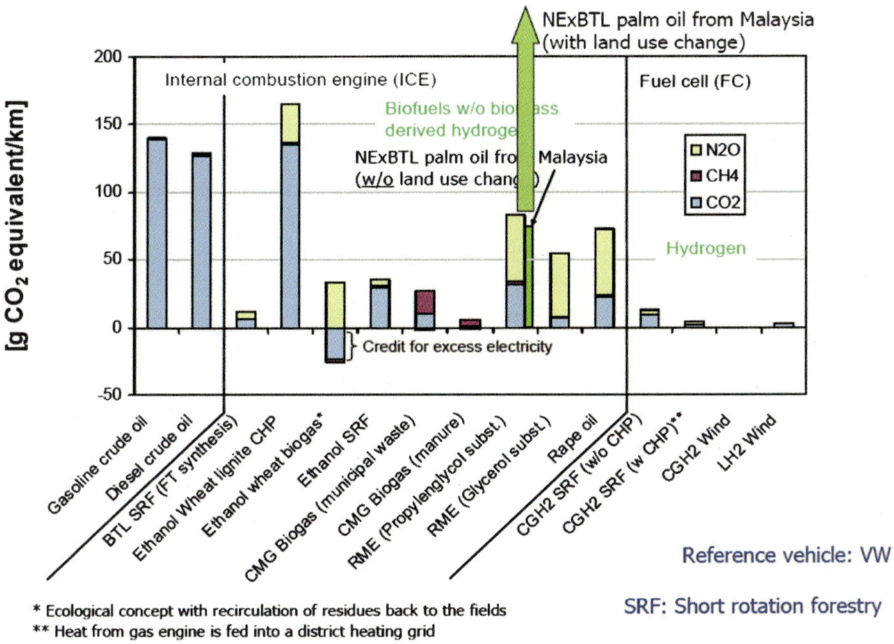

Fig. 14.2 GHG emissions "well to wheel", excluding land use change emissions (except for palm oil: 25 times more emissions). Performance (GJ/ha and year) (Source: Schmidt et al.: 2009)

increase transport GHGs: "The results of several models suggest biofuel-induced land use change emissions might be as large or larger than projected GHG emissions savings from using biofuels as a substitute for fossil fuels" (IEEP 2011: 10). P. Schmidt et al. (2009) reached a conclusion that despite ignoring ILUC, there are AF production technologies which exceed petrol emissions. Yet in the case of Malaysian palm oil, emissions are up to 25 times diesel emissions of CO_2 (Fig. 14.2).

This situation has been produced despite the 2009/28/EC Directive ("Promotion of the use of energy from renewable resources") which establishes sustainability criteria regarding GHG emissions, high biodiversity value areas (forests, wetlands), drained peatland, good agriculture practice, etc. Yet the Commission continues to support agro-fuels: "Biofuels will probably be a main option for aviation, long-distance road transport", and hopes "the market uptake of new bio-energy which reduces demand for land necessary for food production, and increases net greenhouse gas savings" (COM (2011) 85 final). These are the same conclusions reached by a European Expert Group on Future Transport Fuels (EGFTF) report but downgraded via the following statement: "Industrial viability and environmental impacts of producing biofuels from ligno-cellulosic biomass have yet to be demonstrated" (EGFTF 2011: 25).

2 Renewable Electricity – Electric Car

The 2009/28/EC Directive projected that renewable electricity would be one of the energies to contribute 10 % of total road transport fuel consumption in 2020. This poses the problem of defining the mean renewable electricity used per country in the period indicated. In recent years, the vast majority of OECD countries have been fostering deployment of the electric car, with subsidies to: R + D, production, vehicle recharging infrastructure and purchase. Nevertheless, its mass development is hindered by technological immaturity, insufficient recharging structure, and above all by high prices. The latter are justified based on improvement of air quality, although several studies carried out in Germany show that, given the German electric mix, electric cars emit 1.5 to 2 times as much CO_2 as a conventional small car (Hydrogen Mirror 2/2011).

Historically, its main limiting factor was its limited autonomy due to the very limited energy storage capacity of its batteries. Now its proposers expect a rapid increase in the same combined with a rapid cost reduction. Nonetheless, it seems unable to improve sufficiently to become an alternative. Oil, gas and coal have the following energy per kilo respectively: 50, 55 and 20–35 Megajoules (Mj). This ratio reaches 0.1 Mj (and may reach 0.7 Mj) in acid batteries and 0.5 Mj (3.0 Mj in the future) in the lithium ones. Nevertheless, hydrogen has a much better starting point than the battery: 6 Mj per litre (at 70 atm). This explains why manufacturers offer electric vehicles with journeys of only around 150 km maximum, and why the Commission claims "pure electric cars prove more promising in an urban environment given the limited scope of the batteries" (COM (2010)186 final). However, this figure is deceptive, because if we turn on the heating, lights and windscreen wipers, their capacity is drastically reduced. Thus it is no alternative for long journeys. An attempt is being made to reduce this problem by improving batteries and establishing rapid recharging systems, although that requires costly installations. A normal charge requires 4–6 h; however, a rapid charge can be achieved in 15 min, still far from the 3 min required by hydrogen. This problem and the high battery cost could encourage builders to rent them, although this does not seem a good idea, since battery change takes around 10 min. Furthermore, this option would come up against the need for all manufacturers to use standard batteries, which is highly unlikely to happen (Zenz and Johnson 2009). Moreover, we will see in the hydrogen section that the cost of building service station networks with recharging is starting to exceed that of the hydrogen network from 1–2 million cars.

In 2011 and 2012 several automotive companies started selling them; however, the greatest obstacle for deployment is the product cost. In the US manufacturers advertise prices between 30,000$ and 50,000$ whereas in the EU prices range between 30,000€ and over 40,000€, which are two to three times more expensive than the same car types with conventional technology. Moreover, the large subsidies announced by some governments will have little impact on the situation (Woodyard 2010). This is why some manufacturers tend to rent them (at around 500€/month).

We will see further on, that while prices announced for hydrogen cars are similar to those for battery run cars, the costs of the former are expected to drop more quickly. Thus, the electric car will only be used for short journeys, competing with public transport, and if successful will aggravate urban congestion, though the air will be cleaner.

3 Hydrogen

Hydrogen commercialization began in Europe and the US in 1930 for industrial uses (plastics, resins, solvents, fertilizers, steels, etc.). Every year over half a billion cubic metres of hydrogen are produced, i.e. an energy capacity equal to over 10 % of oil consumed. However, its applications are now expanding towards electrical production associated to a fuel cell with a very wide range of uses, particularly in transport. Thus it has become the sole alternative to oil in transport, given the limitations of agro-fuels and battery traction. There are multiple reasons for this: high energy capacity (1 kilo of hydrogen equals 3.5 l of oil); increases in energy safety, because it can be obtained from any energy source; solutions of the intermittence problem of the main renewable energies, since it is the best option for storing large quantities of energy for long periods of time. Furthermore, when associated to the fuel cell its uses are highly efficient. Nevertheless, hydrogen has two serious limitations, i.e. it is a secondary fuel, thus energy must be used to obtain it; furthermore it is be stored at high pressure (especially in vehicles), since it is the most volatile gas. Moreover, it is generally claimed to be dangerous and cited as the reason for the Hindenburg zeppelin accident, which put an end to the development of this technology. For a long time the hydrogen that filled the zeppelin was thought to have exploded; however, subsequent studies state the accident was not due to hydrogen nor did it explode, but due to its high volatility it burnt quickly. Its high safety has been demonstrated through its industrial use for over a century. Furthermore, Ford carried out a study in 1997 that demonstrated hydrogen was much safer than petrol. It would only explode in the event it was ignited in a tight enclosure. Thus "among experts there is no doubt that it (safety) can be achieved" (Altman et al. 2004).

Despite its importance, institutional support has been insufficient and variable, thus in the mid-1990s the previously existing optimism for its rapid growth was doused. This attitude can only be understood due to a lack of comprehension regarding oil shortage. And the accumulated delay means many consider hydrogen could only be an alternative to oil in transport in the medium-long term or longer. However, the increase in investment by major governments is accelerating its expansion.

Most hydrogen is produced from natural gas in industrial processes. Some think photosynthesis is the technique of the future, whilst others go for fermentation, etc. However, the dominating technique is water electrolysis, due to development of the most advanced electrolysers. Electrolysis is already a commercial technology

for industrial hydrogen production. Wind energy is the best prepared among the renewables to be the primary production source of large quantities of hydrogen (in fact the US uses wind energy to produce industrial hydrogen in the MidWest). However, the rapid development of photovoltaic technology is reinforcing its role in decentralized production. Greenpeace Energy (a German energy co-operative, supplying over 110,000 customers) and Gasunie have signed an agreement to inject hydrogen (produced by an electrolyser from the co-operative wind and photovoltaic plants) into the Gasunie natural gas network to supply customers with a mix of both gases. Moreover, the proportion of hydrogen will grow in the future (EHA 2010: 5; Fuel Cell Today, 18 April 2012).

Nevertheless, use of hydrogen from renewable electricity is a very clumsy and inefficient process in many cases (electricity-hydrogen-electricity), which is only justified in the case of an energy surplus. Thus technologies must be developed that enable direct hydrogen production from solar energy. One of the technologies being developed concentrates solar energy to produce steam, which passes through hoppers absorbing oxygen to release hydrogen. The EU is financing development of this technology at the Almeria Solar Platform. After experimenting with a small plant (Hydrosol I), a more advanced Hydrosol II started operating in 2008 and has doubled the capacity of its predecessor, thus taking a necessary technological leap for its commercialization (www.ciemat.es).

4 Fuel Cells

A fuel cell is an electrochemical device combining hydrogen and oxygen to produce electricity and water. A fuel cell unit consists of a stack, comprising a number of individual cells. Each cell has two electrodes and a solid or liquid electrolyte. According to DoE report (2004a) to Congress, "fuel cell technology offers unique opportunities for substantial reductions in energy use and emissions in its application to transport and electrical production". A fuel cell can reach efficiency of up to 90 %, recovering residual heat, and is highly efficient even when not operating at full load. It does not contaminate since it only emits steam. Furthermore, it is a flexible technology, i.e. it adapts easily to all possible uses of electricity and increases demand due to its modular character (Fig. 14.3).

The most significant fuel cell types are, in order of commercial importance: Proton Exchange Membrane Fuel Cells (PEMFC), Direct Methanol Fuel Cells (DMFC), and Solid Oxide Fuel Cells (SOFC). PEMFC and DMFC are similar, as both use a polymer membrane, although the former runs on pure hydrogen, and the latter on methanol. The PEMFC have performances of 40–60 %. It is the most important technology (in 2010, 97 % of units produced were PEMFC, against 2.9 % of DMFC and 0.07 % of the rest). The PEMFC hegemonize the automotive market and a large part of other medium-capacity applications, although absent from the large stationary cell market because manufacturers are concentrating on the most promising applications. The DMFC have a performance of 40 % and

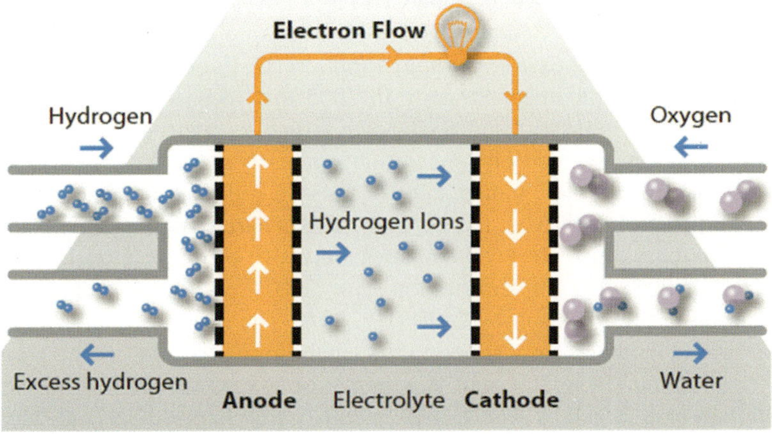

Fig. 14.3 Fuel cell stack (Source: Fuel Cell Today. Industry Review 2011: 6)

are used in mobile devices and transport market niches (boats, motorcycles, to produce auxiliary electricity, etc.). The performance of the SOFC exceeds the previous performance, and they are particularly used for small industrial processes (stationary and auxiliary). The main problems to be solved to achieve massive commercialization of cells are: standardization, cost, durability, reliability and infrastructures. These are solvable and some have already been solved (Callaghan et al. 2010; Fuel Cell Today 2011: 30).

4.1 Strategies for Fuel Cell Expansion

Japan is considered to have the earliest and most coherent promotion of hydrogen. The Roadmap to Fuel Cells of Polymeric Electrolytes was approved in 1993. It aims to manufacture 50,000 vehicles, reach 2,100 Mw capacity in fixed installations during 2005–2010, 5 million vehicles (with 4,000 service stations) and 10,000 Mw between 2010 and 2020. Given the current panorama of the sector, the earliest targets were not reached. Yet government support in recent years has enabled commercialization of fuel cells applied to housing and a commercialization plan for light duty vehicles (Hara 2011).

The USA defined two documents in 2002 constituting their initial strategy: the Hydrogen Vision and Map of the National Hydrogen Energy Route. In 2004 the *Hydrogen Posture Plan* was approved and updated in 2006. Four overlapping stages were established to be carried out at least through 2040. The Plan is centred on road transport and establishes precise targets for its first two stages (DoE 2004a, b; 2006.). At the end of 2010, the US Fuel Cell Council and National Hydrogen Association announced they would join forces to accelerate commercialisation of fuel cell and hydrogen technologies (Fuel Cell Today 2011: 27).

In 2003 the EU Hydrogen and Fuel Cell High Level Group presented the report "Hydrogen and Fuel Cell Energy – Vision of the Future". The Commission subsequently created the *Fuel Cell Technology Platform* (HFP) grouping the main agents and publishing a Deployment Strategy in 2005 that reaches 2020. It establishes four actions areas: transport and recharging structure; hydrogen production and distribution; stationary electricity generation; and early markets. It foresees for this date: 1–5 million cars and annual sales of 0.4–1.8 million cars; electric capacity installed of 8,000–16,000 Mw; in the "early market" area (portable cells, industrial vehicles, auxiliary units, etc.) contemplates a massive commercialization in 2009–2012 (HFTP 2005). The Implementation *Plan* (2007) sets out the strategy in detail based on ten targets to reach by 2020 (IP 2007). The Joint Technology Initiative (JTI) is subsequently created, adopting the legal figure of Fuel Cell and Hydrogen Joint Undertaking (FCH JU), and is a public-private association responsible for meeting strategic targets. The FCH JU establishes a plan in 2009 with targets to be reached by 2015 in the four areas indicated: 500 light duty vehicles and 500 buses; hydrogen price of 5€/kg at service stations; capacity installed 1000 Mw; 14,000 units distributed among the different early markets, 12,000–13,000 units for portable devices and micro-batteries. However, renewable hydrogen dominion and its use for storing energy on a large scale have been put back to 2050. Such targets impede compliance with those of the Expansion Strategy for 2020. Positive aspects proposed are: robust bus development, application to the traction capabilities of lorries, underground railways and trams; likewise, demonstrations of auxiliary units in planes and ships (FCH JU 2009).

Parallel to these plans, in 2007 the HyWays project was developed comprising a strategy (*Roadmap*), and an *Action Plan*, centering almost exclusively on developing lightweight road vehicles (up to 3.5 tonnes). Regarding vehicle market penetration, three phases have been designed. In phase I (until 2010) large demonstration projects will begin. Phase II (2010–2015) is that of early commercialization with 10,000 vehicles and a hydrogen infrastructure that joins the demonstration centers as they are created. Phase III has three sub-phases: 500,000 vehicles (2015–2020); 4 million (2020–2030); 16 million (2025–2035) (HyWays 2007a, b). Thus it exceeds the targets of the FCH JU plan.

Generally speaking these strategies and plans describe a slow expansion rate of hydrogen technologies based on the premise that there will be no shortages of oil or natural gas in the forthcoming decades, thus the renewable hydrogen price is not competitive in the short-medium term. The USA Posture Plan establishes the following cost targets: obtain hydrogen from natural gas at a price equal to 1.5$/gal of petrol and 2.0$/gal from wind energy. Petrol prices greatly exceeded 4$/gal in 2008 and in 2012. HyWays is based on oil barrel price forecasts 26–81$ (average 50$) for 2010 and 54–217$ (average 135$) in 2050 (HyWays 2007a). Whereas the FCH JU (2009) only contemplates diminishing oil in their 2050 vision.

Annual public investment worldwide was around 1,040 million dollars for the period 2003–2005, distributed as follows: Japan (30 %); EU and the US (24 % each), leaving 14 % for the rest of the world. However, the EU has led in public investment since 2009 because the JTI invested nearly 1000 million euro in 6 years on a project called "New Energy World", only half of which will be public. This investment aims for a massive expansion of hydrogen technologies prior to 2020, and as usual to achieve world leadership in the sector, as it already has in transport (EHA 2010: 5).

4.2 Fuel Cell Development

Fuel cells are entering the commercialization phase in some market niches where they offer clear advantages over existing technologies. This occurs particularly in toys and educational kits and is starting to do so in some portable electronic devices for such purposes as emergency electricity production, forklift truck traction, auxiliary electricity, motor-homes and recreational boats; there are also military applications of all kinds; etc.

The number of cells sold in recent years has been increasing around 70 % yearly. Although it was in 2009 when it really started being commercialized en mass, reaching over 150,000 units (almost ten times more than in 2008), even though the vast majority were for toys (power around 1w) and teaching kits. A strong growth trend was maintained in 2010 and 2011. Nevertheless, if we consider the installed capacity (in Mw) we find that although there was a spectacular leap in capacity in 2009, it did not progress in the following 2 years. If we analyse the two graphs from the viewpoint of uses (portable, stationary and transport), we discover the portable use greatly exceeds the other two in number (which accelerates development of fuel cells), yet in capacity its contribution is residual. In this case transport was dominant in 2009 and 2010, while stationary dominated in 2011. This change was due to transport irregularity; since they had not entered the commercial phase, manufacturers produced limited series (around 200 units) intermittently, as per their product development schedules (Fig. 14.4).

4.2.1 Portable

The portable sector has accounted for at least 75 % of total shipments of micro-portable fuel cells every year since 2007, and this ratio increased to over 95 % of total shipments from 2009 onwards. In this section we have: Educational Fuel Cells, Auxiliary Power Units (APU), and Consumer Electronics. Fuel cells are used in the multiple vehicles extant as traction elements and also to produce auxiliary electricity (APU) in motor-homes, recreation boats and lorries. This means over 80 % of cells sold were related to transport (Fuel Cell Today 2011).

Cells applied to portable electronic devices constitute the segment with greatest potential for massive development, since they are highly superior to batteries. The electronic industry demands more and more energy due to rapid increases in the functions of their devices. However, portable fuel cells for consumer electronic applications remain a challenge, due to problems relating to miniaturization and

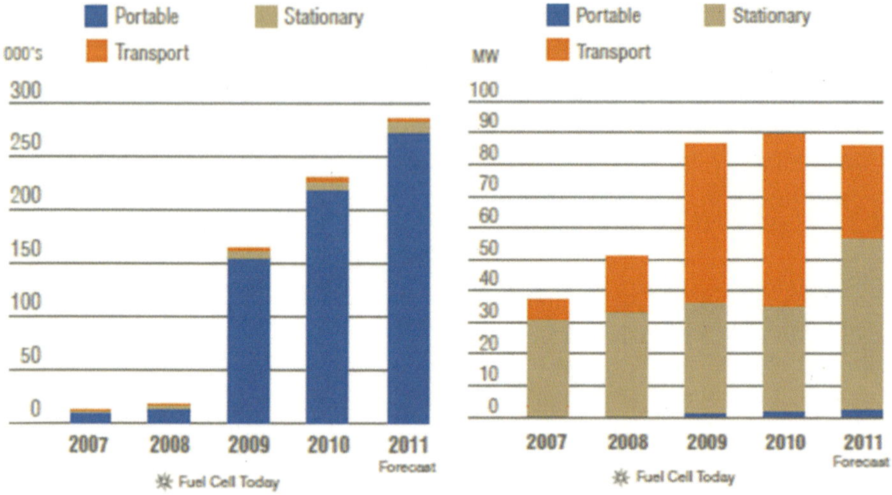

Fig. 14.4 Shipments and Mw by application 2007–2011 (Source: Fuel Cell Today. Industry Review 2011: 13)

system integration. Those problems prevent their full commercialization. As a way round the miniaturization problem, Toshiba started selling chargers at the end of 2009. The experience was successful, likewise in 2012 for the various companies commercializing different varieties. Nevertheless, the integration problem requires a solution and in recent years numerous companies (Antig, CMR Fuel Cells, Nomadic Fuel Cells, Samsung, etc.) have tried without success, since the products are still very heavy. The solution lies with alliances between electronic and fuel cell companies, like the case of Lilliputian Systems and Intel for development of its USB Mobile Power System (Fuel Cell Today 2011: 14, 15).

4.2.2 Stationary

Analysts coincide in the opinion that there is a large potential market for stationary fuel cells. Two groups are defined: small (0.5–10 Kw) and large applications (>10 Kw). There are two application types, i.e. as a reserve device in the event of power outage or to produce electricity and hot water in homes. The former application occurs in banks and telecommunications companies. North America has 2/3 of this market and rapid development is expected. For homes, companies have opted for cells of 1–1.5 Kw, to provide a basic electricity and hot water service. The grid will be used for consumption peaks. In Japan, cumulative shipments had reached 13,500 by the end of 2010. Subsidies for these systems are being phased out, due to cost reductions in the past 5 years. South Korea has developed a similar scheme. There are three market segments for large applications, i.e. 10 Kw batteries (data centres), 250–400 Kw (office blocks, hospitals, etc.) and over 2 Mw to produce electricity for the grid. In recent years, annual sales have stagnated at 50 units, but with capacities rapidly growing. The US dominates this market. A study on Fuel Cells 2000 reached the conclusion that 38 large companies (Ikea, Google, Walmart, Michelin, Motorola, etc.) had installed or contracted over 1000 forklift trucks, 58 stationary units and over 600 units in telecommunications centres (Adamson 2009; Fuel Cell Today 2010, 2012: 17).

4.2.3 Transport

States and automotive companies are centred on fuel cell development for light duty vehicles. Consequently, its application to railway, sea and air transport vehicles is delayed, yet it would be much easier to introduce fuel cells in the vehicles indicated (except aircraft) as they have sufficient space for storing hydrogen. They do not require such demanding features as cars (great autonomy, reduced hydrogen storage space, long-lasting batteries, etc.). Automotive companies are convinced that only hydrogen can guarantee the sector's future. Furthermore electric cars and hybrid car variations are stepping stones towards fuel cell traction, while their infrastructure is under construction. T. Fukui, Honda Chairman in 2007, stated "there is no future for the car industry without fuel batteries" (Hydrogen Mirror 1/2010). While combustion engine performances are generally slightly over 20 %, those with fuel cell can reach 60 %. Thus in a well to wheel energy analysis the fuel cell electric vehicles (FCEV) were much more efficient than petrol ones, since the greater energy consumption in hydrogen production in relation to petrol is greatly compensated by the greater efficiency of the FCEV: it "would represent 50 % reduction in carbon emissions" (Fuel Cell Today 2011: 24). Their commercialization requires a 5 year lifespan (200,000 km) and over 400 km without refuelling. These targets have been reached by manufacturers, however, cost and tank weight are still excessive (Callaghan et al. 2010), although cost is dropping at a strong rate, and is expected to accelerate on commencement of commercialization as shown in the graph in Fig. 14.5.

Since 2004 at least, the main companies have been testing ever larger car series. At first they tested one or several dozen, but in the current pre-sale phase the series are usually 200 units. Honda and Daimler rent cars in California for 600–800$/month for periods of 3 years. Hyundai began doing this in 2012. In 2010, the sector grew 20 % and reached 2,400 cumulative units. 500 cars had already travelled 15 million km and done 90,000 recharges. Thus, "the focus moves from demonstration to commercialization expansion" (Mckinsey and Company 2010; Fuel Cell Today 2012).

The EU aims to build a basic infrastructure by 2017 in two phases. In the first they would create 30 demonstration regions (they are already operative) with nine refuelling stations in each. In phase two they would connect the areas with new stations. However, some States and regions are more advanced. At the end of 2009, a group of energy companies in Germany undertook to build a service station network throughout the country by 2015 and each manufacturer to place 10,000 units on

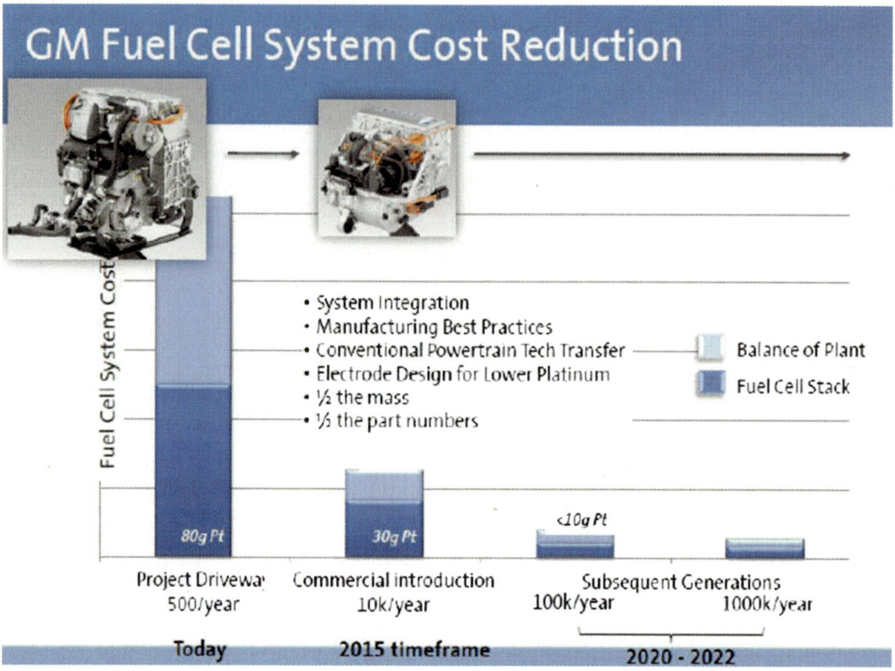

Fig. 14.5 GM FC system cost reduction over one decade (Source: Wurster 2010: 16)

the road. However, in 2012, Daimler decided to bring commercialization forward to 2014, because in 3 years Daimler and Linde will have built 20 more stations. The *Scandinavian Hydrogen Highway Partnership* project aims to join Scandinavian countries with each other and Germany. The aim is to build 15 stations between 2012 and 2015, which will be used by 100 buses, 500 cars and 500 special vehicles. Company groups similar to the Germans are reaching the same kind of agreement in Japan, South Korea, GB, etc. In Japan, a Group of oil and gas companies have announced their intention to build 100 hydrogen refuelling stations: 60 by 2015. California is developing the *California Hydrogen Highway* strategy. A report of the company Pike Research says that, in 2020, one million fuel cells cars will be sold worldwide. 22 new filling stations started operation in 2010 and the total number reached 212 stations (80 each in Europe and North America, 48 in Asia) (EHA 2010; Fuel Cell Today 2011: 24; Hydrogen Mirror 2, 3, 4/2011).

The fleets of captive vehicles (buses, taxis and distribution vans) constitute an early market of great potential, to use centralized fuel recharging and because the excess weight of the system was not an important problem. In 2003 the EU started the CUTE programme where three Daimler buses were each used in ten cities. In 2006 the HyFLEET programme began with 50 new generation Daimler buses, to be used in seven European cities (highlighting Berlin with 14), in Peking and Perth (Australia). Both experiences were a total success, i.e. they used 8.5

million passengers and travelled 2.5 million kilometres. The last project (CHIC-Clean Urban Transport for Europe) has plans, in phase 1, to deploy a total of 26 buses in four cities and regions. The EU aims to put 500 units into operation in 2015. Many large European cities (led by Hamburg, Copenhagen and London) are starting up introduction plans for fuel cell buses. Fuel Cell Today "sees costs decreasing and durability increasing" in the sector (Hydrogen Mirror, 1/2010; Fuel Cell Today 2012: 22, 23).

However, the experience in fuels indicates that 15–20 % of service stations must offer hydrogen to enter into mass vehicle production. The HyWays project estimates the costs of building service station network for at least 25 million vehicles in 2025 at around 15,000 million euros and the entire system (including H2 and transport production) close to 60,000 M€ (HyWays 2007a). The German study mentioned above estimated the cost would be much less, i.e. between 25,000 and 50,000 M€ (McKinsey 2010: 6). This technology and pure electricity have similar infrastructure costs up to 1–2 million vehicles. From there on the quantity of the latter is cheaper. This brings us to the conclusion that massive deployment of rapid charging installations make no sense for a small market fraction that will be used for short journeys. Furthermore the hydrogen infrastructure is valid for all vehicle types (thus its market share would be much greater) and the cost of cell vehicles is expected to drop 90 % by 2020, and electric ones 80 % (McKinsey 2010).

Half of all shipments in the transport sector for 2010 were material handling vehicles, continuing the strong growth in recent years. The vast majority of these were installed in the USA, due to significant government subsidies. This application is expected to grow at an annual rate of 100 % in the next 10 years (Fuel Cell Today 2012: 19, 20).

The application of cells to vehicle traction has, nevertheless, been centred on the car, although this is starting to fragment in the EU at least. The application of cells to trains is one of the clearest examples. Different institutes have developed train and tram prototypes. Between 2000 and 2009 projects were successively started on a mining train, a tram, a military train, a Danish line (57 km) (subsidised by the EU and expected to be inaugurated in 2010) and a train. The aviation sector has also started moving. Boeing successively tested a hydrogen battery run plane with crew in 2008; Airbus did the same in 2009. An Israeli company has designed a 10 passenger plane financed by the EU. NASA is working on reactor propulsion and propeller planes using hydrogen. Several companies are designing aircraft models without crews, a use which is generating growing attention. However, fuel cells are only expected to be used to produce auxiliary electricity in large planes in the medium term. The application for passenger vessel traction is being developed slowly (Adamson and Callaghan 2009; Fuel Cell Today 2010).

Chapter 15
Societies in Energy Emergency

Keywords Societies in energy emergency • Post carbon cities • Transition towns • Community sustainable development • Vulnerability evaluation • Contingency plans

In this chapter we study the following issues: the origin, development and characteristics of societies in energy emergency (SEE); a comparative analysis of the two organisations which form the movement; a study of the process phases in the design and implementation of transformative strategies; and an evaluation of the SEE movement.

1 Origin, Development and Characteristics

SEE are societies that consider peak oil and climate change to represent a very serious threat, and to cope with them they undertake to build a sustainable society. There are three types of SEE: States, federal states and regions, and municipalities. Normally there is no coordination among them, though there is feedback. Some societies become references for others, as we will see in this chapter. Two movements have been created in recent years. One is known as *Post Carbon Cities* (PCC) and is located mainly in North America, although the PCC has been extended to other Anglo-Saxon countries. The second is known as *Transition Towns* (TT), and although it began in Ireland, it soon had a rapid development in Great Britain, and is now spreading into many countries.

Although municipalities have been pioneers in the promotion of sustainable development (as in the cases of Local 21 Agendas or in the field of climate change), some States and federal states have been leaders in taking into account the risk of peak oil. But they have not developed adequate policies to avert risks. In this section we will do a chronological description of the development of SEE.

R. Bermejo, *Handbook for a Sustainable Economy*,
DOI 10.1007/978-94-017-8981-3_15,

In 2003, The Western Australian Government (2003) approved a strategy for sustainable development (elaborated with extensive participation of government representatives from many areas, experts and communities through many workshops) which lacked measurable goals, but which warned of the proximity of peak oil. In following years the government declared that it was aware of the vulnerability of WA, and that they were taking some measures in order to reduce it. A. MacTiernan (2004), Minister for Planning and Infrastructure, declared in 2004, during the *Sustainable Transport Coalition's Oil: Living with Less* Conference, that "the Deputy Prime Minister (. . .) acknowledged that global oil production may peak in the next few years", and that "our government recognises that our continued dependence and reliance on oil makes us vulnerable to future shocks (. . .) We must build resilience into our system. We have started this process": introducing "measures to diversify our fuel sources", "policies that aim to reduce the transport task", and "investing heavily to bring about behavioural change that reduces our dependence on the motor vehicle". But, in 2007, the following Prime Minister made an Institutional Declaration (*Making Decisions for the Future: Climate Change)* that proposed "a long-term goal to reduce (. . .) greenhouse emissions by 60 % of 2000 levels by 2050" (Carpenter 2007: 9).

The Prime Ministers of Sweden and France, at the beginning of September 2005, made two declarations about the peak oil problem (after the impact on the oil sector produced by hurricanes *Katrina* and *Rita*, which happened some days before the declarations). The French Prime Minister stated that we had entered the post-oil era, and proposed a strong impulse for energy efficiency and for renewable energies. The second stated that peak oil was very near, and that it was necessary to design a strategy for phasing out oil consumption by 2020. In 2005 the New Zealand Labour Party issued a statement which acknowledged the proximity of peak oil, and consequently the need to change the energy model. In April 2006, the Prime Minister stated that the world had either reached peak oil or that it was very close. But in December 2006, the Ministry of Economic Development published the *New Zealand Energy Strategy Draft* which stated that "while there will be a peak of cheap oil from conventional sources, the world has plentiful sources of fossil-based oil", that are unconventional (2006: 19). With the exception of the Swedish government (although the result was disappointing), the rest of the governments mentioned did not put their words into practice. On the contrary, in 2005 the Australian Queensland government appointed the *Queensland's Oil Vulnerability Task Force* Commission "to address concerns that future world supplies of oil for energy may diminish (. . .) and that *peak oil* may be a world-wide phenomenon". In October 2007 the government approved the Commission's report (McNamara 2007).

On the other hand, some other governments and parliaments have showed their concerns about the phenomenon, but we have not been able to see the consequences of these statements. Next we will describe some additional information concerning these issues. The Parliament of Southern Australia (2008) appointed a "Select Committee to inquire into and report on the impact of peak oil on South Australia". Later in the year, the Select Committee issued a statement with 23 recommendations. The second one was "that the State Strategic Plan be updated to recognise Peak oil and

Table 15.1 Declarations dates about the peak oil of selected societies

Societies	Date	Population	Societies	Date	Population
Willits (California)	Dec-05	15,000	Connecticut	Nov-07	3,400
Franklin (New York)	Dec-05	2,000	Berkeley (California)	Dec-07	101,000
Burnaby (Canada)	Jan-06	203,000	Bloomington (Indiana)	Dec-07	72,000
South California Association of Governors	Mar-06	18,000,000	Sunshine Coast (Australia)	Dec-07	260,000
San Francisco (California)	Apr-06	744,000	Hervey Bay (Australia)	Dec-07	52,000
Metropolitan area of Portland Portland(Oregon)	Apr-06	1,400,000	Darebin (Australia)	Feb-08	128,000
Hamilton (Canada)	Apr-06	505,000	Spokane (Washington)	Feb-08	201,000
Isla de la Palma (Spain)	Apr-06	86,000	Maribyrnong (Australia)	Apr-08	63,000
Portland (Oregon)	May-06	537,000	Minnesota	May-08	5,198,000
Bloomington (Indiana)	Jul-06	72,000	Whatcom County (Washington)	May-08	186,000
Brisbane (Australia)	Aug-06	1,858,000	Bellingham (Washington)	May-08	78,000
Oakland (California)	Oct-06	402,000	Cleveland (Ohio)	Jun-08	438,000
Marrickville (Australia)	Feb-07	76,000	Somerset County (UK)	Jul-08	518,000
Alachua County (Florida)	Mar-07	240,000	Chapel Hill (North Carolina)	Oct-08	55,000
Austin (Texas)	Sep-07	743,000	Nottingham (UK)	Dec-08	289,000
Bristol (UK)				2009	400,000

Source: Elaborated by the author

incorporate adaptive strategies". Japan and Hong Kong have asked for reports about peak oil, and both reached the conclusion that this phenomenon constitutes a great risk (http://postcarboncities.net/node/3568).

In Autumn 2005, some North American municipalities began to approve declarations warning about the great risk that the proximity of peak oil poses, and stating that they will take adequate measures. Many municipalities are currently taking measures to avert the risks, but in Table 15.1 we only cite municipalities with more than 50,000 inhabitants (with the exception of the pioneering towns).

Most of the municipalities mentioned are part of the movement led by the Post Carbon Institute (PCI). The typical dynamic is the following. The first step is to approve a declaration about the mentioned issues (peak oil, climate change, and sometimes also problems with natural gas). The second step is to create a group to carry out a risk assessment and to propose solutions at two levels: short-medium term proposals to avert the most risky issues, and after that the design of a long-term strategy to transform societies. But in the case of the TT movement's dynamics, being very participative, there is not an agreed-upon protocol. But in the end, they all design transforming strategies. However, the cities and large counties included in the TT movement follow the dynamic of the PCC movement, which begins by approving peak oil resolutions, such as in the cases of Nottingham, Somerset County and Sunshine Coast (www.postcarboncities.net/node/4016).

2 Comparative Analysis of the Two Movements

As the first kind of SEE (States, federal states, etc.) does not have any kind of coordination, we will only analyse the two movements (PCC and TT). They are coordinated and promoted by two organisms: the Post Carbon Institute (PCI) and the Transition Network (TN). Both promote the creation of groups to raise awareness among the population and in governments about pressing issues. However, they differ in several approaches: the organisational models, the role of participation, the intermediate objectives pursued, the dynamics, priorities, etc. To study these aspects we have, apart from the practice of each movement, several handbooks published by the different organisations. The most significant are: Post Carbon Cities: Planning for Energy and Climate Uncertainty (Lerch 2007), The Transition Handbook. From oil dependency to local resilience (Hopkins 2008), and Transition Companion. Making your community more resilient in uncertain times (Hopkins 2010).

There is a basic consensus among both movements. They promote sustainable and decentralised economies, with a great level of self-sufficiency, and they contend that citizen participation in the processes of change is a basic requirement to transform them and to create cohesive communities, apart from having an intrinsic value. However, for the TT participation is the primary force for change, and for the PCC participation is only a tool to move authorities to change. Thus, in the first case communities are the protagonists, and in the second it is ultimately the authorities.

D. Lerch defines three goals to be accomplished by local governments: "*Post Carbon Cities: Planning for Energy and Climate Uncertainty* provides guidance and support to local government officials and staff for meeting three critical goals: breaking community dependence on oil, stopping community contributions to global warming, and preparing the community to thrive in a time of energy and climate uncertainty" (Lerch 2007: v). But most of the book is dedicated to describing the activities of the main cities, and from these experiences it infers three key lessons: "Make a clear government statement on the issue" (peak oil); "Engage the issue with both government and the community"; "Organise and lead task forces carefully" (Lerch 2007: 50). It defines five principles of good governance: "Deal with transportation and land use": "Tackle private energy consumption"; "Attack the problems piece-by-piece and from many angles"; "Plan for fundamental changes and make (them) happen"; "Build a sense of community" (Lerch 2007: 63–66). So its approach is top-down.

Thus the PCI seeks to increase the awareness of societies and authorities of issues concerning peak oil and climate change, in order to move governments to initiate a transformative process with the support and participation of society. And although D. Lerch (2007) states that the PCI works with 200 groups through the Relocalization Network, the 2009 Manifesto declares that its central role is to design solutions to manage "the transition to a sustainable, equitable, post-carbon world": "Post Carbon Institute is dedicated to answering the central question of our times: *How do we manage the transition to a post-growth, post-fossil fuel, climate change world?"*. It further states that "this will require a coordinated effort on the part of those who understand both the problems and the solutions (. . .) The centrepiece of our effort is the development of a select community of *Post Carbon Fellows"*, who "will publish an annual *Roadmap For the Transition*, covering each of the principal issues". Finally "we will develop the framing and messaging of these issues so as to significantly raise the visibility and impact of emerging solutions" (Post Carbon Institute 2009). Thus the PCI's objectives are to promote a greater awareness in societies about the main problems, and about the solutions which this organisation is designing. In 2009 the PCI reached a collaboration agreement with Transition USA, the national organisation of the Transition Network. This could mean a certain distribution of tasks: the primary role of the PCI is to work as a think tank, while the TT is the means for moving communities.

The *Transition Towns* movement began in Kinsale (an Irish town of about 7,000 inhabitants), when the town council "adopted in an unanimous vote at the end of 2005" the *Kinsale 2021: An Energy Descent Action Plan,* elaborated by a student group, under the tutelage of Rob Hopkins. This author has become the creator, principal thinker and leader of the movement. And he is the leader of Transition Towns Totnes, a municipality of 8,500 inhabitants located near London, which is the principal reference of the movement (Branwyn and Hopkins 2008: 9).

The Handbook defends the contention that peak oil means the end of an era of abundant energy. So there is a "need to prepare in advance for the inevitable decline in net energy availability". And to describe this scenario it uses the term "energy descent", which is defined as "The continual decline in net energy supporting humanity, a decline that mirrors the ascent in net energy that has taken place since the Industrial Revolution" (Hopkins 2008: 53). But this is not a gloomy scenario: "It also refers to a future scenario in which humanity has successfully adapted to declining net fossil fuel energy availability and has become more localised and self-reliant" (Hopkins 2008: 53). And this society will be happier: "We are only beginning to scratch the surface of a positive vision of an abundant future: one which is energy-lean, time-rich, less stressful, healthier and happier" (Hopkins 2008: 94). On the other hand, the TT groups should lead the processes of change: "The power of the Transition process is its potential to create a truly community-led process which interfaces with local politics, but on its own terms. The role we identify for Local Authorities in this process is to support, not to drive it" (Hopkins 2008: 144).

For this reason Hopkins criticises other movements (like Local Agenda 21 or the Post Carbon Cities) for developing a top-down dynamic: "Local Agenda 21, although it created many interesting initiatives, was in essence a top-down process"; "For many towns in the US, such as Portland and Oakland, the passing by local authority of a Peak Oil Resolution is seen as a key step. This may be the case, but my sense is that the important first steps are to engage the community in awareness-raising and building the energy for the project, rather than disappearing at the early stage into a bewildering world of policy writing and working at the local government level" (Hopkins 2008: 144, 145).

In 2011, R. Hopkins published a new book (Transition Companion. Making your community more resilient in uncertain times). It contains many contributions by other TT members. In November 2011, R. Hopkins also wrote two articles about the last book (an introduction and a summary).

Through these documents we can see new approaches. R. Hopkins justifies the need for a new book based on the broad experience accumulated: "For the first the Transition Handbook (. . .) was pretty much a speculative question, but for this new book we are able to draw from what has, in effect, been a 5-year worldwide experiment, an attempt to try to put the Transition idea into practice". And the book is called a "*Companion* because that is exactly what it is intended to be" (2011: 13). In my opinion, the principal change of the Transition Companion of the two articles is the treatment of government institutions. Although R. Hopkins maintains the focus on building sustainable communities in a bottom-up process, he introduces a different vision about the role of governments and, not only at a local scale, but also at the national and international levels: "While there is a need for support from central government, and the statutory removal of obstacles that stand in the way of communities creating their own responses and initiatives, well organised community groups can do extraordinary things" (www.energybulletin.net.print/59787). Thus the legal framework contains many obstacles for the path to sustainability, and only governments can eliminate them. Hopkins summarises the section "The hierarchy of responses" (included in Chap. 4) in three ideas, which provide a clear vision of the role government institutions have in bringing about changes: "Transition by itself is not sufficient. National and international responses are necessary. Transition can help create a culture where currently unelectable policies can become electable" (www.eenrgybulletin.net/stories/2011-11-18/transition-companion-outline). Thus it can be said that the TT adopts the awareness-raising role, like the PCI, but not the same tools. The PCI promotes it through the creation of a think tank. The TT preserves as the principal tool the theoretical and practical work of grass-roots groups.

The TT has also created the "Guiding Principles of Transition" which were implicitly promoted by the movement, while Transition Companion makes them explicit:

"1. Positive visioning"
"2. Help people access good information and trust them to make good decisions"
"3. Inclusion and openness"
"4. Enable sharing and networking"
"5. Build resilience". In addition to the novelty of this approach, R. Hopkins places great emphasis on the concept, by dedicating Chap. 4 to it.
"6. Inner and outer Transition. The challenges we face are not just caused by a mistake in our technologies but are a direct result of our world view and belief system".

"7. Subsidiarity: self-organisation and decision-making at the appropriate level" (Transition US: 2011: 9)

Despite disparities between the movements, the TT cities frequently adopt the typical dynamic of the PCC: passing a peak oil Declaration, appointing an Oil Task Force to study the problem and to devise alternative policies, etc. For example, the Nottingham Declaration states: "This Council acknowledges the forthcoming impact of peak oil. The Council has to respond (. . .) It will do it by: developing an understanding of the impact of peak oil on the local economy (. . .) encouraging (. . .) sustainable transport (. . .) pursuing a rigorous energy efficiency and conservation programme", etc. (www.postcarboncities.net/node/4016).

However, behind these formal and logical steps the background is very different. Authorities recognise the TT movement's philosophy and greatly value the group's initiatives. And for these reasons, movement participation and control will be a constant characteristic throughout the process of designing and implementing policies. This is the case of Somerset and Leicestershire (GB), which have approved resolutions supporting the TT Group initiatives (www.transitionculture.org/2008/07/28).

The movement is growing very fast. In June 2008 there were 59 official initiatives. Most of them were located in the United Kingdom (54), and the rest in New Zealand (3), the United States (1) and Ireland (1). By the end of 2010 there were 303 official initiatives: United Kingdom (156), United States (74), Australia (30), Canada (17), New Zealand (8), Ireland (3), rest of Europe (12) [Germany (3), Netherlands (3), Denmark (2), Italy (1), Portugal (1), Finland (1), and Belgium (1)], Japan (3) and Chile (1). In September 2012, there were 430 organisations in over 34 countries, and more than 600 *mullers* (organizations which do not meet all the requisites to be official members of the TT, but which are considered part of the movement). Also, seven countries have organized *hubs*. That is, the organisations are coordinated at a national level (Transition Network 2012).

3 Steps in the Process of Designing and Implementing Strategies

There are many factors which make it difficult to evaluate the transformations that have taken place: the range of experiences goes from States to towns; the functioning and power of government institutions varies widely from country to country; it is difficult to evaluate the importance of the objectives approved without the starting point of each society; trajectories are not linear (sometimes big improvements are manifest (Queensland, Vancouver, etc.)), in other periods societies become stalled or move backwards (Sweden, Western Australia, etc.) due to the election of new governments; in many cases (especially in the case of small societies) there is a great lack of information. But, despite these problems the overall situation shows a very positive picture.

3.1 Resolutions and Declarations

There are two types of declarations: typical (rather complex) and short.

3.1.1 Resolutions

(a) *Typical:* This type usually contains many of the following elements:

- Acknowledgement of the strong dependency on oil (San Francisco, Cleveland, Nevada City).
- There is a short analysis of the oil exhaustion processes (it appears in all the experiences mentioned). State and city vulnerability and official reports about the problem are cited (Austin, Oakland, Berkeley, Minnesota).
- There is an explanation of the huge risk which energy crises becomes to societies (San Francisco, Bloomington, Austin, Berkeley, Minnesota).
- The solution of the problem exclusively through new technologies is discarded (Bloomington, San Francisco, Minnesota).
- They frequently emphasise the cities' traditions of taking measures in advance for environment protection (Portland, San Francisco, Oakland, Berkeley, Minnesota).
- They decide to elaborate a vulnerability study (Austin, Berkeley, Minnesota).
- They decide to elaborate strategies to deal with the problem. They describe their main elements, and in some cases also the main objectives are defined (San Francisco, Portland, Oakland, Bloomington).

–There are other, less frequent elements:

- They ask governments to take measures at higher levels. This is the case of Bloomington in relation to the federal state of Washington.
- They emphasise the high costs of oil and natural gas for the cities (Portland).
- They add a little report explaining the previous steps before approving the resolution (Oakland).

After the declaration, cities, regions, etc. usually create *Peak Oil Task Forces* with different mandates: the preparation of reports on vulnerability (which frequently also incorporate an emergency plan); a report on the problems, with recommendations for tackling them. In small municipalities the organisations that have promoted the declaration are usually asked to prepare the strategy.

(b) *Short:* This type of resolution is typical of towns and is very simple: acceptance of the proximity of peak oil and a resolution approving an agreement with a non-profit organisation, which has the mission of leading the transformation process. This is the case of Willits. In Kinsale (Ireland), the resolution is especially short: "Kinsale Town Council supports the efforts of the not-for-profit company Transition Design in its initiative to act as process leaders in Kinsale's transition to a lower-energy future and in developing the concept of a 'Transition Town'; making the transition from fossil fuel dependency to a state of energy independence" (Rooney 2007).

3.2 *Analysis of Vulnerability*

There are many kinds of vulnerability reports. Some declarations describe the main elements of vulnerability. Normally these reports are a prior step before the strategy preparation. In all reports it is emphasised that societies are very vulnerable and that, as a consequence, they have to act without delay and with great intensity. But many reports state that societies not only face a great risk which has to be averted, but that in doing so they have a great opportunity for improving.

The most common elements of the vulnerability reports are: they declare that their States are very depended on oil imports; they state that the impacts will be crushing in the event of not adopting adequate measures; they identify the most vulnerable sectors and groups of people; there not will be alternative fuels ready at the critical moment; governments will have to deal with mounting demands of aid, but they will have less financial resources available due to the economic crisis brought about by the impact of peak oil.

The South California Association of Governors (SCAG) approved the *Regional Comprehensive Plan* (RCP) in 2008. It has the sidebar "*Assessing our challenges"*, which describes internal and external challenges. The internal ones are: "We will add over seven million more residents to our region by 2035"; "making a real dent in traffic congestion is getting tougher and more expensive"; "we face an air quality crisis"; "we are increasing our dependency on imported petroleum, natural gas, and coal"; etc. The main external challenges are: Climate change "threatens to impact all aspects of our communities, whether it is reduced water supplies, habitat loss, increased air pollution, or public health impacts"; "Energy uncertainty. As the peak of the world petroleum production rate is reached, there could be profound consequences to our region's economy. Southern California transportation, agricultural and industrial systems are highly dependent on inexpensive oil" (SCAG 2008: 4, 5).

The San Francisco Peak oil resolution defines the main vulnerability elements of the city: "World oil production is nearing its point of maximum production"; "the availability of affordable petroleum is critical to the functioning of our transportation system, the production of our food and of petrochemical-based consumer goods"; "San Francisco is entirely dependent on external supplies of petroleum"; "and governmental intervention at all levels of government will be required to avert social and economic chaos". And the city has decided to undertake a wide "assessment study in order to inventory city activities and (. . .) evaluate the impact in each area (. . .) with the aim of developing a comprehensive city plan of action and response to Peak Oil" (www.energybulletin.net/print/15086).

The City Council of Burnaby (Canada) published, in 2006, the report *Global Peak in Oil Production: The Municipal Context.* In relation to vulnerability the main elements of risk are:

- "Canadian per capita energy consumption is well above the OECD average".
- "We are heading for a peak in global production".
- "There may be uncertainties about the timing of the peak, but the potential consequences are so enormous as to justify immediate action".
- "To minimize the impact on our economy and society, all levels of government and the corporate sector should begin preparations well before the peak". (Burnaby City Council 2006).

The Report of *Queensland's Vulnerability to Rising Oil Prices Task Force* "concludes that the overwhelming evidence is that world production will peak over the next 10 years". It poses "a big risk, with impacts arising not only for transport but for many key parts of Queensland's industry and the community". And the Report concludes that "Queensland's vulnerability to the peaking of world oil, and to supply disruptions, is particularly acute given our oil supply and demand trends". And although oil scarcity affects many sectors, "transport is undisputedly the most vulnerable sector in Australia, with Queensland more vulnerable than most states", due to the following factors: "A high dependency on personal vehicles, road freight and air travel"; "a disproportionately high (. . .) dependence on diesel generation to meet remote area power needs"; "a fuel intensive tourism sector"; "disproportionate transport infrastructure costs required to maintain connection between townships and economic centres" (Queenland's Oil Vulnerability Taskforce 2007: 24).

The first part of the Report by the City of Portland's Peak Oil Task Force, and the largest, is "Impacts and Vulnerabilities". The Task Force "concluded that peak oil is likely to occur sooner than later, but the actual timing has only a modest effect on the magnitude and urgency of the overall issue" (2007: 8). The impact could resemble current economic and social problems, but "they will be deeper and more persistent (. . .) more systemic and less susceptible to conventional economic analysis and remedies" (2007: 11) The Task Force also "identified four broad areas that would capture the majority of impacts: Economic Change, Transportation and Land Use, Food and Agriculture, and Public and Social Services" (2007: 10). The impacts on Transportation would be: "Automobile use will decline"; "people and businesses will relocate to be closer to each other and to transportation options"; "transportation of freight will become more costly, likely leading to mode shifts from air and truck to rail and boat"(2007: 12–16). The impacts on food and agriculture will be: "The amount and variety of food produced will decrease"; "food will cost more";"households will experience increased pressure to grow, process and handle their own food"; "there will be less food waste and changes in packaging" (2007: 16–19).

The Brisbane City Council Climate Change and Energy Task force report ("Brisbane Plan for Action"), released in 2007, looks at various factors of vulnerability:

- "The Task force advises Council that (. . .) climate change and peak oil present serious challenges."

– Climate change will have the following direct impacts: "Lower average rainfall", "more extremely hot days", "more intense storms", "sea level rise", "more bushfires".
– "Inevitable rise in petrol prices": it will affect "our economy through high freight costs, flowing through to the price of goods" (Brisbane City Council 2007).

In May 2008, "the State of Minnesota Legislature passed a resolution to prepare a peak oil plan which was ultimately vetoed by the Governor" (Post Carbon Cities blog, posted 9 December 2008). The resolution exposes important elements of vulnerability: "world oil production is nearing its point of maximum production"; "the United States has only two percent of world oil reserves"; peak oil "threatens to increase resource competition and geopolitical instability, and lead to (...) global economic crisis"; "Minnesota is entirely dependent on external supplies of petroleum" (http://postcarboncities.net/node/2331). In May 2008, the General Assembly of the State of Connecticut agreed "to create an energy scarcity and sustainability task force". And "no later than January 1, 2009, the task force shall submit a report" (State of Connecticut 2008).

The City of Austin's *Energy Depletion Task Force Report* identified the following four basic sectors which "could be impacted by an energy supply disruption or depletion: Transportation and Land Use; food, agriculture, and water; businesses, economy, and jobs; low-income populations and public services" (2009: 8).

Numerous USA governmental departments published reports on Peak oil in this period. The Government Accountability Office (GAO) issued the most remarkable one, due to the importance of the body, and because the Energy and Interior Departments agreed with it. It states that "world oil production has been running at near capacity in recent years to meet rising consumption, putting upward pressure on oil prices". It shows great concern about the USA's vulnerability: "While the consequences of a peak would be felt globally, the United States, as the largest consumer of oil and one of the nations most heavily dependent on oil for transportation, may be particularly vulnerable". It concludes: "The prospect of a peak in oil production presents problems of global proportions whose consequences will depend critically on our preparedness". And it recommends "that the Secretary of Energy take the lead" to "establish a strategy for addressing peak oil issues" (GAO 2007: 2, 6, 38, 39).

3.3 Contingency Plans

Many strategies contain short and long-term actions. The former actions contain elements of emergency plans. In addition, the vulnerability reports recommend the design of emergency plans to cope with a strong and sudden oil scarcity. The *Report of the City of Portland's Peak Oil Task Force*: The Task Force, "depicts the profound economic and social vulnerabilities that could result as fuel supplies cease

And the main challenge is peak oil: "As the peak of the world's petroleum production rate is reached, there could be profound consequences to our regional economy" (SCAG 2008: 5). The RCP is the result of 3 years of highly participative preparation, including political representatives, public managers, business representatives, experts, ecologists, citizen community representatives, etc. The RCP "lays out a strategy to reverse the current trends and diversify our energy supplies to create clean, stable, and sustainable resources of energy that address energy uncertainty and environmental health". Its principal objectives are:

- "Decrease the region's consumption of fossil fuels by 25 % from 1990s levels by 2020".
- "Increase the share of renewable energy generation in the region to 20 % by 2010, with additional increases to reach 30 % by 2020". In 2007 they reached 15 %.
- Also, the RCP proposes "reinvesting in public transportation, and revising land use, zoning and building codes to optimize the consumption of our energy resources" (SCAG 2008: 72, 76).

3.4.3 Queensland

This Australian state has developed many initiatives which explain its leading role in relation to the acknowledgement of peak oil and consequently, a policy devised to cope with the problem. Brisbane (its capital), Sunshine Coast and Hervey Bay are carrying out transformations to buffer the impact of peak oil. A. McNamara was the leader of the peak oil initiatives at the state level. In 2005 the Queensland government created the *Oil Vulnerability Task force*, and McNamara chaired the Task Force. After its release was delayed by the Premier, the Labour Party changed the government in 2007 and McNamara was appointed as Minister of Sustainability, Climate Change and Innovation, with the mission of coordinating the rest of the Ministry's tasks concerning peak oil. In addition, the government agreed to release the previous report, and asked McNamara to design a general plan to mitigate the impacts of peak oil. Besides, in 2008 he was asked to carry out a study analysing the impacts of a severe oil scarcity. The *Oil Vulnerability Task force* Report includes the chapters "Queensland Alternative Energy Options" and "Conclusion and Recommendations". It recommends the adoption of an "Oil Mitigation strategy and Action Plan as a matter of priority in order to minimise the foreseeable consequences of substantial liquid fuel price rises and supply disruptions". The three cornerstones of it are:

- "Reduce consumption of liquid fossil fuels;
- Encourage the development and use of alternative fuels, technologies and strategies; and
- Prepare for demographic and regional changes." (Queensland Oil Task force 2007: 154).

McNamara announced in 2008 that the Cabinet had asked him "to oversee the production of a Queensland strategy to mitigate the impacts on the state of severely reduced oil supplies". And he stated that "the challenge of this generation is to build a new economy, one that is powered largely by renewable sources of energy, that has a highly diversified transport system, and that reuses and recycles everything. And to do it with unprecedented speed" (McNamara 2008). However, in the elections of 2009 he lost his seat in Parliament and had to resign from the Cabinet. Since then there has been no news of the above mentioned strategy (www.energybulletin.net).

3.4.4 Portland

The Strategy of the *City of Portland Peak Oil Task Force: Descending the Oil Peak: Navigating the Transition from Oil and Natural Gas* (Portland City Council 2007), approved in March 2007, has become a reference for many other cities. It depicts three scenarios: "**Long-term Transition:** The impacts of peak oil are potentially severe, but the decline in supplies and rise in prices will occur at a fairly gradual pace, allowing time to plan for and potentially mitigate some impacts of peak oil"; "**Oil shocks:** The long-term decline of world oil and natural gas supplies is punctuated by sudden disruptions and price hikes, triggering periodic sustained emergencies"; And "**Disintegration:** Whether sudden or gradual, the impact will become so severe that the social fabric begins to disintegrate". The Task Force focused on the first scenario, "with the intent that its recommendations would reduce the likelihood of severe disruption" (2007: 10, 11). Its first recommendation and the most important is to: "Reduce total oil and natural gas consumption by 50 % over the next 25 years". This means an annual reduction of 2.6 %. Taking into account that the population will rise, "cutting total consumption in half will require a two-thirds reduction on a per capita basis". Most of the rest of the recommendations refer to a reduction of transportation needs, the use of renewable energy and the promotion of energy efficiency (2007: 33–41). We will see that Oakland's and Maribyrnong's strategies were inspired by the Portland one.

3.4.5 Oakland

The Task Force was asked to develop an action plan to become oil independent by 2020, but this target was not possible to accomplish. In February 2008 the *Oil Independent Oakland Action Plan* (OIO) was approved. The primary focus of the OIO is on the transportation sector, because it uses 97 % of the oil used in Oakland, and oil consumption amounts to 40 % of Oakland's energy consumption. The Task Force recommends a 3 % annual reduction of oil consumption, which "will lead to over 30 % reduction in oil consumption from 2008 to 2020" (Oakland City Council 2008: 7, 8). The first recommendation is to "redesign the City using the concept of Urban Villages". This means that each neighbourhood "can provide the full range of many or most of its needs", which reduces the need for transportation (2008:18). The second recommendation is to "advance transportation alternatives to the car", such as the implementation of Pedestrian, Bicycle and Public Transit Master Plans (2008: 25). But the Port is a huge element of Oakland economy, and "since the maritime Port and aviation activities are so dependent on petroleum products (. . .) we urge to reduce oil consumption", although their activities are "beyond the influence of the City Council" (2008: 53). Other important recommendations are: "Oakland needs to create an Oil and Energy Team whose mandate is (. . .) to get Oakland on the path for meeting the annual 3 %" target (2008: 34); and "the development of a comprehensive localization strategy that includes targets for food, energy, and vital goods". By 2008 Oakland had a Resolution to reach the target of "thirty per cent local area food production" (2008: 44, 46).

3.4.6 Maribyrnong

This Australian Municipality approved in April 2008 two documents: the *Peak Oil Policy* and the *Peak Oil Action Plan*. The former establishes the objectives and the latter defines a sophisticated and excellent strategy to achieve them. The Council is committed: "to the Oil Depletion Protocol with the commitment to a 3 % reduction in oil use per year in Councils' operations"; "to a reduction target for oil of 50 % reduction by 2025"; "to develop an annual action plan that directly addresses both the long term transition (. . .) and oil scenarios" (http://postcarboncities.net/node/2875). The *Peak Oil Action Plan* "is designed around three scenarios", which are the same as the Portland scenarios:

– "Long Term Transition Scenario, where the decline in fuel supply is expected to be gradual".
– "Oil Shocks, where a long term transition in oil is interrupted periodically by oil shocks".
– "Disintegration, where oil depletion occurs faster than expected" (Maribyrnong City Council 2008).

The Action Plan defines two kinds of actions: *Precautionary Actions,* "that are designed to avoid the most serious impacts of peak oil": and *Crisis Management Actions* "that are designed to plan for potential breakdowns in the provision of services, business continuity and sudden shocks that affect the oil supply". The contingency plan will define actions to be carried out based on a trigger, "for example petrol prices reaching a predetermined amount", which determine a set of activities to be developed (Maribyrnong City Council 2008).

3.4.7 Brisbane

The City Council Climate Change and Energy Task force released the report *A Call for Action* in march 2007, which "included 31 recommendations across eight strategic areas". Based on this report the City Council approved a *Plan for Action* in which the 31 recommendations became actions. However in these actions nearly all of the quantifiable objectives disappeared. The most remarkable actions are: "Council will use its land use planning powers to change the trend in Brisbane's development towards more compact and mixed use development (featuring TODs) that is well integrated with public transport, cycling and walking networks" (2007: 10).

"Increase investment in public transport to achieve frequent, reliable, high quality, comfortable and attractive services. Plan ahead to provide the buses and public transport infrastructure and facilities needed to meet required increases in public transport patronage" (2007: 13).

"It is recognized that Brisbane is considered a leader in disaster management planning. An ongoing and enhanced commitment to disaster management will be an essential component of Council meeting its obligations to protect its residents" (2007: 14). "Build on the momentum of the current drought response to further drought-proof Brisbane for the long term" (2007: 16)

3.4.8 Totnes

In 2010 TT Totnes issued *Totnes 2030, an Energy Descent Action Plan* (EDAP), which comprised the town and the district (on the whole, there are 24,000 inhabitants), which it had been developing since 2008 (Hopkins 2008). The EDAP considers the community resilience concept to be central. It has devised 21 resilience indicators for a community, where the first 10 are: the leadership is representative of the age, gender, and cultural background of each community; the leadership is visionary, shares power and builds consensus; all members take part in significant community decisions; the community has a sense of pride; people feel optimistic about the community's future; people feel integrated in the community; in the community there is a mutual help and cooperation culture; the community is self-sufficient and seeks within itself and among its resources the elements for facing the primary challenges; there is strong support for education at all levels; there is a wide range of community organisations and enterprise electronics development in the community. Some of the most significant objectives to be fulfilled by 2030 are: energy (reduction by half of the demand in 2009; the local renewable energy supply will meet 46 % of the demand, and adding the proportional share of the UK's production supply it will reach 66 % of demand; the creation of a local company for the production of renewable energy); construction (from 2014 all new buildings will have zero CO2 emissions and will be made only with local materials); transport (car ownership will be reduced by 80 %; Totnes will become totally pedestrian; the bus will be the dominant mode for medium distances; goods will be delivered by rail or sea; finance (80 % of local savings will be invested in the area); etc. Since 2007 a local currency system is being developed (the Totnes pound). The Totnes Transitional Trust (TTT), an investment firm, has been created to promote energy efficiency and renewable energies. The Totnes Renewable Energy Company has been created and has built a wind farm (Totnes EDAP 2010; http://totnes.transitionnetwork.org).

4 Final Remarks

The importance of SEE is very great (especially the TT case), because they constitute a living guide for the development of sustainable societies. TT is a movement which is growing and maturing very fast. Its rate of expansion is so great that the TN defines it as viral. There is much coherence between proposals and actions.

There are various factors which have given rise to this phenomenon. It emerges in societies which have a long tradition of working against climate change and in the implementation of environmental plans. Besides, they appear in countries (USA, UK) which have surpassed the oil and natural gas peak. The acknowledgement of these facts constitutes a qualitative leap, because it represents a great and imminent risk for societies. These experiences create a greater awareness than in societies that have not faced the same problems, and this explains that SEE appeared for the first time in these countries. And being English-speaking and strong countries gives visibility to SEE in the rest of the world, which is a very important factor to comprehend the rapid spread of the phenomenon.

On the one hand, we have seen that the TT movement has grown at a very fast rate. On the other, the PCC movement grew steadily until 2008 (for example, "2008 was a big year for governments' responses to peak oil plus ground-breaking legislation in California" (Post Carbon Cities Blog, posted 9 December 2008), but afterwards it started to vanish. We do not know of any convincing arguments to explain this phenomenon. Between 2005 and 2008 many authors thought that escalating oil prices would produce the collapse of many societies in the near future. However, in 2008 a so-called financial crisis occurred and oil prices plummeted, and although oil prices soon began to rise sharply to almost 126€/b, its level in the London market in 2012 has been oscillating around 112–115€/b. This situation is motivated by the exacerbation of the economic crisis, and also by the effect of high oil prices, as we have seen in Chap. 11. Also in this chapter we saw the significant contribution of high oil prices to aggravating the crisis. However, these factors have been ignored by the mass media and most governments. The outcome of these factors has been that oil has lost relevance among people's concerns. But this widespread perception of the crisis causes has not been an obstacle in the development of TT. An element capable of explaining, in part, that this has been thanks to the prior work of awareness-raising in communities.

Chapter 16
Circular Economy: Materials Scarcity, European Union Policy and Foundations of a Circular Economy

Keywords Circular economy • Geological scarcity • Critical metals • Geostrategic causes of scarcity • Elements of hope • Critical elements

Humanity faces not only an oil scarcity followed by the scarcity of all other fossil fuels, but also of many materials. In both cases the problem is aggravated by the fact that extraction has a peak. But the study of minerals is more complex than in the case of fossil fuels, as many materials which we are using exist in the earth's crust, we frequently use a mix of materials, and information is by far less abundant.

In this chapter we analyse the concepts and principles of a circular economy, critical metals, and non-metallic materials, the EU's policy on materials, and ultimately the basis for a circular economy.

1 Concepts and Principles for a Circular Economy

Although it is very important to precisely define the concept of natural resource, there is not a consensus on the issue. The broader definition comprises all functions which the planet provides: resource function (availability of renewable and non-renewable natural resources); drain function (the Earth's capability to sustainably absorb waste flows); cyclic function (recycling planetary materials and renewing biomass); information function (genetic resources, system models); recreational function (aesthetics, rest, etc.). However, a more reduced definition is frequently used: market valued resources. It means the marginalisation of the rest.

Sustainability obliges us to recycle the materials that we use, eliminating the concept of waste. It means that all materials are nutrients: biotic nutrients and technical nutrients. Biotic materials are recycled by the biosphere; this is the metabolic process. Besides, we have to recycle materials from the Earth's crust,

R. Bermejo, *Handbook for a Sustainable Economy*,
DOI 10.1007/978-94-017-8981-3_16,

using them as technical nutrients, feeding the technical metabolism (McDonough and Braungart 2002: 104).

Natural resources are classified into four groups:

- Renewable resources which are not exhaustible (wind, solar radiation, ocean and air).
- Renewable resources which are exhaustible (biodiversity, freshwater, etc.).
- Non-renewable resources which are not exhaustible, like metal building materials, because they can be recycled endlessly.
- Renewable resources which are exhaustible. This is the case of fossil fuels and radioactive materials (Moll et al. 2003).

In order to build a circular economy it is necessary that the uses of each type of resource must fulfil requirements which make their permanent re-use possible:

- Renewable resources which are exhaustible: their rate of use must not be greater than the production rate, and emissions and dumping cannot surpass ecosystems' absorption capability.
- Non-renewable resources which are not exhaustible have to be recycled or replaced by others which are renewable as much as possible. The Basic Plan for Recycling in Japan (based on the Basic Law for Establishing a Recycling-based Society) states: "To secure a material cycle in nature, the State will restrict the use of natural resources such as mineral fuels and mineral resources that are impossible to regenerate in nature, and promote the utilization of recycled resources, and biomass that is collected" (Japanese Government 2002: 23).
- There is no sustainable use of non-renewable resources that are exhaustible, and their use must tend to zero as soon as possible. However, the quality of being exhaustible depends in some cases on the type of use made. For instance, it is possible to make a sustainable use of fossil fuels when they are used to produce materials (like some plastics) that can be recycled endlessly.

In Chap. 6 we saw that in order to build a circular economy it is necessary for societies to imitate the behaviour of ecosystems, making it its guiding principle. This means: building decentralised production systems; changing product designs so that all their materials are technical nutrients; and creating infrastructures capable of recycling them. To be able to do so it is necessary to apply several tools: implementing an ecological fiscal reform; strengthening companies dedicated to recycling technical nutrients; etc.

In a natural ecosystem there are producers, consumers, scavengers and decomposers, "which break down the waste of other organisms (...) Scavengers are animals that eat dead animals or carrion and help break down or reduce organic material into small pieces. Decomposers are organisms that consume dead plant and animal tissue and return the nutrients to the soil to fertilize new growth"(Geng and Coté 2002: 333-4). Scavengers and decomposers have been categorised as either specialists (they depend on only one resource) or generalists (they depend on more than one resource). The human economy has the same types of companies. Scavengers in the industrial system are those companies that feed off the waste resources of other companies: "Scavengers redistribute resources (. . .) back into the system to companies that recycle materials (. . .) Decomposers are those companies that use the waste resources from both producers and consumers (including scavengers), and transform or recycle them back into the system" (Geng and Coté 2002: 336). These companies have had a low status in the product chains, their development has been precarious, and their technological level low. Although this situation is changing, their current status is far from reaching that of manufacturing companies.

Summarising what has been described, the requirement for sustainability is maintaining the status "of the resources as resources", which means to replace the traditional concept of "cradle-to-grave" with another: "cradle-to-cradle". It means mimicking nature: "C*radle-to-cradle design* defines a framework for designing products and industrial processes that turn materials into *nutrients* by enabling their perpetual flow within one of two *metabolisms*: the *biological metabolism* and the *technical metabolism*" (Braungart et al. 2006: 1343). But it means the need to change the traditional approach to production. Manufacturers design products to be applied in any circumstance: "This aim guarantees the largest possible market". But "the attempt to impose universal design solutions on an infinite number of local conditions and customs is one manifestation of this principle and its underlying assumption, that nature should be overwhelmed" (McDonough and Braungart 2002: 30, 31). Thus, sustainability is only possible in decentralised economies, but the dimension of the adequate territory depends on the kind of products. The rule of thumb could be: the more vital a product is (food, energy, housing, etc.) the more decentralised production has to be.

2 Analysis of Critical Metals

We use a great amount of materials, but only a group of them (mostly metals) are indispensable for our societies. For this reason this section is focused mainly on the situation of metals, and especially critical metals. At the end of the chapter we study other materials.

In 2009, the Öko-Institut made the report "Critical Metals for Sustainable Technologies" for the UNEP. These technologies are for renewable energies and energy efficiency. The report reaches the conclusion that indium, germanium, tantalum, PGM (platinum Group metals), tellurium, cobalt, lithium, gallium and rare earths "are the basis for cleaner technology innovation". The PNUMA created the Global Metal Flows Group in order to elaborate six reports: "Metal Stocks in Society", "Recycling Rates of Metals", "Environmental Impact of Metals", "Geological Metal Stocks", "Future Demand Scenarios for Metals", and "Critical Metals and Metal Policy Options". The five first reports form the basis for the last one (UNEP 2010a: 10).

However, today the UNEP has only published the first three. Metals are classified into five groups: nine ferrous metals (iron, manganese, vanadium, nickel, etc.); eight nonferrous metals (aluminium, cobalt, copper, lead, etc.); eight precious metals

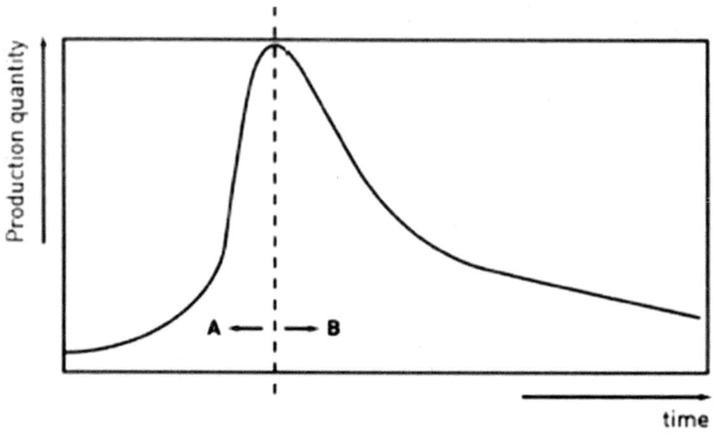

Fig. 16.1 Theoretical graph of materials extraction (Source: Diederen 2010: 51)

(gold, silver, platinum, palladium, etc.); 28 special metals (antimony, cadmium, lithium, indium, tellurium, selenium, tungsten, etc.); and the family of alloy steels and stainless steel.

2.1 Geological Scarcity

The annual consumption growth rate of five strategic metals (copper, zinc, nickel, tin and platinum) has been remarkable as a whole: from 1.8 % (tin) to 4.9 % (platinum), but consumption of the rest of metals has been over 3 %. These growth rates are leading to a growing scarcity of many metals in the short-midterm. However, these trends can be softened if the level of recycling becomes much higher and significant resources are discovered, although the probability of this is low for most metals. Scarcity of the main metals meant that prices tripled in the period from 2002 to 2008. The prices of iron ore and copper rose more than oil in the same period, but also aluminium, zinc and lead became very costly. Although the ensuing economic crisis caused a sharp price reduction at the beginning, the rising trend of prices soon recovered, as in the case of oil. Besides, some exporting countries have been restricting the amount of exports in order to lengthen the resource life and Earth's crust materials also reach peaks (COM(2008) 699 final).

The theoretical extraction curve of materials has a bell shape, as Fig. 16.1 shows, but this is slightly different, because the peak happens sooner than in the case of oil. The first part of the curve represents the extraction of the richer and easier to reach fields. But then they begin the exhaustion process and it is necessary to discover other fields, but these are poorer and poorer and less reachable. Due to this trend and rising oil prices the extraction of metals becomes more costly, reducing the number of marketable fields. The process is similar to the case of oil. The extraction

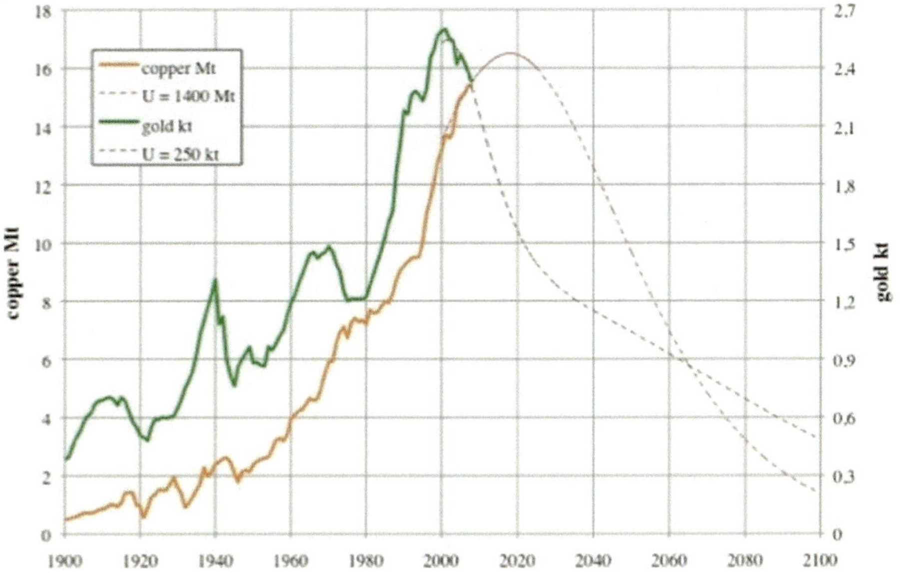

Fig. 16.2 Gold and copper extraction curves (Source: Laherrere (2010), www.theoildrum.com/node/6307)

costs grow because metals and energy become more and more costly. The energy sector is the greatest consumer of metals, especially its uses of steel and copper. So many fields become uneconomical. Although technological improvements can change the tendency, making more fields exploitable than in the past, technologies cannot resolve the scarcity problem (Diederen 2010: 51–55).

We shall show, as an example, the depletion curves of gold (its peak appears in 2000) and copper (its peak could be reached in 2020). U. Bardi and M. Pagani have studied the United States Geological Survey (USGS) statistics, and they have found that the following metals have reached their respective extraction peaks between 1962 and 2002: mercury, tellurium, lead, cadmium, thallium, selenium, zirconium, molybdenum, rhenium, and gallium. The mercury peak occurred in 1962, much sooner than the date of prohibition agreed by the European Union (Bardi and Pagani 2007) (Fig. 16.2).

Diederen (2009) estimates that, without taking into account the effect of escalating oil prices, the elements of Fig. 16.3 will be exhausted in 10–50 years' time, in the event of a 2 % annual consumption growth. It must be underlined that some of the most abundant metals, like zinc and above all iron, shall be exhausted in the next decades. But it must be considered that their consumption has grown strongly. The exhaustion of many metals which are critical for the deployment of renewable energies and fuel cells (like rare earths and gallium, tellurium, indium, copper, etc.), that could happen in the next few decades, could jeopardize the highly necessary energy model revolution towards sustainability (Diederen 2009; 2010: 54, 55).

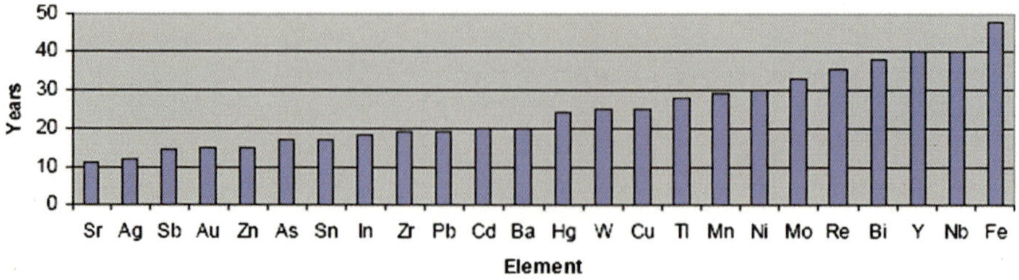

Fig. 16.3 Years which remain before exhaustion with an annual consumption growth rate of 2 % (Source: Diederen (2009))

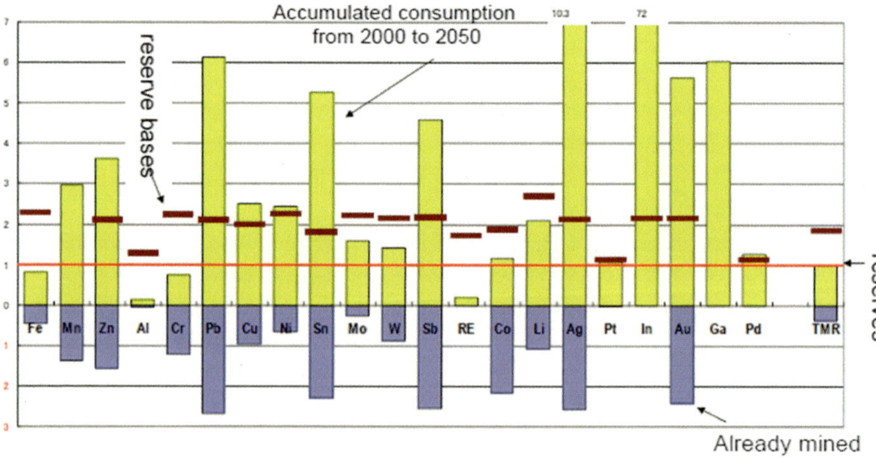

Fig. 16.4 Metal reserves needed by 2050 (Source: Halada (2007))

However, we shall see that other authors do not believe that such a level of scarcity could be possible. K. Halada (2007) has classified metals into three groups depending on their scarcity, as shown in Fig. 16.4. Bars below the horizontal axis show extracted metals, and bars above this axis, future consumption until 2050. The little red lines indicate the estimated level of each metal resource. Iron (Fe), molybdenum (Mo), wolfram (W), cobalt (Co), etc., will not be depleted by 2050. The consumption of nickel (Ni), lithium (Li), indium (In) and gallium (Ga) would surpass by far current known reserves. But the author does not consider that gallium could be scarce, because he estimates that reserves will grow strongly in the future. On the contrary, consumption of copper (Cu), lead (Pb), zinc (Zn), gold (Au), silver (Ag) and strontium (Sn) could widely surpass their reserves.

Ultimately, the LBST (Ludwig-Bölkow-Systemtecnik) consultant, which made numerous reports for EU institutions, has published a report about the geological scarcity of elements. It reaches the following conclusions: "Based on their historical production profile, probably cadmium, chromium, copper, gold, lead, nickel, silver, tin and zinc are close to or at peak production already. Others like bismuth, boron, germanium, manganese, molybdenum, niobium, tungsten, and zirconium might experience peak production within the next two decades"(Zittel 2012: 77).

In conclusion, there is a broad consensus, at least, about the acute scarcity of gold, silver, copper, lead, zirconium, zinc, cadmium, nickel, strontium, and mercury.

2.2 Other Scarcity Factors

Geological scarcity being the most important factor in the evaluation of the critical state of resources, there are other factors which can aggravate the evaluation significantly: great economical importance; reserves highly concentrated in a few countries; policies aimed to restrict exports; and low level of recycling.

Due to the accumulated access problems and mounting prices, some governments are promoting numerous reports in order to evaluate the problem's magnitude and identify the resources which present the greatest problems and have the greatest importance for the economy: that is, the most critical elements.

Most reserves are concentrated in very few countries, and their governments frequently decide to slow down exports in order to lengthen the life span of reserves. In other cases, political instability in those countries leads to the same outcome. Many OECD countries have a very strong dependency on imported resources, so they have a high level of vulnerability. China is the leading exporting country of 16 metals: it exports 97 % of rare earths (17 metals which are so named because they are in the earth's crust in tiny quantities (richness is below 0.01 %)), 81 % of tungsten, 50 % of indium, and 45 % of tin. Besides, China possesses 75 % of global reserves of germanium and tungsten, 62 % of indium, 38 % of rare earths (the difference between exports and reserves is because the USA and Australia do not want to extract them, due to their great environmental impacts), and it has a dominant position in the supply of antimony, gallium and magnesium. Brazil extracts 90 % of niobium and controls the supply of tantalum together with the Democratic Republic of Congo, which has a dominant position in the supply of cobalt. South Africa extracts 79 % of rhodium and platinum, and 42 % of chromium (Ad-hoc Working Group 2010: 34; Diederen 2010: 55).

The EU identified in 2008 450 restrictions of element imports, above all from China, Russia, Ukraine, Argentina, South Africa and India. In the case of nonferrous and precious metals these restrictions have produced a 40 % supply reduction, while exports of secondary raw materials have increased by 125 % in recent years. And these exports are often illegal, because they do not meet the Basel Protocol requirements on hazardous wastes from OECD countries to Non-OECD countries. In 2009, the EU, USA and Mexico raised complaints about China in the WTO due to export restrictions on yellow phosphorous, bauxite, fluorspar, coke, magnesium, manganese, silicon carbide, zinc, etc. However, in 2010 China drastically reduced rare earth exports. This measure multiplied prices by six. In the previous year it satisfied 97 % of the demand, despite possessing only 38 % of global reserves,

Table 16.1 Critical metals

Timeline	Metal
Short-term (within next 5 years)	
+ **rapid** demand growth	Tellurium
+ **serious** supply risks	Indium
+ moderate recycling restrictions	Gallium
Mid-term (till 2020)	
+ **rapid** demand growth	Rare earths
and	Lithium
+ **serious** recycling restrictions	Tantalum
or:	Palladium
+ moderate supply risks	Platinum
+ moderate recycling restrictions	Ruthenium
Long-term (till 2050)	
+ moderate demand growth	Germanium
+ moderate supply risks	Cobalt
+ moderate recycling restrictions	

Source: Öko-Institut (2009: 93)

due to its low labour costs and not taking into account the strong environmental impacts of its mining. The reduction was due to the fact that the Chinese government estimated that their reserves were being rapidly depleted. However, raw materials are not the only resources with problems related to access. In the EU the use of secondary resources (metal scrap) has risen from 40 to 60 % of total resources over the last decades. And the EU faces a twofold problem to access needed supplies: growing domestic demand; import restrictions; and mounting exports, most of them illegal, as we have already mentioned (Ad-hoc Working Group 2010: 20; yxiao@bloomberg.net).

We analyse two reports which apply only non-geological criteria for defining critical resources, as examples of other evaluation methods. One is elaborated by Öko-Institut and was requested by the European Commission. The other was produced by the Ad-hoc Working Group on Defining critical raw materials, and it was created by the Commission. The first has a partial approach, because it only studies metals which are critical for the development of sustainable technologies. It establishes three premises: fast growth of consumption; economic importance; and restrictions to recycling in the mid and long term. While the second report only establishes two factors: economic importance and supply risks. The first report's conclusions are summarised in Table 16.1. In the short term (within the next 5 years) tellurium, indium and gallium will be critical, because of a rapid growth in demand, serious supply risks (>90 % of reserves controlled by three countries, physical scarcity and technical scarcity), and moderate recycling restrictions. In the midterm (until 2020) rare earths, lithium, tantalum, palladium, platinum and ruthenium will be critical, but they are classified into two sub-groups. Rare earths, lithium and tantalum are suffering a rapid growth in demand, and serious recycling restrictions (high level of dissipative applications, physic-chemical limitations, lack of technologies and infrastructures, etc.). While palladium, platinum and ruthenium are also experiencing a rapid growth in demand they suffer moderate supply risks and recycling restrictions. And in the long term (until 2050), germanium and cobalt will be the only critical materials, and their critical characteristics are moderate (Öko-Institut 2009: 6, 7).

The concern about the lack of a secure supply of raw materials led the EU to approve in 2008 *The Raw Materials Initiative*, which put forward several policies to tackle the problem, also creating the Ad-hoc Working Group to define critical raw materials to be assessed. The Initiative states that there are "*high-tech metals* such as cobalt, platinum, indium, titanium and rare earths". But the dependency on platinum and indium is considered "particularly critical: first, they have a significant economic importance for key sectors (. . .) second, the EU already experienced supply risks (. . .) and third, there is currently a lack of substitutes" (COM(2008)699 final: 3).

In 2010, the Ad-hoc Group issued the "Critical Raw Materials for the EU" report, which defines several critical materials based on two factors: risk of supply and great economical significance. In the first case the Ad-hoc Group only takes into account that the risk is caused by exporting countries which limit trade due to political reasons and to reduce the environmental damage associated to extraction. The report begins by strongly stating that there is no geological resource scarcity: "Over the past 50 years, the extractive industries sector has succeeded in meeting global demand and the calculated life time of reserves and resources has continually been extended further into the future" (Ad-hoc Working Group 2010: 16). It claims that technological development will allow much deeper mining in the Earth's crust than today. It seems to not take into account the huge consumption growth of emerging countries, mounting extraction costs due to ever-deeper mining and rising oil prices.

However, the report is contradictory, because it describes many serious problems and only analyses the next 10 years, due to considering the future to be too complex and uncertain to make any further predictions. With this time span it is impossible to design a long-term strategy regarding materials, and it is precisely something that is sorely needed. The report shows many examples of metals which are by-products of the carrier (gallium, germanium, tellurium, selenium, indium, platinum group, rhenium, etc.): "For example, gallium is found in bauxite (. . .) germanium and indium typically with zinc, and tellurium with copper and lead ores". And, "for example, it would not be economic to raise zinc production just to meet an increase in germanium demand". The report cites the special case of rhenium: "it is produced as a by-product from molybdenum, which in itself is a by-product of copper" (Ad-hoc Working Group 2010: 17–19). The report defines three groups of minerals, and only one shows risks, because two problems converge: supply uncertainty and economic significance (between moderate and high). In this Group the most important are rare earths, the platinum group, niobium, manganese, germanium, indium, tungsten, etc. It considers that silver, lithium, copper, etc., have low factors. And rhenium, tellurium, iron, aluminium, zinc, nickel, etc., have a high economic value, but a very low supply risk (Ad-hoc Working Group 2010: 34).

2.3 Recycling

The number of metals that we are using is increasing steadily, we frequently use metals in tiny quantities, and product chains are increasingly longer. All of these reasons and many more that are not mentioned, make it impossible to build a circular economy.

Apart from the resources located in the Earth's crust, others exist which were previously extracted by man and are located in infrastructures, buildings, machines, durable products, etc.

Besides, many discarded products are dumped into landfills. Although it is likely that a resource scarcity would oblige us to *mine* them in the future, in this section we only analyse metals which are part of the technosphere, because they are the most important ones. In developed societies a huge amount of metals has accumulated over the centuries. In the United States, the accumulated copper has risen from 73 kg/person in 1932 to the current 240 kg/person. In OECD countries the main metals used are known with certain precision. Among metals for engineering we have: iron (7,000–14,000 kg/person), steel (7,085), aluminium (350–500), copper (140–300), zinc (80–200) and lead (20–150). The scale of special metals used is rather low: manganese (100 kg/person), cadmium (80), chromium (7–50), titanium (13) and mercury (10). The amount of precious metals is much less: gold (35–90 g/person), silver (13), palladium (1–4), platinum (1–3) and rhodium (0.2). However, there is a great lack of information about most metals used in alloys (chromium, manganese, nickel and tin) (UNEP 2010a: 17–25; 2010b). And without enough data it is impossible to design policies for recycling.

The recycling rates are low, and only a few metals (such as steel and platinum) have indexes of 50 % or more. There is a great variability between sectors. In heavy industry up to 90 % can be reached, but in the automotive sector the rates are around 50–55 %. On the contrary, in cases where many metals are used in tiny quantities the recycling rate is very low. In the electronics sector rates are between 5 and 10 %. And in renewable energies (photovoltaic, wind turbines, fuel cells, batteries, high efficiency bulbs, etc.) the rate is around 1 %, although it is improving rapidly as technologies become mature and companies implement recycling plans, as is the case of photovoltaics in OECD countries. The metals involved are lithium, tellurium, lanthanum, indium and gallium (UNEP 2009: 6–19).

Some of these metals could suffer in the future explosive increases of demand. The use of indium (the main consumption of which is produced by the electronics sector) could increase from 1,200 tonnes (in 2010) to 2,600 in 2020. But it is a by-product of zinc mining. In the EU, scrap amounted to 9.3 million tonnes in 2005. Normally, products are dismantled at a national scale and copper, aluminium and ferrous metals are easily obtained. But the complex elements (integrated circuits, batteries, mobile phones, etc.) use many metals in tiny quantities (mobile phones contain around 60 materials), so there is a need to build costly recycling plants of high technological complexity. Although only a reduced number of countries have built recycling plants (Japan, Germany, Sweden, etc.) capable of recycling indium, tellurium, ruthenium, etc., most of the e-waste is exported to Non-OECD countries. This dynamic is risky for companies, because the sector uses 80 % of indium and ruthenium, and 50 % of antimony, apart from using essential materials for renewable energies and fuel cells, such as selenium, tellurium, indium, platinum or ruthenium. These plants also recycle materials for other sectors, but soon their capacity will be exceeded. Ultimately, there is no technology capable of recycling metals in tiny quantities, like tantalum, rare earths, lithium, gallium, germanium, etc. However, recycling metals greatly reduces energy consumption. Recycling aluminium requires twelve times less energy than primary production, and six times less in the case of steel (UNEP 2009: 6–19, 2010b).

3 Analysis of Non-metallic Materials

3.1 Material from Construction and from Demolition

Construction materials represent the great majority of all materials used in a developed country. The kinds of materials used are very diverse, because it depends on their availability in the surrounding areas and customs. They are classified in relation to their possible treatment once they become waste:

- Recyclable and/or reusable: metals; wood and other biotic waste; glass; plastics; fabrics, paper and cardboard
- Exclusively reusable: stone (natural or artificial), which usually must be crushed to be reused
- Only reusable, because it is found mixed with other materials: mortar (Glinka et al. 2007).

Besides, construction generates many other materials, which do not form part of the elements constructed: earth and rocks as a result of excavation; plastic bags; paper and cardboard packaging; and also other types of broken elements (bricks, paving stones, bulbs, etc.). The most frequent treatment is burying them (EPA 2002).

The buildings' design determines (as happens with all other products) the structural materials, roofing, piping, insulation, painting, etc. And these are critical for the durability, recycling and re-usability of materials, in the cases of demolishing or reshuffling buildings. The building's structure must be made of steel, because it is easily recyclable and consumes less energy. The design has to be modular, that is, the building is constructed by assembling pieces which have been produced by manufacturers. And the demolition is carried out by dismantling the parts. These elements can be used again. Some European Nordic countries show recycling percentages a little under 100 %. However, some of the recycled materials are not used at the same level as before. For instance, they are used to build road bases, once they have been ground down (EPA 2002).

3.2 Materials and Substances from the Chemical Industry

The Chemical industry produces large quantities of products, substances and waste. In many cases all of them are hazardous. The sector uses oil, minerals, metals and biomaterials to manufacture tens of thousands of substances which are used as biocides (disinfectant agents, fungicides, pesticides, wood preservatives, etc.); paints, inks, varnishes, etc.; and polymers (plastics). The EU produces 30 % of the world's chemicals, and the chemical sector is the second largest sector in Europe. In the EU there are more than 100,000 substances registered, of which 10,000 are produced in quantities above 10 tonnes/year and about 20,000 in quantities between 1 and 10 t/y. The EU acknowledges that it knows very little about the other 70 %, which are products and substances marketed before 1981 (when there was no toxicity control), and which represent 99 % of the weight. Between 200 and 300 new products, which are produced in significant quantities, are marketed every year. Only 3,000 have been tested, and it is known that over 800 of them are carcinogenic, mutagenic or endocrine system disruptors. World purchases have multiplied by nine since 1970 and they grow by 3 % annually (Eder and Sotoudeh 2001; EEB 2004a: 23).

A sustainable strategy for the chemical sector must meet two premises: urgent detection of product families which are the most hazardous, and their rapid elimination; the development of a green chemistry, which means a radical change of the raw materials, reactants and production technologies used.

Studying all products and substances case by case would be extremely lengthy and costly, and could never be completed, due to the many combinations of products and substances that there are. This is the Royal Commission on Environmental Pollution's opinion: "The current system is overloaded because of the difficulty of applying an expensive testing and assessment regime to the very large number of chemicals already in the market" (RCEP 2003:9). Work can be facilitated by analysing product families, because the great majority of hazardous products belong to a few families. The organohalogen compounds constitute by far the most important family, and especially the organochlorine sub-family. It has about 11,000 marketed products. Another important family is aromatic hydrocarbons (benzene, benzopyrene, etc.). The RCEP (2003: 9) "recommend that we go beyond the assessment of individual chemicals and extend the process to include monitoring data to target the limited assessment resources onto those chemicals that might already be causing environmental harm".

The second element of a sustainable strategy, and a transcendental one is the substitution of conventional chemistry with a green chemistry. It is considered that this chemistry has 12 requirements: maximising the incorporation to the final product of used products; to use and generate substances which have little or no toxicity; to prefer renewable materials instead of non-renewable ones; to design products which can be decomposed; to prefer reactions which need a catalyst instead of those that do not use one; etc. This is a chemistry that imitates nature: "Many forces give rise to chemical pollution, but there is one overarching scientific reason why chemical technology pollutes. Chemists developing new processes strive principally to achieve reactions that only produce the desired product. This selectivity is achieved by using relatively simple reagent designs and employing almost the entire periodic table to attain diverse reactivity. In contrast, nature accomplishes a huge range of selective biochemical processes mostly with just a handful of environmentally common elements" (Collins 2001). In 2001, 200 scientists and policymakers held a conference to assess the state of the art in green chemistry, and a key finding was: "Many technologies that meet green chemistry objectives already exist and offer immediate opportunities to reduce environmental burdens and enhance economic performance" (Ritter 2001).

3.3 Materials from the Petrochemical Industry

The petrochemical industry has a 20 % share of all oil consumption and it consumes a significant quantity of natural gas (IEA 2008), in order to manufacture many types of products (polymers-plastics, textile fibres, lubricants, solvents, tensoactive agents, etc.). In many cases recycling is not possible: the diffuse use of tensoactive agents makes recovery impossible; lubricants pass to the environment from lubricated chains and outboard boat engines, chainsaws, etc.; many types of plastic are not recyclable.

Plastics are by far the most important materials made by the petrochemical industry. Splitting molecules from raw materials makes it possible to obtain simple molecules, which are combined to produce a great amount of polymers (plastics). The global production of plastics has been growing by 9 % annually, going from 1.5 million tonnes in 1950 to 260 Mt in 2007, though in 2008 it dropped to 245 Mt, due to high oil prices and the economic slowdown. In Europe production was 60 Mt (25 % of the total), a little less than the previous year (Plastics Europe 2009).

There are two types of inorganic plastics: thermoplastics and thermosetting polymers. Thermosets can melt and take shape once; after they have solidified, they stay solid, so the chemical reaction is irreversible. Thermoplastics have a flexible molecular structure because they do not undergo a chemical change, so they can be heated and moulded again and again. More than 80 % of plastics are thermoplastics. In the EU, in 2008, packing used 38 % of thermoplastics, construction (21 %), car industry (7 %), electrical and electronic appliances (6 %), and the rest of uses 28 % (Plastics Europe 2009).

The plastics used the most are: polyethylene (PE, 29 %), polypropylene (PP, 18 %), and polyvinyl chloride (PVC, 12 %). Also used widely are polyurethane (PU), polystyrene (PS), polycarbonate (PC), ABS, and polyethylene terephthalate (PET)

The annual growth rate of each of those plastics varies between 0.5 and 7.5 % (Plastics Europe et al. 2008). PE and PP are cheap, recyclable and very versatile, and for this reason they can substitute the rest of plastics, which do not have these qualities. PET is used in packaging and has few additives. PU is used especially in insulation and to produce foams. In its manufacture hazardous products are used and it produces hazardous waste. PS is used to produce insulation materials and in applications which need hard materials, like some toys. It is a compound which contains many hazardous elements. ABS is applied in piping, toys and bumpers, all of which need hard materials (http://archive.greenpeace.org/toxics).

X. Ren states that "plastic products are difficult or non-recyclable in an economically viable way" (2002: 28). And the most complex plastics are the least recyclable. The only way to recycle thermoset plastics is chemically, by converting polymers into the original monomers. The recyclability order corresponds with the order of exposing types of polymers. PE and PP do not pose significant problems to be recycled, because they need few additives, but they normally need the addition of virgin plastic in order to obtain their previous characteristics. PET follows, in order of recyclability. Intermediate plastics (PU, PS, PC, ABS) have a low recyclability, and PVC is not at all recyclable. Thus, PE and PP ought to substitute the rest of the plastic families, because they are versatile, recyclable, cheaper, and their environmental and health impacts are low (Plastics Europe 2009).

The recycling rate of waste is low. In the EU mechanical recycling is growing steadily due to Directives which impose recycling targets.

In 2002, only 38 % was recovered, and most of it was incinerated (23 %); the remaining 15 % was recycled (Plastics Europe 2003). In 2008, recovery reached 51.3 % (50 % in 2007), with 21,3 % recycled (20,4 % in 2007) (Plastics Europe 2009). Thus, plastic recycling is increasing slowly and steadily, though almost half of all plastics go to landfills. Recycling saves 70 % of primary used energy, while the energy recovery rate from incineration is very low. The packaging sub-sector is particularly able to recycle, because it uses only six kinds of plastics; they do not have harmful substances and the waste materials are young, due to the short life span of the products. In 2008, the percentage of recycled waste was 63 %, due to the mentioned characteristic and, above all, to the packaging Directive. However, an increasing amount of plastic waste is being exported, mainly to China (Plastic Europe et al. 2008; Plastics Europe 2009).

3.4 Agro-materials

Numerous reports state that there is a great development potential for biotic materials. The GB Government-Industry Forum on Non-Food Uses Crops (GIFNFC 2002: 35) has found a high level of growth potential in the packaging industry, in graphic arts (solvents, inks and so on), etc. The Interactive European Network for Industrial Crops and their Applications (IENICA) estimates that there is an important potential for development in organic oils to produce lubricants, paints, plastics, tensoactive agents, detergents, textiles, paper, surface covering, soaps, polymers, inks, cosmetic products, perfumes, glues, adhesives, medicinal products, in several applications for the car industry, etc. (Christou 2010).

The low or non-existent recyclability of most plastics obliges us to work in two directions: substituting non-recyclable plastics with other recyclable ones, and with agro-plastics. Substitutions are applied in cases when it is the better option: "Introduction of biodegradable plastics is mainly driven by the situation under which plastic products are difficult or non-recyclable" (Ren 2002: 28). The introduction of agro-plastics has been difficult in many applications due to low oil and natural gas prices, and because of the high efficiency of oil refineries, contrasting with the low efficiency of agro-material processes. But escalating oil prices in the period from 2002 to 2008 gave a great boost to the production of agro-plastics, going from a market share of 0.1 % to almost 1 % in 2006 (85,000 t/a) (GIFNFC 2002: 23; http://en.wikipedia.org).

Agro-plastics are used especially in short-lived products: food packaging (for fruit, vegetables, dairy products, fruit juices, etc.) catering (trays, dishes, cutlery, cups, etc.). They are also used in durable applications, likes fibres for the textile industry, in the car industry and in buildings (insulation). Several reports predict a 17 % annual growth for agro-plastics. They would potentially mean 1.5 million tonnes/year in 2011. An EU report predicts 2 million t/y in Europe, for the following applications, in order of importance: catering; packaging; packaging for vegetables; wheel manufacturing; agriculture; bags for organic waste; etc. (http://en.wikipedia.org).

It is not convenient for agro-plastics to degrade under environmental conditions, because it would force a design of plastics with different rates of degradation, depending on the durability of each good. Normally they are degraded in the environment of a compost facility. But there are exceptions. In the case of being used in agriculture they are decomposed by soil microorganisms (Ren 2002).

Synthetic solvents are being replaced by organic ones, especially in Graphic Arts. In the EU 280,000 tonnes of agro-solvents are being used annually, most of them in paints and metal cleaning products, and they are gradually substituting synthetic products, because in many cases they offer better results and also for environmental reasons. 50 % of the synthetic solvents used in the world are dumped in natural environments. In GB agro-solvents have 75 % of the market share in uses such as chain saws, and it is a common prediction that they could reach 90 % of the share. In the EU-15, uses in paints and artificial surface coverings are increasing. Organic inks are dominant in Belgium (market share of 80 %), Netherlands, Germany and Scandinavia (GIFNFC 2003: 20, 27; IENICA 2006: 33–36).

Many companies are producing great quantities of agro-plastics. Cargill produces in Nebraska 140,000 tonnes/y. In Brazil, Dow has built a plant capable of producing 350, 000 tonnes/y, and in Braskem another 200,000 tonnes/y. But the raw material is usually produced by genetically modified plants (http://nz.theoildrum.com/node/4458).

In the EU, natural fibres (hemp, flax, sisal, jute, cotton, etc.) are used in the car industry, fundamentally to substitute fibreglass. This industry has passed from zero use in 1994 to more than 28,000 tonnes in 2000. 60 % has been used in Germany. It is estimated that production will have reached 100,000 tonnes by 2010. Natural fibres are used because they offer energy efficiency (their production needs 80 % less energy than fibreglass production), they reduce vehicle weight, save energy, etc. But some car manufacturers have begun to use agro-materials as car parts (GIFNFC 2002: 28, 52).

Hemp has the potential to be applied in many uses. It has almost all the applications of the rest of the plant matter mentioned: textile fibre, paper, solvents, lubricants, inks, surface coverings, hygienic products; its oil can be used as food and for animal feed. There is no need to use pesticides because it does not have predators, nor herbicides, because its growth rate is very fast. It protects soils because its roots create a complex network, and it absorbs heavy metals. Hemp does not need chlorine in paper manufacturing. In the USA, during the 1930s, hemp was used massively to produce paper. But Du Pont, which had the technological monopoly on paper-making from wood, obtained the ban on hemp due to its narcotic properties (as happened in Europe). From 2002 onwards a recovery process in the world (especially in China, North America and Europe) has led to occupy a surface of 23,000 Ha in 2003, with a production of 70,000 tonnes in this year, 30,000 tonnes of them in the EU. Hemp can substitute cotton, the cultivation of which uses 26 % of all pesticides used in the world. Hemp and flux are the most significant textile fibres in Europe, and their cultivation surface grows rapidly, although their market share was only 1.5 % in 2005. Despite new applications of hemp the surface dedicated to it is stagnant in Europe, at about 16,000 Ha. Ultimately, a GB report estimates that the land which agro-materials will use will be minor in comparison to agro-fuels. Agro-materials need less than 1 % of the land dedicated to produce food in order to reach 5 % of the market quota (GIFNFC 2003: 13 and ss.; IENICA 2006: 54–69).

4 The EU's Broad Policy on Materials

The EU acknowledges that it has a special dependency on importing metals, because it only extracts 3 % of the global total. Besides, having only 15 % of the world's population, it consumes 56 % of the oil, 60 % of the natural gas, and over 50 % of non-energetic raw materials. This data does not take into account the resources embedded in the products imported from China and other emerging countries. The EU and the rest of the OECD countries are facing growing import restrictions. It is because emerging countries possess most of the reserves of some resources, and they want to lengthen the life span of these resources. On the other hand, India and China are changing the way of accessing foreign resources. In the traditional way large corporations reach commercial agreements with exporting countries (mainly through nationalised companies) on prices and quantities, but always with the diplomatic backing of the superpower. Under the new model China and India agree with exporting countries a long- term broad policy of cooperation, which includes investments in infrastructures, in technological development and, as compensation, the emerging countries obtain long-term contracts for resource exports (Ad-hoc Working Group 2010: 20). This behaviour is a factor, though not the only one, which shows that in an era of resource scarcity the market begins to be marginalised as a tool for obtaining resources.

The European Commission apparently does not believe in resource scarcity: "From a global geological perspective, there is no indication of imminent physical shortage of the majority of raw materials in the world". It defends this conclusion with the argument of inadequate investment: "the speed and scale of the increase in demand is generally expected to result in supply lagging behind demand" (COM(2008)669 final). However, the adjective "imminent" is too restrictive, because national security determines the need for designing mid and long-term plans to cope with resource supply scarcities, because adaptation to scarcity can't be obtained over short periods.

However, the risk that a strong resource dependency poses has led the EU to multiply initiatives to reduce it. The Raw Materials Initiative (COM(2008)699 final) defines the following "3 pillars": "Access to raw materials on world markets at undistorted conditions", "Foster sustainable supply of raw materials from European sources", and "Reduce the EU's consumption of primary raw materials". And the Waste Framework Directive (2008/98/EC) defines on page 11 the following objectives for 2020: to re-use and recycle at least a minimum of 50 % of household paper, metal, plastic and glass, and non-hazardous construction and demolition waste to a minimum of 72 % by weight.

The Waste Framework Directive also establishes the following hierarchy of activities: prevention, re-use, recycling, recovery and disposal.

Prevention means "measures taken before a substance, material or product has become waste, that reduce: (a) the quantity of waste, including through the re-use of products or the extension of the life span of products; (b) the adverse impacts of the generated waste on the environment and human health; or (c) the content of harmful substances in materials and products". Re-use means "any operation by which products or components that are not waste are used again for the same purpose for which they were conceived". Recovery means "any operation the principal result of which is waste serving a useful purpose by replacing other materials which would otherwise have been used to fulfil a particular function, or waste being prepared to fulfil that function". Recycling means "any recovery operation by which waste materials are reprocessed into products, materials or substances whether for the original or other purposes". Disposal means "which is not recovery even where the operation has a secondary consequence the reclamation of substances or energy" (European Parliament and Council 2008b: 10).

Besides, the EU has changed its vision, as the "Europe 2020. A European Strategy for Smart, Sustainable, and Inclusive Growth" shows, which states on page 15 that resources are scarce: The EU has "to prosper in a low-carbon, resource constrained world" (COM(2010) 2020 final). This means that the cheap resource era has ended: "Trends show, however, that the era of plentiful and cheep resources is over" (COM(2011) 571 final: 2). The White Paper on transport calls for abandoning oil dependency as soon as possible (COM(2011) 0144 final).

The "Roadmap to a Resources Efficient Europe" (COM(2011) 571 final) defines several "milestones" to be reached by 2020. The most important are:

> "Citizens and public authorities have the right incentive to choose the most resource efficient product and services, through appropriate price signals and clear environmental information".

"Market and policy incentives that reward business investments in efficiency are in place".

"Waste is managed as a resource. Waste generated per capita is in absolute decline. Recycling and re-use of waste are economically attractive options for public and private actors due to widespread separate collection and the development of functional markets for secondary raw materials".

On the other hand, the EU has conceptual contradictions which constitute additional hurdles to design a strategy for a circular economy. For instance, there is not a clear-cut distinction between sub-products and waste. If a substance is considered to be waste, its treatment pursues the minimisation of its environmental impacts, despite of being recyclable in some cases. If it is considered a resource, its subsequent productive processes will not be hampered. The problem is to determine where the limits are between them. There have been many conflicts and the European Court of Justice has ruled on the subject, although the conflict continues because the resolution has a low degree of applicability: "Waste means any substance or object which the holder discards or intends or is required to discard" (European Parliament and Council 2008a, b: 9 and 10). Thus using or dumping waste depends on the company's behaviour, which depends on markets. And prices can mean that a secondary resource becomes a by-product [which is "a production residue which is not waste" (European Commission 2007c)] or remains as waste.

Ultimately, EU advances towards a circular economy are undermined by economic globalisation, which produces longer and more complex product chains. In Chap. 6, we saw that in nature the optimal recycling rate is obtained when the circuits are short. On the contrary, the international market has been growing more than GDP. Besides, ever-increasing product complexity means that product elements are being manufactured in a growing number of countries. In this situation it is very difficult to implement resource recovery policies, and energy consumption tends to grow without limits. Lastly, competition between companies leads to an accelerated production of *new* goods, which determines planned obsolescence.

5 Basis for a Circular Economy Strategy

It is impossible to build a circular economy in a globalised world, which is why we must strengthen decentralised economies. But we also need to reduce the number of materials which are used. Thus, to reach a circular economy, economic structures have to be simplified. Other requirements are: minimising the use of critical materials; substituting scarce resources for more abundant ones; reducing recycling speed, by making more durable products; banning most hybrid materials, because they are not recyclable; substituting abiotic materials for biotic ones, but without being an obstacle for producing enough food.

Although recycling permits a rapid reduction of resource consumption, it is unable to avert exhaustion, because of the great number of resources used, some of them in applications which make recycling impossible, like in alloys. These reasons

Table 16.2 Elements of hope, and frugal and critical elements

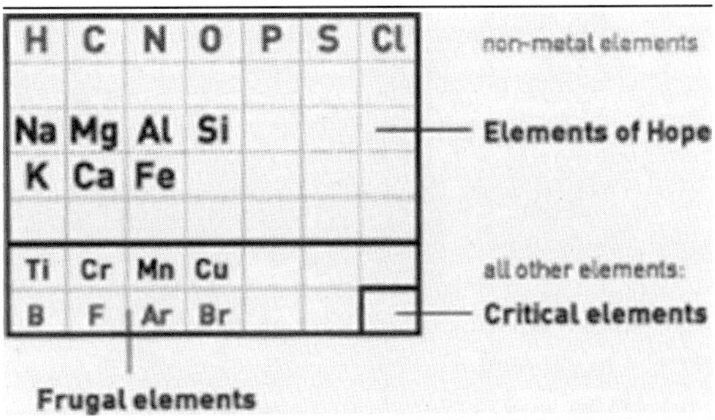

Source: Diederen (2010: 91)

make it impossible to surpass the 90 % recycling rate. Thus, apart from reaching the highest possible recycling rate we need a drastic reduction of the amount of resources used. Table 16.2 shows a proposal by Diederen. Above all, our economy has to employ the "elements of hope", which are those used by nature. However, our economy needs other materials, like "frugal elements" and "critical elements". Frugal elements are those which do not have substitutes for indispensable applications, like chrome in stainless steel. Critical elements must be used only in indispensable applications and without substitutes, like rare earths in renewable energies. Among the table's elements there are no heavy metals (Diederen 2010: 91, 92).

These measures have guidelines to be applied in each sector. As an example application we propose some measures for the following sectors: construction, chemical, petrochemical, and for biotic materials.

In construction, concrete must be replaced by steel, and the structural elements of buildings have to be modular, which in the case of demolition can be reused. In the chemical sector hazardous families have to be banned and traditional chemistry must be substituted with a green chemistry. In energy it is necessary to stop burning fossil fuels, and conventional energies must be replaced by renewable ones. Fossil fuels can be used to produce renewable raw materials, especially renewable plastics. We have seen that plastics obtained from the polyethylene and polypropylene families are renewable, and that they can substitute the other families in all applications.

Biotic materials are beginning to replace abiotic materials in many applications, although in small quantities. This change has sometimes reduced environmental impacts, like in the case of lubricants, textiles, inks, medicines, perfumes, single-use products; in some uses of plastics (agriculture, food packaging, bags for waste food, etc.) and so on. But they are limited by the available land, which varies from country to country.

Ultimately, in the next two chapters we shall provide a broader picture of a circular economy, with an analysis of industrial ecology and a description of its main applications, as well as a design of the basis of an integrated product strategy, also adding the main applications.

Chapter 17
Industrial Ecology

Keywords Industrial ecology concept • Material flow accounting • Material flow analysis • Industrial symbiosis • Material recycling • European metabolism

1 The Industrial Ecology Concept

There are several issues which are debatable in IE: its definition, its status as a science or not, as a normative field or not, whether it is related to sustainability or not, whether they are mutualistic or cooperative relationships, etc. But before looking at these issues, we will provide a short historical overview. Two events took place in 1989 which are the point of departure for IE. During that year Robert Frosch and Nicholas Gallopoulos (General Motors Research Vice President, and Director of engineering respectively) published in Scientific American the article "Strategies for Manufacturing. Waste from one industrial process can serve as the raw material for another, thereby reducing the impact of industry on the environment", in which they defend the creation of productive systems which function as natural systems do: "The industrial ecosystem would function as an analogue of biological ecosystems. Plants synthesize nutrients that feed herbivores, which in turn feed a chain of carnivores whose wastes and bodies eventually feed further generations of plants. An ideal industrial ecosystem may never be attained in practice, but both manufacturers and consumers must change their habits to approach it more closely if the industrialized world is to maintain its standard of living and the developing nations are to raise theirs to a similar level without adversely affecting the environment" (1989:144). Besides, the same year the Kalundborg industrial ecosystem was uncovered, in a Danish town, and has become the world's reference in this field and is the best known case.

This chapter is dedicated to analysing the concept of Industrial Ecology (IE), and its two main fields: Material Flow Accounting and Industrial Symbiosis (IS).

R. Bermejo, *Handbook for a Sustainable Economy*,
DOI 10.1007/978-94-017-8981-3_17,

IE began to be developed rapidly based on these facts. In 1991 the US National Academy of Sciences held a symposium on the subject. At the beginning of 1994 the National Academy of Engineering published "The Greening of Industrial Ecosystems". In this book the leading articles on the subject of this period appeared (Garner and Keoleian 1995: 4). In 1997 the Journal of Industrial Ecology began to be published (Erkman 2002: 35). In 2001 the International Society for Industrial Ecology was created, and adopted it as its official journal. The Journal of Cleaner Production also regularly publishes articles about IE. It has become a common subject for teaching, research and collaboration with enterprises in many universities.

However, IE has not become the "science of sustainability" as Allenby stated (1999: 40). Instead of taking into account such a broad endeavour, IE has only been working on two fields: Material Flow Accounting and Industrial Symbiosis or Industrial Ecosystems. A study carried out by Ehrenfeld (2007) on the articles published in journals dedicated to IE shows that a minimal percentage of them had not dealt with issues about the two fields. The social metabolism studies the material and energy flows in societies, industries, etc. IS studies and promotes material and energy exchanges between companies. This field has normally been developed by engineers, while the social metabolism has been a field for economists. The discussion about what is science is too complex to be analysed in depth here. And although the term symbiosis reflects the interchange of materials and energy which takes place in nature and is studied by biology, its translation to the interchanges produced between industries cannot be considered the creation of a new science. And as it is a field of engineering, and engineers always strive to put a solid science-based foundation under itself (Ehrenfeld 2007: 74).

On the other hand, there is a discussion about whether IE should be normative, and this is deeply related to the above discussion. Allenby (1999) rejects this idea: "Industrial ecology should not become a normative tool", because it will be an "imposition of ideological absolutes on complex real-world systems". But many authors claim that IE is a normative field due to being connected to the concept of sustainability. Boon and Roome (2001: 49) urged "researchers in the field of industrial ecology to reflect upon its normative aspects".

Some authors argue that IE comes from the existing analogy between nature and industrial relationships. Frosch states that "the idea of IE is based on a straightforward analogy with natural ecosystems. In nature an ecological system operates through a web of connections in which organisms live and consume each other's wastes [...] The system structure of a natural ecology and the structure of an industrial system, or economic system, are extremely similar" (1992: 800). From this perspective, industrial systems, despite being formed by a network with many interconnected elements, are not complex systems. For this reason, the present generation of models and tools are based on assumptions of linearity and stability. They do not take into account the system's evolution, the sociological and organisational processes that are involved when theories and technological models must be put into practice. However, some authors are beginning "to account for changes and material stocks over large periods are now being incorporated into the framework for analysing material flows in large and long-lasting systems".

And this change goes further and deeper. More and more authors think that the present linear models limit the scope of this field and its potential for change, because "industrial ecology can be a powerful new frame for thinking about and acting towards sustainability" (Ehrenfeld 2009: XVIII, XIX). This author, who is the director of the International Society of Industrial Ecology, kicks off the debate about the two approaches of IE: analogy or metaphor (or normative vision), declaring that "my own normative vision for industrial ecology is based on the metaphor of ecosystems as flourishing or sustainable (. . .) Metaphor differs from analogy in that it can be generative, producing new visions of the world as it might be" (2007: 76).

The book Dynamics of Regions and Networks in Industrial Ecosystems shows that the paradigm is changing. We will see the opinion of three authors: M. Chertow, B. Davidsdottir and M. Ruth (2009: 224, 225). These last two authors state that "industrial ecosystems are complex adaptive systems". M. Chertow states that "looking at industrial ecosystems as complex adaptive systems greatly helps to explain observed dynamics" (2009: 13).

The last debatable issue to be analysed is the use of the ecological concept of mutualism in IE. Chertow and Ehrenfeld state: "When symbiosis benefits both species (which exchange materials, energy or information), it is called mutualism" (2012: 15). This affirmation leads to applying mutualism to industrial symbiosis: "Mutualism can be seen to lead to new economic structures- reconfigured relations among economic actors enabling new interaction". For example, in Kalundborg, companies, in addition to exchanging natural resources, "also become willing to share other assets for mutual benefits, such as personnel, equipment, and information" (Chertow and Ehrenfeld 2012: 15). However, we have seen that ecology states that there are six kinds of interplays between two or more species in ecosystems: competition, predation, parasitism, commensalism, cooperation, and mutualism. The first three relationships are negative and the other three positive. Commensalism is a simple kind of positive interaction. It benefits one of the actors, and the rest (another actor or more) are not affected. Cooperation means that all species obtain benefits from the symbiotic relation, though they are not vital. Mutualism is a vital or very necessary relationship for the survival of the species that take part (Odum 1992: 116). In IE the relationships are positive for all the actors, but it cannot be ascertained that it is a kind of mutualism, but rather it must be considered as cooperation, because the benefits produced in the relationship are not vital.

2 Material Flow Analysis and Accounting of Socioeconomical Systems

2.1 Methodology

During the last two decades there has been great progress in learning about the metabolic processes of many societies. This knowledge is vital to move them towards sustainability, because we can comprehend the material and energy flows of the societies, due to the diverse economic activities, how these flows move across the economic system, and how they are transformed and treated at the end of their life.

This metabolic analysis can take place at different levels: the analysis of the resource flows and environmental impacts associated to the elaboration of a product or substance (Life-Cycle Analysis-LCA); the study of materials and energy associated to an area, like an industrial plant, a city, a region or a country (Material and Energy Flow Analysis (MEFA)); the knowledge of the flows of dangerous substances (SFA); the study of the material or energy flows and environmental impacts produced by the consumption model of a region (Ecological Footprint, Carbon Footprint, etc.); the study of those flows in the exchanges between countries; etc. Material Flow Accounting (MFA) is based on the principle of matter conservation: the quantity of matter which is moved through a Biosphere is constant. The MFA indicator is normally used as an account of all materials, including those used to produce energy, such as fossil fuels. For this reason, we use it in this broad sense.

The MFA, as a tool for understanding the relationship between Anthroposphere and Biosphere, has been developing since its creation in the late 1960s, and today it is considered a consolidated methodology. During the 1990s, institutions such as the Institute for Interdisciplinary Research and Continuing Education of the University of Vienna, the Wuppertal Institute for Climate, Environment and Energy or the World Resources Institute, joining efforts with governmental organisms and agencies from Germany, Italy, Japan or the Netherlands, gave the MFA a strong push which has endured until today. The metabolic analysis of our socioeconomic systems gives a systemic overview of them, through the study of the physical flows of natural resources from their extraction to their end of life treatment, covering production, use, recycling and deposition processes, and taking into account the resources lost through them. The use of these tools is related to the concern about the Earth's capacity to provide enough resources and to cope with the environmental impacts, without suffering greater harm (Adriaanse et al. 1997). From the MFA several indicators have been created to give a general vision of the physical foundations of our economy, which is necessary to know its level of sustainability. It can also measure the material and energy flows between countries. The difference between inputs and outputs causes changes to the physical stocks of a country. The MFA-derived indicators show the scale of resource consumption of a country, its tendencies, and relationship with GDP. So we can evaluate whether an economy follows a process of materialisation or dematerialisation. The MFA gives the indispensable complement in physical units (tonnes or Petajoules) to the monetary terms of the National Accounts (Bringezu et al. 2001: 20, 21). Thus, "the metabolism of an economy can hence be summarized in terms of a material balance" (Moll et al. 2005: 33).

Most of the studies which have been carried out about regional and national physical accounts have applied the MFA methodology. This methodology provides several indicators of material flows to obtain a vision of the pressures caused by an economy on natural systems. In general, these indicators have been designed to obtain a broad picture of the socioeconomic metabolic processes, as a first step in order to deepen the analysis in posterior studies (Bringezu et al. 2004) (Fig. 17.1).

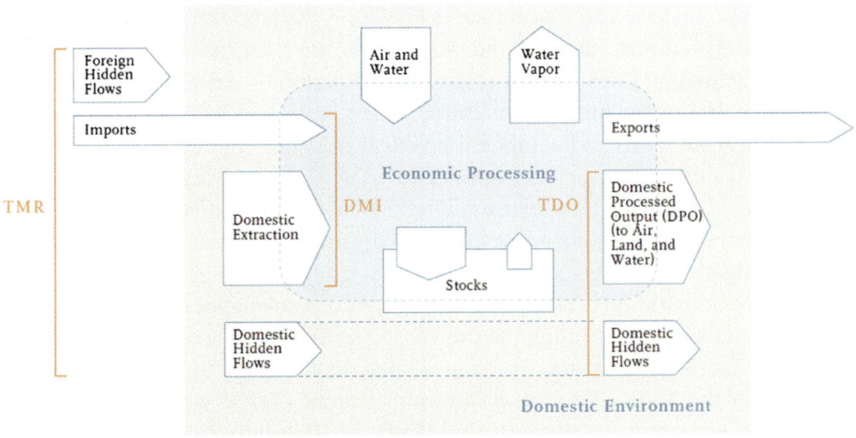

TMR (Total Material Requirement)=DMI+Domestic Hidden Flows+Foreign Hidden Flows
DMI (Direct Material Input)=Domestic Extraction+Imports
TDO (Total Domestic Output)=DPO+Domestic Hidden Flows
NAS (Net Additions to Stock)=DMI–DPO–Exports
DPO (Domestic Processed Output)=DMI–Net Additions to Stock–Exports

Fig. 17.1 Material cycle (Source: Matthews et al. 2000: 5)

From the perspective of the MFA several kinds of materials are treated separately: materials which are used directly in the economy (Direct Material Input-DMI), and materials which are removed from the Earth, but are discarded due to being non-economic materials (Hidden Flows-HF). The DMI includes biotic and not biotic (or abiotic) materials. The abiotic material comprises minerals (ores, sand, stone, etc.) and fossil fuels. Hidden flows are generated at the harvesting or extracting stage of the material cycle. They have two components: ancillary flows (fish captures, plants and forest biomass that are removed from the soil, in order to obtain logs and grains, but later are separated from the economic materials), and excavated and/or disturbed materials. In this case, in the mining sector some materials of the Earth's crust are removed in order to access the ore, and also in the building sector there are excavating needs. On the other hand, industrial agriculture produces soil erosion. HF is associated with domestic activity, and with imported materials. In the current globalised economy this factor has a growing relevance for OECD economies, and especially for the European Union, which is the most dependent economy in the world on raw material imports. This situation shows a huge asymmetry between OECD countries and the rest of the world. The former import a large portion of the natural resources that they consume, while the exporting countries suffer great environmental impacts. However, the accounting of the hidden flows from imports is only partial. They are only taken into account in the case of importing raw materials, but not those associated to the manufactured goods, because it is impossible to know the hidden flows associated to them, due to the growing complexity of the product chains. But, if all of the hidden flows could be accounted for, many aspects would be left out, like biotic diversity losses, river pollution, etc. The DMI plus the HF measure the total material requirements (TMR) for a product, service or economy, without taking into account air and water. This physical indicator is measured in tonnes per capita and year, and it represents all materials extracted from nature for an economy (Bringezu and Schütz 2001c).

The TMR and DMI indicators are physical measures of economic activity, and complement other monetary measures like GDP, thus showing the most complete and integrated vision of an economy. The TMR also gives an approximate measure of the overall pressure imposed on the Earth by an economy. However, this measure is an aggregate of many elements, so it is too broad and general to be precise. Frequently, industrial processes generate dangerous substances, and it is necessary to analyse their impact through the production and consumption processes.

There are other indicators, which give specific visions within the overall picture provided by the TMR. The Total Domestic Output (TDO) takes into account the total processed output and domestic hidden flows. This indicator represents the total quantity of material output to the domestic environment caused by an economy. This indicator accounts, on the one hand, for the domestic HF and, on the other, for the materials processed and emitted to the environment (emissions, waste, etc.). Domestic Processed Output is the result of taking away from the DMI the Net Additions to Stocks and the Exports (Matthews et al. 2000: 5). Direct Material Consumption (DMC) represents apparent consumption of a region calculated by subtracting the exports from the IMD (domestic extraction + imports).

Net Additions to Stocks (NAS) measures the net quantity (weight) of materials added to the stock of an economy. For example, the net materials added to the stock of buildings and other infrastructures, incorporated into new durable goods as cars, industrial machinery, household appliances, etc. Thus, NAS "measures the physical growth of the economy". PTB (Physical Trade Balance) defines the physical trade surplus or deficit of an economy. It is calculated by taking the amount of imports in physical units minus exports in physical units. PTB may also be applied to account for single material categories (e.g. for fossil fuels), indirect flows associated to imports and exports, and to measure the land use associated to agro-materials trade. PTB gives a broad picture of the raw material dependency on imports, mainly in OECD countries. The MFA also includes a Material Balance (MB) which compares material inputs and outputs. The MB is based on the first law of thermodynamics: Total Inputs equals Total Outputs plus Net Addition to Stocks (NAS). This set of material flow indicators offers a full metabolic vision of an economy, supplying very useful information for the design of different strategies for sustainability (Moll et al. 2003: 25, 26).

2.2 *European Union Metabolism: Material Flow Accounts*

At the turn of the last century many metabolic studies for OECD countries and some regions were developed. Several studies were carried out by the Wuppertal Institute (WI) for the European Union. The data given here comes from two reports

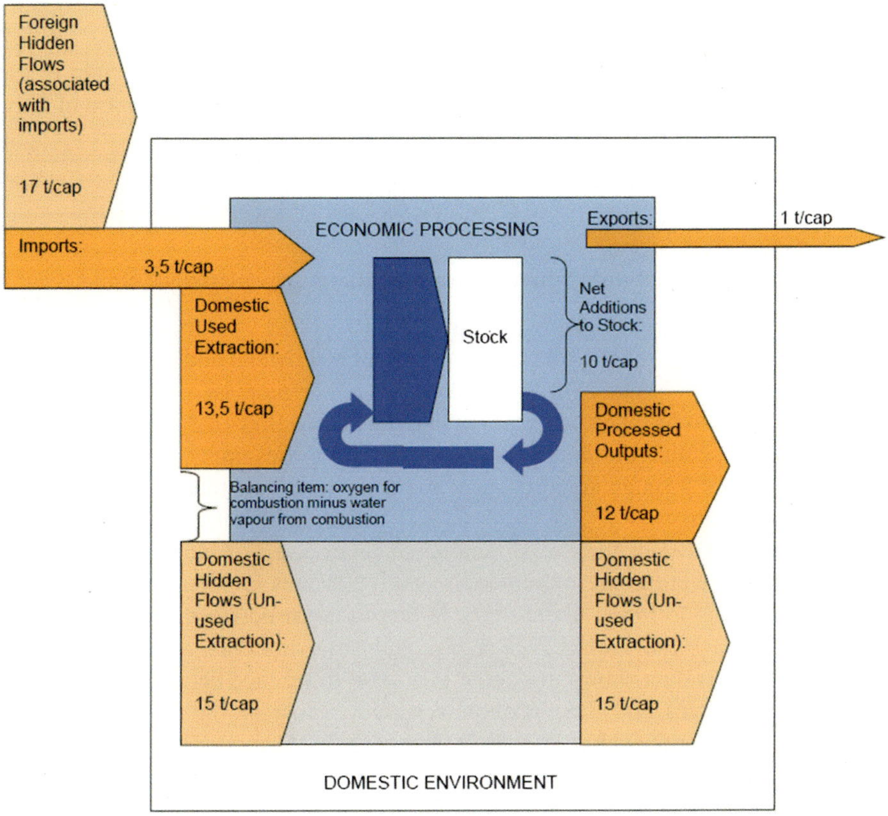

Fig. 17.2 Estimated economy-wide material flows in the EU, on a per capita and yearly basis and for the second half of the 1990s (Source: Moll et al. 2005: 31)

carried out by WI experts (Bringuezu and Schutz 2001; Moll et al. 2005). Besides, Eurostat developed a methodology for MFA, with the technical advise of the WI. So the data obtained for each country could be equally compared. And studies carried out during different periods of time can give an evolutionary vision. We analyse here a report elaborated by the WI to study the MFA for the EU-15, and for each of the member States. Figure 17.2 gives a synthetic vision of the first study. Its main conclusions are:

In 1997, the TMR of the EU-15 amounted to more than 50 tonnes per capita. This is an average level among OECD countries: well under USA's level (80 tonnes per capita) and a little higher than Japan (45 tonnes per capita). This difference is due to Japan's low per capita energy consumption. About 28 % of the EU's TMR was determined by fossil fuels. Metals constituted the second biggest component with some 23 %, followed by construction minerals (18 % of the TMR). Related to the latter are excavation and dredging which accounted for 6 %. Biomass accounted for 12 %, and erosion for 9 %. About 88 % of the TMR was comprised of non-renewable resources. Great differences appeared among countries, oscillating between Italy's TMR (32 tonnes per capita) and Finland's (100 tonnes per capita). These differences can be explained by the hidden flows and by the weight of the primary sector. Finland has a strong primary sector, and this is not the case of Italy. On the whole, differences are due to several factors:

- The level of economic development (GDP per capita).
- The production structure (the weight of each sector).
- The technological level.
- The domestic availability of natural resources (land, raw materials, climate, etc.).

In 1997, the up-stream resources extraction associated with imports of the EU-15 was more than 20 tonnes per capita, 40 % of TMR. Metal and energy resources amounted to 55.9 % and 27.8 % respectively, while biomass only amounted to 1.9 %. But it has to be taken into account that the hidden flows associated to imports are much greater than the domestic ones. The EU's consumption of resources is increasingly met through imports, and this has caused increasing hidden flows associated to them.

In 2000, the DMI of the EU-15 amounted to around 17 tonnes per capita and year (13 tonnes were domestic materials and 3.5 tonnes imported materials), and was composed as follows. About 40 % is due to construction minerals, followed by biomass and fossil fuels (25 % each). Industrial minerals amounted to only around 9 %. This data shows that there is a great difference between the physical weight of resources and their economic weight. Most of the EU-15's DMI is comprised of non-renewable resources: about 74 %. Around 60 % of the annual DMI input increases the material stocks of the EU-15 (about 10 tonnes per capita). Approximately 40 % is discarded in the form of air emissions, waste, waste water and dissipative losses. It amounts to about 12 tonnes per capita and year.

In the period from 1980 to 2000, resource consumption grew slightly; DMI went from 15.9 tonnes per capita in 1980–17.5 tonnes per capita in 2000. However, there were significant differences among the various kinds of resources and a strengthening of imports. In the first case, a strong increase of the metal component of the TMR is observed, a steady increase of the construction component, and a steady decrease in fossil fuel weight. This last tendency is due mainly to the replacement of coal with natural gas to generate electricity. The foreign part of the TMR has been increasing steadily since the late 1980s from around 30 % to nearly 39 % in 1997, in the EU-15, and during the period from 1995 to 1997 it grew by 11 %. To a large extent this tendency was due to the domestic demand of precious metal ores, and also to finished products such as jewellery, gold, silver goods, and plated ware. These imports grew by 51 % during the period from 1995 to 1997, and the collateral effect was a strong increase of foreign hidden flows, which grew from 14.7 to 16.5 tonnes per capita from 1995 to 1997. The imports of metal resources caused 17 times higher hidden flows than domestic extraction.

3 Industrial Symbioses

3.1 The Concept and Overall Perspective of the Field

In the nineteenth century and at the beginning of the twentieth century there were strong symbiotic exchanges between various industries (in particular the meat industry provided others with waste which was raw material for them). After this period these symbiotic relations declined sharply, but at the end of the past century, a new process of recovery began and an extraordinary range of experiences have been developed. And, as we have seen previously, these experiences constitute a part of a new field of knowledge: Industrial Ecology.

From the beginning industrial ecologists tried to give a practical meaning to the analogy between natural ecosystems and industrial ecosystems. They used terms like "industrial food chain" and "industrial food web" and emphasised that a mature industrial ecosystem had four key actors: "producers, consumers, scavengers and decomposers" (Chestow 2009: 7). The same tools as those used by ecology to study the dynamics of ecosystems must also be applied by the industrial ecologists: potential, connectedness and panarchy. For example, Van Berkel studies connectedness or "symbiotic intensity", and has measured it at several locations: Gladstone (Australia) (3 symbiotic projects between 6 companies); Kwinana (Australia) (47 symbiotic projects between 22 companies); Guitang (China) (5 symbiotic projects between 5 companies); Ulsan (Republic of Korea) (9 projects between 12 companies); and Kawasaki (Japan) (14 symbiotic projects between 9 business) (2009: 484). These accounts enable researchers to define physical indicators to quantify and compare the intensity of the symbiotic experiences. However, "even with consensus and standardisation on physical indicators, the quantification of symbiotic relationships in a given industrial area will remain dependent on several factors", such as the "level of knowledge of the industrial system" or the "industrial organisation and firm boundaries in the industrial system" (Van Berkel 2009: 485).

Many authors describe IS as an exchange of resources among enterprises. For Chertow and Ehrenfeld it means a "cooperative management and exchange of resource flows – particularly materials, water, and energy- through clusters of companies" (2012: 13). Here the core message is "cooperative management and exchange". Lombardi and Layboum's definition describes the objectives that the firms want to obtain: "IS engages diverse organisations in a network to foster eco-innovation and long-term cultural change. Creating and sharing knowledge through the network yields mutually profitable transactions for novel sourcing of required inputs, value-added destinations for non-product outputs, and improved business and technical processes". Here "profitable transactions" obtained through "eco-innovation and cultural change" is the main idea. Ashton describes industrial symbiosis "as partnerships among firms in a region to physically exchange and share in the management of resources such as energy, water, and input and output materials, as well as to pursue broader strategies for sustainable industrial development" (2008: 35). This definition adds the important dimension of sustainability. But there are other significant aspects which are not manifested in the definitions, and that are shown in the following explanations. For example, Chertow states that "central to the field of industrial ecology is *biological analogy*" (2009: 6). Chertow and Ehrenfeld state that "looking to science, industrial symbiosis reflects the notion of biological symbiotic relationships in nature" (2012: 15). As a conclusion, there is not a single definition which describes the multiple dimensions of the IS, and it is not clear that most authors could agree on a definition which includes all the aforementioned dimensions.

There are several factors which drive the development of industrial symbioses. Boon and Speklink define five factors which provide the enabling conditions for the development of symbiotic experiences: "(1) the need for a learning process and a strategic vision"; (2) "diversity of involved actors and its consequences for connectance and interdependency"; (3) "trust and ways to foster it"; (4) "presence of anchor tenants or coordinating bodies"; (5) "an enabling context" (policies, regulations, other institutions, culture, etc.) (2012: 62). But we must add other factors: scarcity of resources and its consequence of rising prices, proximity and isolation factors. Proximity lowers the transactional costs, and reinforces other factors, like trust or interdependency. Isolation is mentioned as one of the factors that facilitated the development of the Kalundbourg ecosystem. The term "anchor tenant" is used to describe enterprises (like chemical manufacturers, breweries, energy generators, etc.) which produce a large and relatively constant flow of waste or by-products that can be used by other companies, paving the way for the development of many symbiotic processes.

3.2 Terms Used in the Application of Industrial Symbiosis

As is usual concerning similar experiences that have been developed in very different countries and circumstances, such as the case of exchanges among firms, many terms have been used to describe systemic relationships, but with a significant range of differing meanings. At the beginning the following terms were used: industrial ecosystems, eco-industrial parks, industrial eco-parks, virtual eco-industrial parks, eco-towns, networks of industrial symbiosis, regional networks of industrial symbiosis, and many more. As an example of the most extreme variance of meaning, the Asian term 'industrial eco-parks' was used to name an industrial park which promotes reductions of environmental impacts, without taking into account symbiotic exchanges. However, at the end of the last decade, a broad agreement was reached concerning the meaning of the most-used terms: eco-industrial parks (predominant in Asia and less so in North America), eco-industrial networks, and industrial ecosystems (used in Europe and its use is growing in North America). Chertow and Ehrenfeld state that "we use the terms 'industrial ecosystem' and 'networks' and also 'eco-industrial networks' in a similar fashion throughout this article as the manifestations of industrial symbiosis" (2012: 25). And Geng et al. claim that an eco-industrial park "emphasizes the establishment of an industrial symbiosis network composed of varied industries" (2009: 16). However, although the term 'industrial systems' can be used in all cases (parks, eco-towns, regional networks, etc.), the term 'eco-industrial' is too narrow (although it is the most frequent application of industrial symbioses) to describe other spatial applications. In these cases the terms used are 'regional networks' or 'regional eco-industrial networks'. Last of all, the term 'industrial park' has a different meaning and potential for fostering symbiotic relations in Asia and in North America or Europe. Frequently, in Asia an industrial park is owned by an authority (like a municipality), which appoints a manager: "An industrial park is land reserved by a municipal authority for industrial development; it includes an administrative authority for making provisions for management and enforcing restrictions on tenants and detailed planning with respect to sizes, access, and facilities" (Geng et al. 2009: 16).

On the other hand, there is a debate about the minimum requisites needed to distinguish industrial symbiosis from other types of exchange. Chertow states that "my colleagues and I have adopted a '3–2 heuristic' as a minimum criterion. Thus, at least three different entities must be involved in exchanging at least two different resources to be counted as a basic type of industrial symbiosis". The author defends that this requisite draws a line between "complex relationships" and "linear one-way exchanges". This criterion has some problems. It is debatable whether to consider the 3–2 heuristic as a case of complex relationships. Thus it appears that two-way relationships between companies (two or more) must be considered cases of industrial symbiosis, but not as complex systems. Another problem is to decide the level of complexity which has to be reached to catalogue the symbiotic exchanges produced in a case as an industrial ecosystem or eco-industrial park.

A review of the literature on industrial symbioses shows that the most frequent process in the creation of a network of symbiotic relationships has three stages: sprouting, uncovering and embeddedness, and institutionalisation. In the sprouting stage firms begin to exchange waste, energy, services, etc., but the relationships are not stable. But little by little the level of exchanges grows and becomes stronger. The whole process is developed without external influence, and frequently the actors do not have an idea of the whole level of exchanges occurred. In the second stage, an external actor often studies the existing exchanges and *discovers* their level of maturity and complexity. And at this point, the stage of embeddedness and institutionalisation begins to develop. In addition to the self-organising previous process, "further expansion of the network becomes intentionally driven by an institutional entity", "tying together the actors and interpenetrating whatever other structures may influence their behaviour" (Chertow and Ehrenfeld 2012: 19–21).

There are many industrial ecosystems which were "uncovered", frequently by academic studies. The Kalundborg industrial ecosystem became known when high school students made a scale model of all the connections. The Kwinana case (Western Australia), a study of the large volume of waste from mineral processing in the context of the Mining, Minerals and Sustainable Development Project backed by the World Business Council for Sustainable Development, revealed more than 100 exchanges. Schwardz and Steininger studied the large province of Styria (Austria), and uncovered a highly diverse and complex network of symbiotic exchanges. In Finland, Korhonen and his colleagues realised that the city of Jyvaskyla had a co-production of heat and electricity in which industrial waste was used as fuel. In North Carolina staff members of the six-county Triangle J Council of Governments discovered that their project of creating an industrial ecosystem had a developed base, after making an inventory of 182 business inputs and outputs. They found that 36 % of the firms had been making exchanges. Similar cases have appeared in Gladstone (Australia), in Guayama and Barceloneta (Puerto Rico), in the industrial parks of Burnside and Alberta Heartland (Canada) (Chertow 2007).

3.3 The Role of Government Policy in the Promotion of Industrial Symbiosis

Government institutions have promoted two kinds of policies to develop industrial symbioses: creating an enabling framework for fostering private initiatives, and promoting the development of industrial ecosystems from scratch in specific sites. Here an overview of the main governmental policies (in some cases, strategies) in this field is developed.

Recently the European Union has issued several documents which support IS. The most significant has been the Europe 2020 Strategy. Its Resource Efficiency Flagship Initiative recommends that promoting resource efficiency through IS should be a priority for Member States. The strategic document "Sustainable Industry: Going for Growth and Resource Efficiency" issued by the Directorate-General (DG) Enterprise and Industry (Laybourn and Lombardi 2012: 11) goes in the same direction. However the "priority" has not been made concrete in a specific policy to promote industrial symbioses in Europe, with objectives and tools to meet them.

The U.S. President's Council on Sustainable Development, formed during the Clinton administration, promoted 15 so-called "eco-industrial parks", being at different stages of development. 10 years later, six were open (but only two as industrial parks, the rest had changed the concept or this was uncertain), and nine never emerged or failed (Chertow 2007: 15, 16). The Dutch DBT program was aimed to stimulate industrial symbioses by providing subsidies for specific projects and monitoring their improvements. The DBT programme obtained positive results, but failed to meet expectations. It was implemented during the period from 1999 to 2004 (Boon and Spekkink 2012: 67, 66). The Japanese Eco-Town Programme was launched in 1997. It aimed to promote the recycling sector and in particular to renew cities with an aged industrial infrastructure. The programme contained voluntary initiatives and financial support from the central government. It was further supported through the approval of the Basic Plan for Establishing a Recycling-Based Society, which was established in 2003. The Eco Towns promotion has three main objectives: the promotion of environmental industries, waste treatment and community development. The programme has been successful: "This application of eco-industrial principles, legislation and support funding has been instrumental in rejuvenating Japanese industry" (Fergurson 2010: 35, 36).

In 2005 Korea initiated a plan to create many "eco-industrial parks" over the course of 15 years, with the period divided into three stages. The leading organism is the newly formed Korean National Cleaner Production Center (KNCPC). The first phase (2006–2010) is dedicated to identifying possibilities for industrial symbioses, and to develop two eco-industrial parks. The second phase (2011–2016) seeks to create 20 eco-industrial parks. The third phase (2016–2020) would review the flaws and seek to solve them (Mathews and Tan 2011: 446).

China is the country with the most consistent strategy for promoting industrial symbioses, and it has the greatest potential for development. The State Environmental Protection Administration (SEPA) has been the leading organism in promoting IS, and its main policy is the National Demonstration EIP Programme. Since 2007 SEPA has been collaborating with the Ministry of Commerce and with the Ministry of Science and Technology. And since 2009 the EIP Programme has had the umbrella of the Law for the Promotion of the Circular Economy. It has to be highlighted that China is the first country in the world "to make circular economy a national strategy" (Mathews and Tan 2011: 436). The EIP Program has three stages: planning, implementation, and accreditation of a National Demonstration EIP. Due to the fact that the least successful experiences are those developed from scratch, China has focused its policy on transforming existing parks into eco-parks. In 2011, three ministries approved 60 National Trial EIPs: 48 are mixed industrial parks, 11 sectoral industrial parks, and only one is a resource recovery park. In 2011, 15 industrial parks obtained the accreditation, and they have become references for all other initiatives. Thanks to the preferential public policy, better developed planning, the hundreds of qualified researchers and professionals working in the field of EIP planning, "most of the 60 National Trial EIPs have brought about rapid real estate development" (Shi et al. 2012a: 8–10). Despite this success, there are many barriers to be removed: "the EIP programmes remain a primarily government-led initiative"; there is a "lack of effective financing mechanisms"; "the lack of pertinent hard and soft technologies"; "the annual performance evaluation (...) is mostly based on self-reported data from the national trial EIP" (Shi et al. 2012b. 391, 392). Last of all it must be taken into account that China has developed the largest number of IS initiatives, and its potential for growth is huge. China has at least 1,568 national and provincial industrial parks, and they account for an increasing proportion of national GDP (Shi et al. 2012a: 8).

There is a debate about the government's role in the promotion of IS. We have seen the poor results of the U.S. Presidential Council on Sustainable Development programme, although Chertow (2007: 15–17) has studied it and considers that some of the initiatives were not properly defined. However, other government policies have obtained better results, especially in some European States and in Asia. Some authors believe that the different outcomes are due to the weak culture of public intervention that exists in North America, while in Europe and Asia it is stronger. On the other hand, the actors involved tend to pursue quick results, not realising that frequently the maturation pace of IS initiatives is very slow. It seems plausible that a collaborative government can make feasible projects which private actors only by themselves cannot develop. This kind of governance can improve three dimensions:

- "Relational capacity enables actors to engage in risky transactions that would be too costly in the absence of strong personal and professional relationships".
- "Knowledge capacity enables the actors to collect timely and relevant information about feasible symbiotic linkages"
- "Mobilisation capacity enables actors to target and involve the actors that are necessary for symbiotic exchanges, to influence policies and regulations that are relevant to these exchanges, and to attract external resources" (Boons and Spekkink 2012: 63).

3.4 *Industrial Ecosystem Taxonomy*

Chertow and Ehrenfeld (two of the leading researchers in the field) state: "We do not have enough knowledge about industrial ecosystems to understand fully their life cycles and the categories into which they fall" (2012: 18). However, Chertow (2008) proposed five categories. Type 1: "Through waste exchanges". Normally an agent carries out the main action: the collection and separation of waste. And they become raw materials for many companies in the region. Type 2: "Within a facility, firm, or organisation" (holding). Type 3: "Among firms collocated in a defined eco-industrial park". Type 4: "Among local firms that are not collocated" in an eco-industrial park. Type 5: "Among firms organised virtually across a broader region". This taxonomy emphasises the spatial dimension of the symbioses. For this reason we classify industrial ecosystems (as a tentative approach) into five categories; three are based on the kinds of exchanges, and only the last one follows a spatial criterion: agro-industrial; based on the waste produced by companies from one sector; complex and multi-sectoral ecosystems; urban-industrial ecosystems; and regional ecosystems.

3.4.1 Agro-industrial Ecosystem

Frequently the industrial processes which use biomass generate a great quantity of waste. 92 % of the barley used in a brewery is waste. The most productive biomass-to-good process is the production of paper from wood: 80 % of the biomass is wasted. The same problem appears in harvesting: 99.8 % of the biomass generated to produce coffee beans has little value. The ZERI Institute, promoted by Gunter Pauli, has been developing several industrial ecosystem models, all based on the

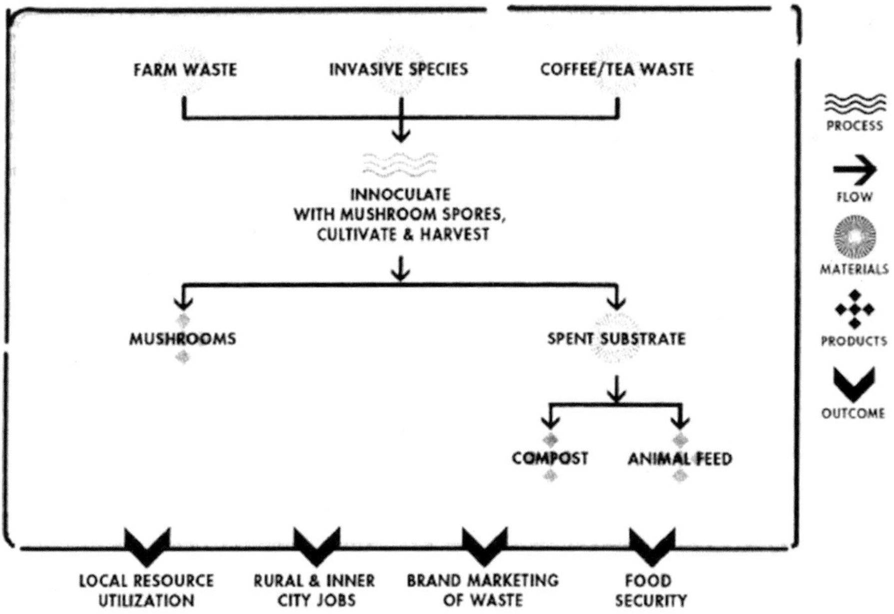

Fig. 17.3 Pulp-to-protein cascade (Source: Pauli 2010: 93)

same process (called the "Pulp-to-Protein model"), to create "value for 100 % of the nutrients". He states that this "model could generate a staggering 1.5 million times more revenue than coffee produces today". As the picture shows the process begins with the inoculation of mushroom spores in the cultivated and harvested biomass. Two products are obtained: mushrooms (mostly reishi and shiitake varieties which have high nutritional and medicinal qualities) and spent substrate. From this, compost and animal feed is obtained (Pauli 2010: 92, 94) (Fig. 17.3).

This cascading process has been tested and fully applied from Colombia to Zimbabwe, from San Francisco to Germany. In the Colombian province of El Huila there are more than a hundred coffee-to-mushroom companies, and this number is growing steadily: "Economies that cascade like ecosystems achieve the goal of providing multiple benefits for diverse partners": jobs, cash flow, and profits (Pauli 2010: 93).

The Guanxi Autonomous Region in China produces 40 % of China's sugar. The Guitan Group located in the region has transformed the traditional activity. The Group reached an agreement with the provider farmers to produce organic sugar cane. Besides, it produces ten by-products from the two basic by-product flows: molasses (refined sugar residue) and bagasse (fibrous waste). The complex includes an alcohol plant, two pulp and paper plants, a toilet paper plant, a calcium carbonate plant, a cement plant and a power plant. The Group has created a cluster of firms that reuse the by-products. This experience "has inspired the town of Guigang to adopt a 5-year plan to become an eco-city" (Geng et al. 2006; Chertow 2009: 17).

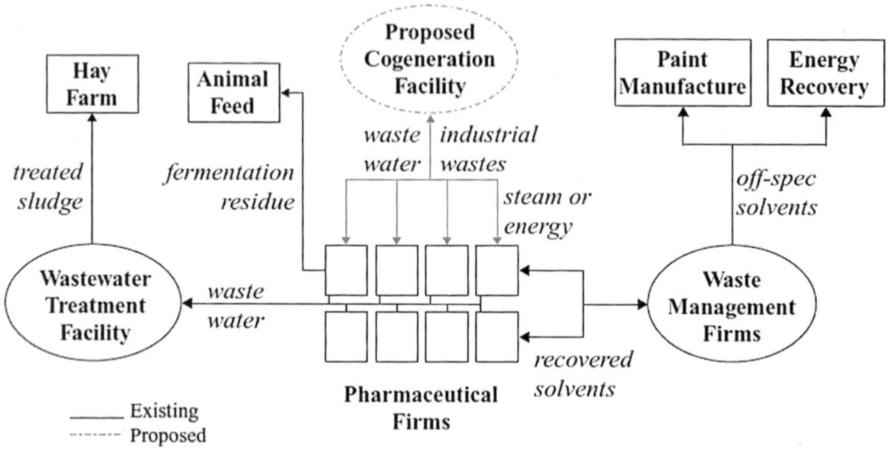

Fig. 17.4 Existing and proposed industrial symbiosis linkages in the Barceloneta pharmaceutical cluster (Source: Ashton 2009: 39)

Ecosystems based on the waste produced by companies from one sector. In this case one company (Chertow calls them "anchor tenants") generates a large quantity of waste, which is used by other firms as raw material. Enterprises of this kind are power stations, refineries, petrochemical plants, etc. Refineries have become the biggest sulphur providers, which is frequently used to produce fertilisers (Ayres 2002: 44 and ss.). In the area of the Houston (Texas) canal there are many refineries and petrochemical companies which have developed a dense network of exchanges. Value Park (Germany) integrates petrochemical plants, plastic producers and service companies. Power plants have been developing symbiotic projects in Guayama, Barceloneta (Puerto Rico) and Londonderry (New Hampshire). Puerto Rico has one the highest concentrations of pharmaceutical companies in the world. Since the late 1970s, a group of industrial companies, located in Barceloneta, began to develop several projects (promotion of a common infrastructure, service sharing and by-product exchange). During the period from 1978 to 2006 eight pharmaceutical firms formed the Wastewater Advisory Council of the Barceloneta Regional Wastewater Treatment Plant with the aim of receiving the water in return for paying for a large part of the plant management, which also treats domestic wastewater. The sludge was used by an adjacent hay farm, and the hay was sold to livestock farmers in the region. In the mid-1970s, a waste management company moved to the area to collect and recycle used solvents from pharmaceutical facilities. Companies have been discussing with the relevant authorities the creation of a cogeneration facility to use solvents and hazardous waste to generate energy, though without results. But "the Wastewater Advisory Council and local solvent recycling facility ceased functioning at the end of 2006, due to a combination of economic and regulatory changes, in favour of internal management of waste treatment and recycling activities" (Ashton 2009: 240) (Fig. 17.4).

3.4.2 Complex and Multisectoral Industrial Ecosystems

The Naroda Industrial Estate (NIE) is one of the largest sites in the world in which symbiotic exchanges are being developed. The estate comprises some 700 firms with 35,000 workers. Companies from many sectors (chemical, pharmaceutical, dye and dye intermediates, engineering, textile and food production) are located there. Under the leadership of the NIE and with the technical assistance of the University of Kaiserslautern a general project for the creation of an "eco-industrial network" in this area is under way. At the beginning four projects were identified as feasible: recycling of spent acid, recycling of chemical gypsum, recycling of chemical iron sludge and the re-use or recycling of food waste. Four chemical companies planned to use spent acid to produce ferrous sulphate. The chemical gypsum was dumped into landfills, but four firms developed a common plant to dry the gypsum. They recycled 300 tonnes per month of dried gypsum. Producers of dye and dye intermediates produced large quantities of iron oxide, quite hazardous waste. Several companies joined forces to develop a cleaner production solution to reduce the amount of ferrous acid. Many small food companies generated great quantities of waste and they collectively designed a process to use it. After these four projects, 15 firms in the ceramic industry developed the fifth project: assuring the purity of their materials (Lowe 2003: 347, 348).

The Canadian Burnside Industrial Park, located in Halifax, was established in the early 1970s. Approximately two thirds of the 1,200 ha have been developed. The municipal government is responsible for the park, and has been strengthening the environmental awareness of the companies, and among other actions symbiotic exchanges have been developed. Some 90 % of the businesses are SMEs from several sectors: manufacturing (10 %), sales and services (48 %), industry construction (11 %), distribution and warehousing (8 %), retail (8 %), and professional, financial and other business (14 %). The most successful waste exchanges have been wooden pallets and metal and packaging materials. Some firms are also involved in province-wide exchange programmes which include the recycling of computer hardware, chemical exchanges and food waste recovery (Coté and Crawford 2003: 323–327).

3.4.3 Municipal-Industrial Ecosystems

Cities need food, materials, energy and water, and generate a great amount of waste. In order to build a circular economy of materials in the municipality and to, at least, improve energy efficiency, both the cities and their industrial systems must be integrated in a unique symbiotic entity. This case is very frequent in Asia, although it is extending to all the continents. Japan is becoming a leading country in this field, due to facing huge problems with waste management, as landfills had nearly reached their capacity, and also due to the ageing of many industrial sites, which generated

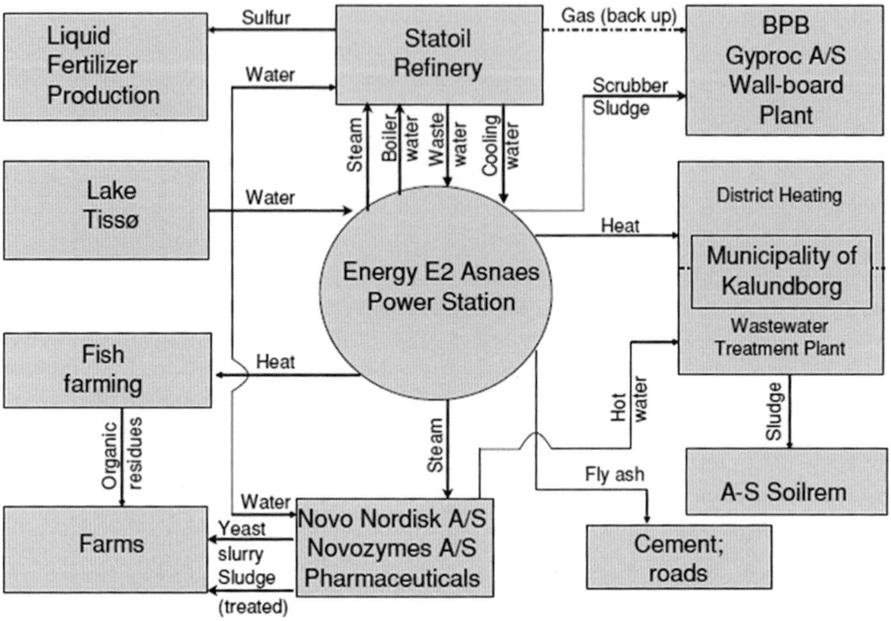

Fig. 17.5 Selected industrial symbioses in Kalundborg (Source: Chertow 2007: 17)

severe environmental impacts. For these reasons, the Ministry of International Trade and Industry (MITI) promoted the so-called eco-town project by providing technical and financial support to local authorities that wished to establish eco-towns "through various recycling and industrial symbiosis efforts" (Morikawa 2000: 10).

The Danish town of Kalundborg (20,000 inhabitants) is the most well-known and best analysed case of a self-organised industrial ecosystem, and was the first case to be uncovered. Due to these facts, Kalundborg stands as a benchmark case of industrial symbiosis. The town is predominantly industrial and its economy is based on four basic industries: a coal-fired power plant (Asnaes), an oil refinery (Statoil), a pharmaceuticals and enzymes producer (Novo Nordisk, with two plants: Novo Nordisk and Novo Enzymes), and a plasterboard manufacturer (Gyproc). The municipality also provides various shared facilities (mainly a wastewater treatment plant) and services. The most significant kind of symbiosis is the cascading use of water and energy. After being uncovered, the Symbiosis Institute was launched, as part of Kalundborg's industrial development agency, to accelerate the quality and number of symbiotic exchanges. Mathews and Tan (2011: 445) state that recent studies show "that the industrial symbiosis exchanges have been upgraded from time to time from generally low-value by-product exchanges (. . .) to high-value by-product exchanges" (Mathews and Tan 2011: 445; Chertow 2007: 19, 20) (Fig. 17.5).

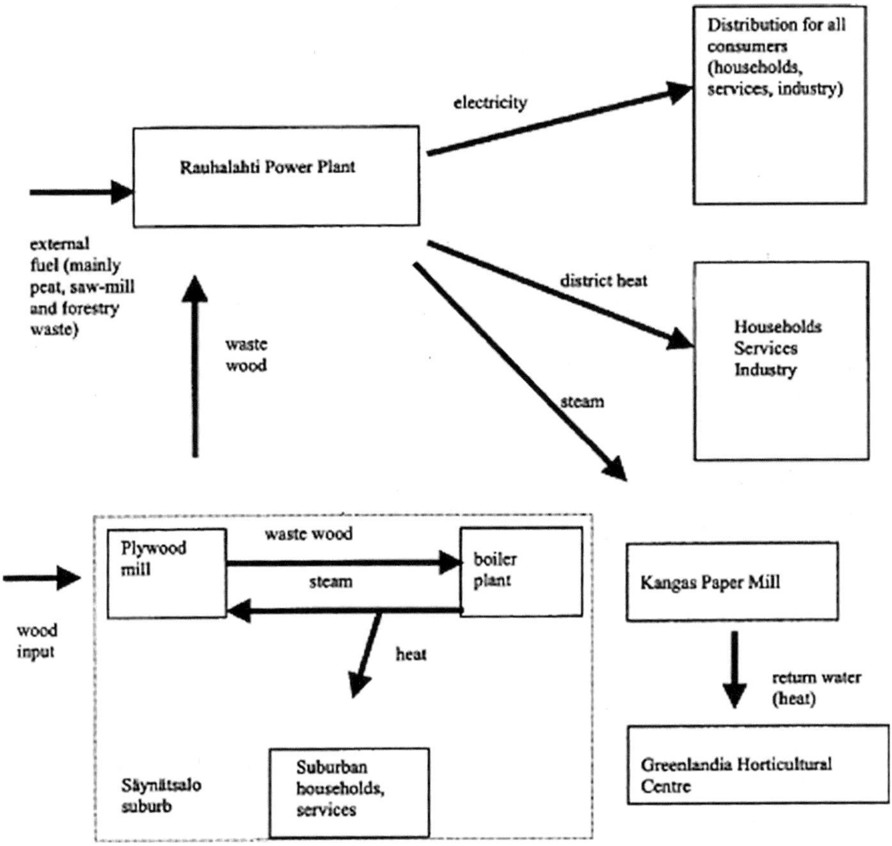

Fig. 17.6 Jyvaskyla industrial ecosystem (Source: Kohronen 2001: 65)

3.4.4 Regional Ecosystems

We have commented that in Jyvaskyla (Finland) and Styria (Austria) regional and self-organised regional ecosystems have been developed, and that they are rather simple. Finland, Denmark and the Netherlands are the only three countries in which the energy supply is organised to a large extent by co-production (generation of heat and power) (CHP). The surplus heat produced by a power station is used to heat buildings or in production processes. We have seen that in the Kalundborg ecosystem the waste heat from Asnaes is used to heat the town buildings and many industrial firms in particular by the Statoil refinery. In the Jyvaskyla ecosystem there are exchanges of waste energy and industrial and forestry waste between the actors of the region (Fig. 17.6). The energy is used in a cascading fashion, and the wood waste is used as fuel. The ashes are collected and moved to the earth, so there is a recycling of matter. The whole symbiotic process is described in Fig. 17.6. The main actors are a power company, a power plant, a paper mill, a plywood mill, forest companies and a horticultural centre. These actors are located within a distance of 50 km. The waste energy from electricity production is used mainly to satisfy the heating needs of the Kangas paper mill. The paper mill provides the local horticultural centre Greenlandia with warmed water. The plywood mill in the Saynatsalo suburb of Jyvaskyla provides the power plant with waste wood and receives electricity. The Rauhalahti plant receives waste wood from regional saw mills, and also from the forest residues (cuttings). The Saynatsalo plywood mill also uses wood waste, generated locally, in a suburban boiler plant from which it obtains steam, the boiler also providing the households and buildings in the immediate vicinity with heat. The waste ash from the Rauhalahti power plant is currently used by the horticultural centre Greenlandia, which is building a model for green gardening and green construction in the 'Green Land' city project (Korhonen 2001: 61).

The INES (Industrial Ecosystem Project), started in 1994 with 69 participating industrial firms, is located in the Rotterdam harbour and industrial complex (the Europort/Botlek area, near Rotterdam). It is a project driven by the interest of the companies involved, through their organism, Deltalinqs. The chemical companies were dominant, because the main cluster of this industry in the Netherlands is located in this region. The first INES project took place during the 1994–1997 period. Fifteen symbiotic projects were defined. Three of them were selected to analyse their feasibility. One steam exchange project was put into practice and it involved eight companies. The INES project was followed by the INES Mainport project, which was developed in the period from 1999 to 2002, by a platform of representatives from industry, government, a university, and an environmental NGO. Its principal aims were the development of a long-term strategy and to design the elements for the third stage (Baas and Boons 2004: 1078–1080).

The promotion of the Rhine-Neckar region (50 × 50 km) started from a university-driven research project to test whether an industrial ecosystem could be built in the industrial area of Pfaffengrund. The study proved that the project was feasible. In the first phase four types of exchanges were implemented: direct cascading processes; companies opening disposal paths; joint transportation of pellets; and informal communication. After that phase, they tried to extend the model to the whole region, where companies such as Roche, BASF, ABB, etc. are located. Two actions have been implemented: the creation of a network structure (UKOM) as a means "for the discussion and preparation of coordinated actions" among the actors (firms, university and authorities); "a waste management information system to facilitate data exchange between participants" (Adamides and Mouzakitis 2008: 20, 21). In the Swedish region of Landskrona an industrial region is being built driven by two forces: a declining economy and regulatory pressure to improve the poor environmental quality of the region. The project is funded by the Swedish Business Development Agency. A total of 21 firms, belonging to the sectors of metal works, chemicals, printing, auto parts, etc., are participating. After identifying the potential synergies, the project is limited to the following activities: district heating using waste heat, collective waste management, using unprocessed waste, cooperation in transport, and use of renewable energies (Adamides and Mouzakitis 2008: 23–25).

Chapter 18
Basis for an Eco-effective and Integrated Product Strategy

Keywords Integrated product policy • Extended producer responsibility • Integrated product strategy • Product-service systems • Intelligent material pooling

In this chapter we analyse the so-called Integrated Product Policy (IPP) and its limits, and propose the basis of an Integrated Product Strategy (IPS), which in our opinion has by far the greatest potential for transformation.

1 Integrated Product Policy

1.1 Overview

At the turn of the century the Netherlands, Denmark, Sweden, Finland and Austria were developing an integrated approach regarding products, and were pressing the European Union to adopt it. Although the European Council formally adopted this approach in its *Sustainable Development Strategy* (2001) (the "IPP is an integral part of the EU Sustainable Development Strategy") (COM(2003)302 final), and the Commission approved a Green Paper on the subject that same year, the EU was not able to produce a White Paper, previously agreed to be issued in 2002. Instead of it, the Commission approved in 2003 the Communication *Integrated Product Policy* (COM(2003) 302 final).

The aforementioned countries had gone through three previous phases before reaching the integrated product policy: (1) "policies for managing a growing number of product-related wastes"; (2) "policies which aim to generate and make accessible environmental information about product-systems"; (3) "policies specifically aimed at stimulating innovation and market creation for green products". The IPP sought to integrate the three previous policies (EC-DGXI 1998: 5).

R. Bermejo, *Handbook for a Sustainable Economy*,
DOI 10.1007/978-94-017-8981-3_18,

The *Green Paper on Integrated Product Policy* defines it as "an approach which seeks to **reduce the life cycle environmental impacts of products** from the mining of raw materials to production, distribution, use, and waste management". The Commission considers that the "IPP seeks to minimise the environmental impacts of products by looking at all phases of their life-cycles and taking action where those impacts can be reduced best and most cost-effectively" (SEC(2009) 1707 final).

But the IPP is not a tool to seek "concrete actions for specific products but rather to provide a conceptual framework, guidance for a large variety of policies and actors" (COM(2009) 693 final). Due to the great variety of products and impacts there is not a single policy to cope with them. For each product a combination of appropriate instruments has to be selected (legislation on product design, emissions ceilings, fiscal measures, etc.). In order to get a better result from these instruments, "a large variety of stakeholders needs to be involved, including policy makers, business and consumers" (COM(2009)693 final).

On the whole, the IPP has the merit of being a systemic approach, which is needed to cope with the problems created by the production system. But we will see that the application of this approach presents many shortcomings.

The Commission states that the EU is introducing the IPP principles into its initiatives: "The IPP principles are embedded in many initiatives" (COM(2009)693 final). The most important among many actions are: The *Action Plan on Sustainable Consumption and Production/Sustainable Industrial Policy* which "combines IPP instruments into a coherent policy, packaged towards *greener* products and smarter consumption"; the *Eco-design for energy-using products Directive* which "was a direct transportation of IPP principles into product design legislation"; the new *Framework Directive* "makes several IPP elements legally binding"; the *Thematic Strategies on Sustainable Use of Natural Resources* and on the *Prevention and Recycling of Waste* "have endorsed the life-cycle thinking and the call for continuous improvement" (COM(2009)693 final). Also, the Commission organises bi-annual IPP Regular Meetings with Member State representatives and key stakeholders to report and inform Member States and stakeholders on IPP-related activities carried out by the European Commission (SEC(2009)1707 final).

In October 2003, the European Council supported the IPP principles in its "Conclusions", but the Parliament regretted that "it provides only limited guidance on how to move society in the direction of truly sustainable systems of product development and design" (Resolution (2004) 0349. Besides, many shortcomings are hampering the potential of these policies: "Most product legislation addresses only specific aspects of a product's life-cycle" (COM(2008)397 final). For example, the Eco-design Directive is focused on providing eco-design requirements for energy-using products.

Sometimes in EU literature important initiatives are proposed, like differentiated taxation or the company's obligation of providing relevant information about the life-cycle of their products. However, most of these proposals have been rejected. For example, the Green Paper proposes that a cornerstone of the IPP "is *differentiated taxation* according to the environmental performance of products", and a first step could be "to apply *reduced VAT rates on products carrying the European eco-label"* (COM(2001) 68 final). However, the following Communication on IPP states that, due to the opposition of Member States, "the Commission will not apply reduced rates to products bearing the EU eco-label for the time being" (COM(2003)302 final). The Action Plan informs us that "the Commission is examining *inter alia* options for revising the energy taxation framework" (COM(2008)397 final). However, "the EU's role in the promotion of fiscal measures for sustainability has been rather small to date" (SEC(2009)1707 final).

Also there has been no significant improvement in informative transparency. The Green Paper declares that "a possible instrument to increase the generation and availability of information is to **oblige and/or encourage producers** to *supply key data along the product chain and consumers*" (COM(2001) 68 final). But there have not been any important improvements regarding this issue, as we will see in the next chapter.

The IPP is developing within the current economic framework, and this system generates tendencies which run against sustainability. And although it is gradually improving the environmental quality of products made in the EU, the overall effect of many structural factors means that the negative trends outweigh the positive ones: The quantity and variety of products placed in markets are growing rapidly; product chains are becoming longer and more complex; a growing number of products sold in the EU are imported from Non-OECD countries; etc. The Commission acknowledges that "the situation is worsening due to the ever increasing number of products being consumed in the EU and globally" (COM(2009)693 final).

Due to the analysed shortcomings the IPP concept is fading in the EU. Most Member States include IPP principles in their national Sustainable Consumption and Production Strategies. Only some Member States (Finland, Sweden, Belgium, and Poland) have specific national strategies for IPP and only two of them (Poland and Belgium) specifically use the term IPP (SEC(2009)1707 final).

1.2 Contribution to the IPP of Extended Producer Responsibility

Extended producer responsibility (EPR) implies, in a broad sense, that the responsibilities which were assigned to authorities (like waste management), are shifted to the producers. The Waste Framework Directive (2008/98/EC) gives legal status to this concept. Article 8 states that "Member States may take legislative or non-legislative measures to ensure that any natural or legal person who professionally develops, manufactures, processes, treats, sells or imports products (. . .) has extended producer responsibility". Among such measures the same Article includes "an acceptance of returned products and of the waste that remains after those products have been used, as well the subsequent management of the waste and financial responsibility for such activities". In fact, this vision means only a take-back system at the end of a product's life-cycle. The OECD gives a broader perspective, but goes beyond today's reality. EPR "is the concept that manufacturers and importers of products should bear a significant degree of responsibility for the environmental impacts of their products throughout the product life-cycle, including upstream impacts inherent in the selection of materials for the products, impacts from the manufacturers' production process itself, and downstream impacts from the use and disposal of the products" (1998b: 8). However, in most cases, the extension of the manufacturer's responsibility means shifting part, or all of the responsibility for the end-of-life management of products from tax payers, waste management authorities and conventional waste dealers, to manufacturers.

Over the 20 years since the EPR concept was first conceived in Sweden, and the famous German packaging take-back law was passed, the EPR concept has become an established instrument of environmental policy in many countries. Although EPR means different things to different people, as the level of responsibility varies greatly: "A core characteristic of any EPR policy is that it places some responsibility for a product's end-of-life environmental impacts on the original producer and seller of that product" (OECD 2006: 7).

There are three types of responsibility: economic, physical and informative. Economic responsibility means that "the producer will cover all or part of the expenses, for example, for collection, recycling or final disposal of the products he is manufacturing". Physical responsibility means that the manufacturer is involved in the physical management of "the product and/or their effects". In some cases the producer retains the ownership of their goods throughout their whole life-cycle, but this is a private option, as we will see in the case of Product Service Systems. Informative responsibility means that the manufacturer has to supply "information on the environmental properties of the products" (Lindhqvist 2000: 37, 38), and also information about the environmental impacts of its manufacturing processes has to be added.

1.2.1 Physical and Economic Responsibility

The EU and other governments imposed by law, on the producers of several types of goods (cars, electric and electronic devices, batteries, etc.), that they take back their products at the end of their life in order to treat them, assuming the costs. The treatment brings about a growing rate of recycling in a certain period. They are also obliged to eliminate hazardous substances from new products. These policies mean different levels of physical and economical responsibility.

Over the 20 years since the EPR concept was first conceived in Sweden, and since the famous German packaging take-back law was passed, the EPR concept has become an established instrument of environmental policy in many countries. Although EPR means different things to different people, as the level of responsibility varies greatly: "A core characteristic of any EPR policy is that it places some responsibility for a product's end-of-life environmental impacts on the original producer and seller of that product" (OECD 2006: 7).

For the International Institute for Industrial Environmental Economics (IIIEE) an effective EPR implementation must have two main environmental goals: "(1) *Design improvements of products* – the EPR system should provide incentives for manufacturers to improve products and systems surrounding the life cycle of products (2). *High use of product and material quality through effective collection and re-use or recycling*" (van Rossem et al. 2006: v).

The German government adopted the Ordinance on the Avoidance of Packaging Waste in 1993, which set the requirement that producers had to take back the packages of their products to be recycled. Companies created an enterprise to put into practice the requirements, and they had to pay for this service amounts of money proportional to the volume of waste delivered. The recycling rate grew from 12 % in 1992 to 86 % in 1997. The plastic recycling rate was multiplied by 19. Packaging became lighter: weight was reduced by 17 % during the period from 1991 to 1997 (Gardner and Sampat 1999: 54; Lindhqvist 2000: 44).

In the case of the EU, the following Directives are among the most important EPR initiatives: On Packaging and Packaging Waste (94/62/EC); on Waste Electrical and Electronic Equipment (WEEE) (2002/96/EC); on End-of-Life Vehicles (2000/53/EC). These Directives oblige companies to recycle a growing proportion of cars and electric and electronic devices and to eliminate the use of hazardous substances, such as several fire-retardants, in new products. Despite having positive results, many obstacles remain, especially in the case of electric and electronic devices. The WEEE Directive has achieved a poor outcome. Besides, there are widespread illegal exports of both groups of goods. At the end of 2008 the Commission decided to revise the WEEE Directive (COM(2008)397 final).

Other policies, besides the end of life product take-back, may be considered as EPR, as they oblige manufacturers to improve the energy efficiency of products or forbid the production of inefficient goods. The first field includes, for example, Directive 2009/125/EC establishing a framework for setting Eco-design requirements for energy-related products; the Directive 2009/33/EC on the promotion of clean and energy efficient road transport vehicles, Directive 2006/32/EC on energy end-use efficiency and energy services; or Directive 2002/91/EC on energy efficiency of buildings. In the second field, we have prohibition of the incandescent lamp or of mercury (SEC(2009)1707 final).

On the other hand, there is a debate about the EPR's repercussion on the design change of products. The OECD states that "it also appears that a limited form of design for the environment has taken place in many instances – reductions in material use and product/packaging downsizing – in response to policy" (OECD 2006: 35). But a IIIEE report reaches the conclusion (after analysing the repercussions of the EPR policy in industrial sectors of Sweden and Japan) that the "EPR law has been central for specific design changes for the products investigated (…) Upstream measures in design, both in terms of reduction of hazardous substances and enhancement of source reduction of material use, re-use and recycling, have been undertaken in both industry sectors in Sweden and Japan respectively" (2006: 52).

Additionally, extending producer responsibility to other life cycle phases in addition to the end-of-life stage may encourage new business approaches where instead of selling products, they hire them with associated services.

1.2.2 Informative Responsibility

It is necessary to make it clear that EPR and so-called Corporate Social Responsibility (CSR) are different concepts. CSR means that corporations assume social and environmental responsibilities, and that they are publically transparent about their behaviour. But these actions are voluntary. In 2001 the Commission presented the Green Paper *Promoting a European Framework for Corporate Social Responsibility* (COM(2001) 366). In the subsequent consultation process about the GP "enterprises stressed the voluntary nature of CSR", and the Commission's definition of CSR makes this aspect clear: "CSR is a concept whereby companies integrate social and environmental concerns in their business operations and in their interaction with their stakeholders on a voluntary basis". However, consumer organisations "underlined the importance of trustworthy and complete information about the ethical, social and environmental conditions in which goods and services are produced and traded" (COM(2002) 347 final). On the contrary, the informative responsibility of businesses in the context of EPR means that it is imposed by the authorities.

However, this type of broad responsibility is not frequent. The USA is a partial exception. It is the leading country in community right-to-know legislation due to the Emergency Planning and Community Right to Know Act (1986). Its main tool is the Toxic Release Inventory (TRI), which "gave the public information about what some factories in some industrial sectors have emitted to air, land, underground water supplies, rivers, publicly owned treatment plants, disposal sites such as incinerators, and even external recycling plants" (Thorpe 1999: 15).

Based on TRI data many communities have been empowered by the TRI to set up dialogues with their neighbouring industries, and have achieved emissions reductions, the development of pollution prevention plans, and an improvement of safety measures.

Governments also use the data to devise aspects of their fiscal policy. For example, the state of Massachusetts financed its toxic substance use reduction programme with company fees based on the number of chemicals that they used.

Then in 1997, after 8 years of applying the law, the state obtained the following improvements:

- Companies generated 41 % less toxic waste.
- Companies reduced their use of toxic chemicals by 24 %.
- Companies achieved an 80 % reduction in emissions released to the environment.
- Net savings to industries and the state amounted to $15 million without considering the environmental or public health benefits (Thorpe 1999: 15, 16).

On the contrary, in Europe information responsibility only appears as a *by product* of the application of the EPR concept, above all in the case of *take-back* systems: "A manufacturer may also be required to *label* the material composition of components and to provide information to recyclers regarding the content and structure of their products. Recyclers must meet certain treatment standards" (van Rossen et al. 2006: 4). But some Directives have established information requirements for enterprises. For example, the Reach Directive is applicable to chemicals, and the Commission requires information about the existence of GMO in food.

2 Inputs for an Integrated and Eco-effective Product Strategy

2.1 Overview

What we define as an integrated product strategy (IPS) comprises elements of the IPP, in a deeper and more integrated way, and other new elements which are needed in a strategy for sustainability. In the first place, the basic approach has to change: the concept of eco-efficiency must be replaced with that of eco-effectiveness. As we have explained the concept of eco-efficiency is incremental and has a narrow range of application, such as reducing waste or recycling it to be used in less important forms. On the other hand, sustainability requires waste to be recycled without losing quality as a resource.

However, the concept of eco-effectiveness has a narrow scope because it is focused on the activity of an enterprise during the life span of its products. The scope must be broadened, to make it an element to be implemented with others: the creation of an information system which provides essential product and life-span data; reinforcement of extended producer responsibility; implementation of strong fiscal burdens on resource consumption; subsidising enterprises which are radically improving the ecological quality of their products, etc. These policies can only be implemented by governments.

Besides, selective policies are needed, especially for products and services which are responsible for most environmental impacts, for the producer groups most inclined to change, and so on. Governments should also be leaders in the adoption of SPS in their purchasing policies. Many reports like the EIPRO ("Environmental Impacts of Products"), by the European Commission, have reached the conclusion that private consumption "is dominated by housing, transport and food and drink" (EEA 2010: 9). This means that not only an improvement in products and services is required, but that also the consumption model should be changed (diets with less animal protein, collective transportation instead of the intensive use of cars, collective housing instead of individual housing, etc.). Due to the implementation of advanced policies in seven European countries, the EU's green public procurement amounts to 45 % of the total value of public procurement. The Dutch government seeks to reach 100 % by 2010 (COM(2008)693 final). However, these data must be looked at from an adequate perspective, because the EU eco-labelling system has to be improved considerably.

Although these measures can be considered utopian, we must remember that the scarcity of very significant resources will lead to the implementation of very radical measures in the short to mid term, in order to abruptly reduce the level of resource consumption. In the context of growing scarcity eco-effective firms will be reinforced, because they will be able to offer cheaper products and also have a greater capability for improvements, due to the acquired knowledge and expertise.

2.2 *Product-Service Systems*

Physical producer responsibility is greater than in the case of the EPR experiences, at least in the most advanced experiences. And in order to advance towards sustainability these experiences have to be generalised. Shortcomings in the way to reach this goal are: the experiences are voluntary, and there is a lack of adequate governmental support for the development of these systems.

At the end of the 1990s some Nordic governments (and especially the Dutch government) began to promote Product-Service Systems (PSS). In 1998 this government issued the *Dutch Policy Document on Environment and Economy,* which offered favourable prospects for sustainable development, and described a policy plan to evaluate the economic and environmental potential of these systems. Later three consultant enterprises were commissioned by the government to analyse this potential. In March 1999 they presented the report *Product Service Systems, Ecological and Economic Basics*, which reaches the following general conclusion: “We consider PS systems to be of interest for business, policy makers as well as Non-Government Organizations” (Goedkoop et al. 1999: 12: 101).

From this starting point, a growing number of experiences shows that Product Service Systems (PSS) have a great potential to improve the sustainability of the production system, constituting an important element of an integrated product strategy. But, as usually happens, the contribution to sustainability of many experiences is frequently exaggerated.

2.2.1 Concepts

Ten years ago this field was frequently defined as “new product uses”. However, currently there is broad agreement on the term product service system (PSS), although some authors use terms like servicing, eco-efficient services or remanufacturing (Cook et al. 2006: 6). The US Environmental Protection Agency (EPA) uses the terms “servicizing” and “green servicizing” (in this case to emphasise the environmental dimension), although it acknowledges that “the PSS concept encompasses

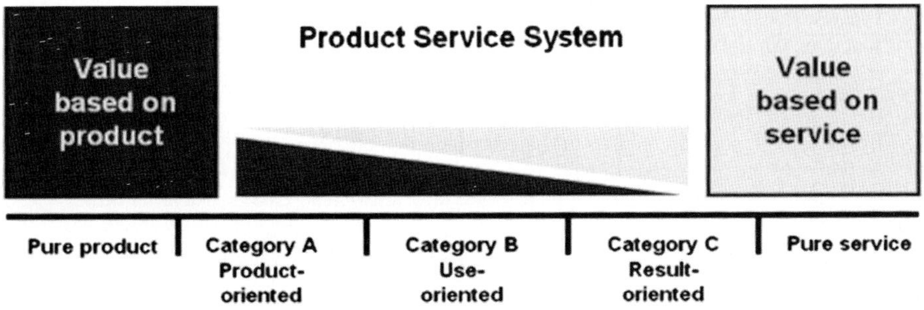

Fig. 18.1 Product service system (Source: Sundin et al. 2009: 33)

a broader range of economic activity", and that it is "by far the dominant term of art" (EPA 2009: 6). There are many definitions of PSS. In some of them only the dimension of the producer-client interface appears, but in others economic and environmental aspects appear. A Commission report (2008: 9) defines this concept in summarised form: "Tangible products and intangible services designed and combined so that they are jointly capable of fulfilling specific customer needs". This definition is taken from A. Tucker and U. Tischner (2006). U. Tischner et al. (2010: 95) defines it as "a system of products and services (and related infrastructure) which are jointly capable of fulfilling client needs or demands more efficiency and with higher value for both companies and customers than purely product solutions". The definition is more complex and adds in the factor of efficiency (but without an explanation), though in essence it represents the same approach. I. Omann (2007: 3) places the focus on environmental impacts: "PSS are then eco-services resulting in less negative impacts on the environment through partial or full substitution of tangible material components". It does not mean that previous authors do not take into account the environmental dimension. In most cases this dimension appears in the subsequent explanations, but it is a shortcoming of the definitions. For example, U. Tischner states after the aforementioned definition that a PSS could "decouple the creation of value from consumption of materials and energy and thus significantly reduce the environmental load of current product systems" (2010: 96).

2.2.2 Typology and Transformational Potential

There is a broad consensus on three types of PSS. But practice shows multiple combinations of products and services, and sometimes it is difficult to establish limits between them, as the picture below reflects (Fig. 18.1):

> Product-orientated PSS (PO-PSS): "The ownership rights of a material artefact are transferred to the customer and a service arrangement is provided to ensure the utility of the artefact over a given period of time". A typical example is a maintenance contract (Cook et al. 2006: 4).

Use-orientated PSS (UO-PSS): "The product is owned by the service provider, who sells functions instead of products by means of modified distribution and payment systems; for example, sharing, pooling and leasing" (Sundin 2010: 33).

Result-orientated PSS (RO-PSS): "While ownership rights of material artefacts are retained by the service provider, similar to use-orientated PSS, the consumer purchases a utility as an outcome and not the use of a *product* over a given period of time". For example, instead of leasing a washing machine, the customer purchases clean clothes delivered through a washing service (Cook et al. 2006: 4).

Resource productivity is the highest in category C and the lowest in A. In theory, PSS provides opportunities to improve resource productivity by a factor from 4 to 20 (Cook et al. 2006: 4). But Tukker and Tischner (2006: 4) consider that the first and second types only have the potential to reach a factor of 2, and the third type has a potential factor of 10. However, we will see later that other authors attribute higher factors to the first category. It is due to the fact that when the producer retains ownership "there is a financial incentive to direct innovatory activities to produce more durable goods and the producer has responsibility for disposal" (Cook et al. 2006: 5). And it represents "the more popular interpretation of the features of a PSS" (Baines et al. 2007: 1547). A European Commission report states that the application of PSS in some cases "has led to 20 % reductions in waste and 65 % increases in recycling – with financial gains" and it estimates that the most promising models "have markets worth €5bn or more per year" (European Commission 2008c: 3).

In relation to PSS development there is a broad agreement on these points: its application is growing; the level of development is limited; the potential is very big. But there is a lack of precise information about what trends there are, although in the next section, where we will analyse the three models, we can add further information. Visnjic and Van Looy (2009) state that "manufacturing firms diversifying their offerings to include services has been a rising phenomenon over the last decades". Colen and Lambrecht (2010: 8) reach the conclusion that "in certain industries more and more contracts are performance based and less based on the actual use of labour and material". But, on the whole, it is "clear that the market for PSS is developing slowly and faces several barriers" (European Commission 2008a, b, c, d, e, f, g, h, i, j: 47).

This report identifies three main barriers which justify this dynamic: "Uncertain rewards; lack of knowledge of the potential benefits"; and "new infrastructure requiring investment". So it is necessary to set up a public policy which should: "Examine the areas of potential"; audit existing regulations (. . .) to remove legal barriers to PSS"; and "create economic policy that supports growth of these models – internalising environmental externalities" (European Commission 2008a, b, c, d, e, f, g, h, i, j: 47).

These recommendations are insufficient and they have to be completed with others, as we have seen in previous chapters. They also lack the necessary integration. Reviewing the proposed initiatives put forward in this text, we find the following: fiscal policies which charge the use of resources and polluting emissions; public requirements which mean new responsibilities for producers; extension of the implementation of more demanding EPR systems; subsidies for their development;

information, education and technical advice services (the Swedish Environmental Protection Agency offers these services), especially for small and medium-sized companies; the creation of institutes for the improvement of these policies, and of promotion programmes (the Austrian government has developed the *Factory of Tomorrow* programme); information and promotion campaigns directed to the population; adoption of these systems by public administrations; etc. (www.serviceinnovation.at).

2.2.3 Product Oriented PSS (PO-PSS). Remanufacturing

Although the most frequent application of PO-PSS is adding services to an existing product (Sundin et al. 2009: 41 calls it "service integration"), in our opinion the most interesting application is in the case of upgrading products until they reach an *as-new* status. There are many terms which are used as if they were synonymous with remanufacturing (although they are not the same): refurbishment, reconditioning, repair, rebuild, etc. Besides, other terms are used in specific activities, like retread (remoulding of tyres) or rewound (rewinding rotating electric equipment). Also, some authors and organisms use the term "remanufacturing" as synonymous with SPP. Last of all, in some countries there are terms used which have the same meaning; Inverse manufacture (Japan); Product recycling (Germany); Renovation (France). This situation makes it difficult to know "the extent to which remanufacture is practised" (Gray and Charter 2007: 8, 9).

However, many authors and organisms give definitions which contain the same central meaning. Gray and Chapter (2007: 7) give a short and precise definition: "Remanufacturing is a process of recapturing the value added to the material when a product was first manufactured". The EPA definition is cited frequently: Remanufacturing is "the standard term for the process of restoring used, durable products to a like new condition". But for the European Commission remanufacturing is synonymous with reuse: "Remanufacturing is a strategy to bring used products back to the market" (European Commission 2008a, b, c, d, e, f, g, h, i, j: 21). So in these definitions ownership is not taken into account.

There are two kinds of remanufacturing and component remanufacturing. Reports about remanufacturing show very high ratios (factors) of natural resources saved. Sundin (2010: 42) considers that ratios of energy used in remanufacturing "are in order of 4:1 and 5:1". But Calleja et al. (1999: 60) reach the conclusion that, in the case of the car remanufacturing industry, the energy saved is 84–91 %, and 88 % in the case of materials. The EPA reaches similar conclusions: "A limited survey estimated that the entire remanufacturing industry recoups between 85 % and 95 % of the materials and energy" (2009: 64). There have been two main drivers for remanufacturing: cost saving and compliance with environmental laws (EPA 2009: 65).

In the USA remanufacturing was historically dominated by heavy and/or durable machinery and machined goods such as automotive parts, construction equipment, refrigeration components and textile machinery. But lately other products have appeared: consumer electronics, office furniture, carpets, etc. The US Department of Defence is the largest single remanufacturer (EPA 2009: 64, 65). The European Union states that the most successful sectors are automotive parts, computers, laser toner cartridges; photocopiers, cleaning equipment and single-use cameras (2008: 86). But there are other successful sectors: aerospace, trains, marine engines, etc. On the other hand, in some of the aforementioned sectors the company policies are situated in the field of RO-PSS, as we will see. A 1996 survey estimated that there were 73,000 firms in the USA selling approximately $53 billion worth of goods. ASML is a leader in providing microlithography systems for the semiconductor industry. It sells them, but at the end of their life-cycle ASML takes them back to be remanufactured (EPA 2009: 64).

Lufthansa Technik is dedicated to aircraft maintenance and remanufacturing. The service is given to Lufthansa aircraft and to other companies, and it has become the world's leading company, with a market quota of 11 %. The Swiss company SR Technics has a quota of 5 %. General Electric Medical Systems remanufactures not only its products but also the products of competing companies. General Motors, Caterpillar and Volkswagen remanufacture their own engines and sell them in the secondary market (Guide and van Wassenhove 2002: 487). Dell and IBM remanufacture computers and sell them to customers who do not require the latest generation performance. In the USA many third party remanufacturers sell remanufactured phones to less developed markets (EPA 2009: 66).

2.2.4 Use Oriented PSS (UO-PSS)

Personal Use

The most frequent experiences in this field are renting and leasing. Renting usually means a short-term hiring of goods. From the point of view of sustainability renting has positive effects, because it reduces the amount of purchases and the existing goods are used intensively. Leasing is a mid to long-term hiring system and it facilitates buying goods, because the accumulated payments are deduced from the good's price in the event (very frequent) the user decides to buy it. In Germany, more than half of the total leasing transactions are vehicle leasing (Behrendt et al. 2004: 30). This system has a negative effect on sustainability, as it facilitates purchases. Besides, it reduces the useful lifespan of goods, because within the renting system a continuous offering of new models to clients is also included.

Collective Use

In this field we briefly describe two kinds of experiences: car-sharing and cohousing.

Car-sharing

There are the systems which provide car services to non-owners: carpooling, car rental and car-sharing. In the first case various car owners reach an agreement by which they use a single car each time, but the service is provided by all at agreed times. In the car rental system enterprises sell short-term services provided by a large variety of vehicles. There is not a broad consensus on a car-sharing definition. For M. Gossen and G. Scholl (2011) car-sharing is defined "as the organized collective use of passenger cars". This definition integrates carpooling. S. A. Shaheen et al. (2003) define it as access to "a fleet of shared-use vehicles on an as-need basis". The consultant Frost & Sullivan (2010) gives a broader definition of car-sharing: It is "a mode of transport where vehicles are owned by a separate firm or an organisation and shared between a number of different people at different times". This definition means that the users do not own the enterprise or organisation which provides the service. But in the case of cooperatives the members own the enterprise. It is the case of Mobility, the biggest car-sharing organisation in Europe. So there is not a car-sharing definition able to describe the various experiences usually termed as car-sharing.

Members of car-sharing organisations pay for short-term access to a shared-vehicle fleet, and pay monthly on a per-hour or per-mile/kilometre basis. Vehicles are usually distributed throughout cities. The system can be cost-effective for people who can use other alternative modes for daily mobility. The system reduces car-ownership and energy use: "Previous research has shown that people joining car-sharing organizations reduce their personal driving and vehicle ownership, which translates into reduced energy use" (Matin and Shaheen 2011: 2095). In Europe, sharing cars register up to 15–20 % lower CO2 emissions than private cars because the former are smaller (Loose 2011). According to several studies, a car-sharing vehicle reduces the need for 4–10 private cars in Europe, and between 6 and 23 cars in North America (Shaheen and Cohen 2007: 3).

Although after World War II several car-sharing experiences appeared in Europe, successful organisations began in Switzerland and Germany in 1987 and 1988. A decade later, car-sharing began to be developed in North America. Across Europe there were almost half a million car-sharing members in 2009, and a Frost & Sullivan report predicts 1.5–5.5 million by 2015 (EEA 2011: 47). Switzerland has developed the world's most successful car-sharing experience. Its participation rate is over 1 % of its population, while in other successful countries such as Germany, the Netherlands, and Sweden the rate is one seventh of the level of Switzerland (Loose 2011). In North America car sharing had grown to nearly 640,000 members in 2011 (Martin and Shaneen 2011: 2095). But in Europe car-sharing organisations and cooperatives predominate, while in North America private companies are dominant. The Swiss Mobility Cooperative had more than 100,000 customers by the end of 2011 (www.mobility.ch). The American company Zipcar had 72 % of the market share in 2011. It operated in the United States, Canada, and the United Kingdom, and had about 650,000 members (Boston University Student Research 2011).

Costumers obtain positive outcomes through the implementation of PSS: they receive more services, which are personalised; the services are flexible, better adapted to their changing needs; the chosen product is adapted to their needs; they are free from the inherent responsibilities of ownership; they do not have to make a large payment and in a short period of time. On the whole, "customers prefer to form long-term relationships with a small group of reliable and reputed suppliers in order to handle information asymmetries (. . .) Focusing on a limited number of suppliers leads over time to more efficient transactions and lowers the overall transactional costs" (Visnjic and Van Looy 2009: 10). However, the scientific literature presents "few real-life examples" (Baines et al. 2007: 1548).

However, these positive factors frequently also mean hurdles: changes in the organisational principles, structures and processes (there are big differences in managing product and service-related activities); there is a need for organising services separately from products; sales networks have to change and be decentralised; employees need new skills; the responsibility for taking care of the products throughout their lifetime runs counter to the traditional culture; the same applies to the intensity of the relationships with clients; changes carry costs. These hurdles can be overcome through a gradual implementation of the new system. For example: multinational industrial equipment has been developing its "service portfolio over the last years and has gradually embraced more intensive models" (Visnjic and Van Looy 2009: 16, 17).

Many authors have too narrow a vision: "The PSS concept tries to solve sustainability problems (almost) entirely by changes in a business-client interaction along the value change" (Tukker and Tischner 2006: 7). Accepting the contribution of PSS-RO to sustainability, it should be emphasised that there are many structural problems, which cannot be solved at the market level. Nor can States do it in many cases. Thus, to begin solving the problem international cooperation between all stakeholders is needed.

Manufacturing Enterprises

Frequently it is difficult to know which firms are applying the PSS-RO concept, because many authors and official reports account as cases of manufacturing experiences of the three types indicated. Besides, the term remanufacturing is used to define the broad field of take-back artefacts in order to be renewed, without taking into account their ownership. So we will cite the companies which maintain the ownership of their products.

Xerox is the best-known case, and perhaps the most significant one. The company rents photocopiers, printers, scanners, and besides provides several services to customers. Xerox takes the machines back after five years or more, depending on the intensity of their use. This policy was established in the early 1990s. As the company maintains the product ownership, it is interested in manufacturing durable, easy to repair, and recyclable products. It has designed photocopiers with a modular structure, formed by seven parts, which are easy to remove and assemble. A study by the Xerox Australia division shows that in the case of modular photocopiers they have saved materials by a factor of 1.9 and energy by 3.1. In 2002, 10 years after the start of the policy, the company prepared a report detailing the results obtained. Savings in costs amounted to 2 billion pounds sterling, and waste generation was reduced by 64 million tonnes. In 2003, 95 % of the waste produced was recycled (175,000 tonnes). Remanufactured output accounts for 25 % of the company's total output. It is typical in this kind of companies that the remanufacturing task is outsourced (Gray and Charter 2007: 36, 37). Canon and Océ and have also developed a "pay per copy" system. Siemens and Electrolux have put into practice PSS-RO programmes for home appliances (Baines et al. 2007: 1547–8).

In the aerospace industry some companies have found it profitable to develop PSS-RO models, especially in the case of engines. Airlines, instead of buying engines and a service package, rent them and pay a fee for every hour an engine runs, what is often referred to as "power-by-the-hour": "The manufacturer must provide ongoing real-time monitoring and through-life service of the engine in order to deliver the availability of power" (Johnstone et al. 2008: 525). JetCo and Rolls-Royce are the leading companies offering these services. In the Rolls-Royce TotalCare programme, company engineers monitor the performance of 3,500 engines flying at any given moment. And this programme has become the main source of its revenue: "Spare parts and long-term servicing operations have overtaken new sales to account for 63 % of the engines division total revenues" (www.designcouncil.org.uk).

Interface, DuPont, Collins & Aikman, etc., have been implementing a carpet PSS-RO. Carpet modules are placed and changed in the areas which suffer more wear, such as corridors. Customers pay a fee for the use of the carpet. However, the case of Interface is a special one. It has devised seven policies (zero waste, renewable energy, closed-loop recycling, etc.) in order to "bring us to the summit of Mount Sustainability and our goal of becoming the prototypical company of the 21st century", in a process lasting 25 years until the year 2020 (Anderson 2003: 22–27). By 2010 Interface had reached about 60 % of their objectives, and is a business success: "Interface has generated substantial business value in its brand and reputation, saving costs of $405 million" (Anderson et al. 2010: 97). But its mission goes beyond being a business success; it is to build a community of employees, suppliers, costumers and communities: "It will be strongly connected to its employees and constituencies – with engaged communities, engaged customers, and engaged suppliers who have bought into the vision. What we believe will emerge is our own *ecosystem* of connected constituencies, with cooperation replacing confrontation" (Anderson et al. 2010: 27). IBM, Siemens-Nixdorf, Sony, Hewlett-Packard, Océ, Canon, Parkersell, Herman Miller and Steelcase (the two leading furniture manufacturers in the USA), etc., have developed PSS-RO programmes in their respective sectors (Baines et al. 2007: 1547).

Chemical Product Services

Many Chemical companies (Olin, Castrol, Ashland, Henkel, Dow, DuPont, BASF, etc.) are departing from their traditional business model (products are sold in combination with some basic services) towards new models characterised by offering more services. But the cases that we analyse here are only the PSS-RO ones. Wikipedia defines it as "a business model in which the chemical company supplies a substance for a specific service (for example dyeing or powder coating), but retains the ownership of the chemical" (http://en.wikipedia.org/wiki/Chemical-leasing). Only a few types of chemicals are involved in this practice. A report by the European Joint Centre reaches the conclusion that the most frequent chemicals are paints and cleansing and degreasing solvent applications. Adhesives, lubricants, tanners and water treatment chemicals have also entered this market, "but apparently to a lesser extent" (Kortman et al. 2006: 8). The overall reduction of environmental impacts varies broadly from case to case. In car body painting this impact varies from 15 to 25 %. In the case of metal cleaning the impact varies from 5 to 70 %. These variances are due to the application of PSS-RO and the cases with consumer ownership. Undoubtedly the best reduction is obtained by the PSS-RO programmes. Large corporations, like GM, outsource the chemical management to specialised enterprises (Kortman et al. 2006: 9). The main driver of these companies' change of culture is public regulation: "Numerous examples from the chemical sector show that regulations were the primary source of this favourable development" (Mont and Lindhqvist 2003: 7).

Volkswagen, Daimler, Volvo, Ford, Toyota and GM had begun adopting the CMS policy contracting the paint, but are now moving to contracting all chemicals. GM reports a 30 % reduction in its chemicals, and has now implemented CMS in nearly 90 % of its manufacturing facilities (Yachnin and Associates 2010).

In some cases, products have been eliminated because they are not needed or are hazardous, and because they can be replaced by other products which are innocuous or less harmful. A decade ago, in the USA the penetration by sectors of so-called Chemical Management Services (CMS) was: Automotive (75–80 %); automotive suppliers (30–40 %); heavy equipment (15–25 %); aerospace manufacturing (25–30 %); air transport maintenance (40–50 %); electronics (30–40 %); steel manufactures (20–30) (EPA 2009: 41). Between 2000 and 2004 CMS application had grown from 4 to 11 sectors. Europe lags behind the USA, but CMS is to a large extent applied within the automotive and aerospace industries. In the electronics sector Intel and Motorola lead the way in the application of CMS. In the EU this kind of business amounts to 14 % of all sales in the chemical sector. In painting the system is used by 40–70 % of the automotive industry, and in the aerospace industry the rate is 30–45 %. The reduction of consumption varies between 5 and 30 %, due to the product types. UNIDO (United Nations International Development Organisation) and the International Working Group on Chemical Leasing are collecting data on the subject (European Commission 2008a, b, c, d, e, f, g, h, i, j: 68, 69).

2.3 Design of EC-Efficient Products by Means of the "Intelligent Material Pooling" Methodology

After analysing the concept of PSS, and especially that of PSS-RO, the "cradle to cradle" principle, designed by McDonough and Braungart (2002) constitutes a further development of PSS-RO. They have designed a framework for the full development of the cradle-to-cradle principle, which needs the development of PSS-RO experiences, thus enabling enterprises to build a circular economy. The framework defines various steps to reach this objective: to create an upstream infrastructure which permits suppliers to deliver adequate materials and substances, and to develop a downstream infrastructure to transform end-of-life products into (technical) nutrients.

However, most companies are not capable of knowing the quality of the materials and substances delivered by their suppliers, nor do they have control over the treatment of their products at the end of their life-cycle. So it is difficult or impossible to develop a cradle-to-cradle system. To be successful enterprises must create an *Intelligent materials pooling* (IMP), which "is a framework for the collaboration of economic actors within the technical metabolism which allows companies to pool material resources, specialized knowledge and purchasing power relating to the acquisition, transformation and sale of technical nutrients and their associated products". However, in order to reach the full potential of the IMP it is necessary to create a "*materials bank*". It leases chemicals and materials "to participating companies, who in turn transform them into products and provide them to consumers in the form of a service scheme. After a defined use period, the materials are recovered and returned to the materials bank" (Braungart et al. 2007: 1346). The materials bank is responsible for the reprocessing of substances and materials, and their storage to begin a new cycle by leasing them. These authors have developed a collaborative methodology (sharing a list of substances and materials) for defining which substances must be eliminated, and devising lists of preferred chemicals and materials.

The cradle-to-cradle concept has been applied by many enterprises and some organisations in the redesign of their products and services. Some of them are: Ford, Nike, Herman Millar, Rohner Textil, Honeywell, etc., Chicago City, and the Sustainable Development Center of China and the USA. But these companies have not created a *materials bank,* supposedly because it is a tool only adequate for small and medium sized enterprises (SME). But this is not the case; apart from Rohner Textil, the rest of the companies are so large that they have the power to impose their requirements on their suppliers (Braungart et al. 2003: 146).

Chapter 19
Sustainable Consumption

Keywords Consumerism causes • Motivation systems • Systems of provision • Systems of access • Systems of certification • Eco-labels • Green public procurement

In this chapter we study, on the one hand, the structural causes of the current high-level consumption model through the lens of motivation, provision and access systems, following the UNEP methodology. On the other hand, we describe the policies which must be adopted as part of a sustainable consumption strategy.

1 An Unsustainable, Unfair and Pathological Model of Consumption

All ancient religions and philosophies reject (with slight differences) the accumulation of wealth. Aristotle defined the concept of *pleonesia* as an "insatiable desire for more". This phenomenon was an exception in the ancient world, but in capitalism it has become a goal pursued by most people. Veblen denounced conspicuous consumption and Galbraith our effluent society (Jackson 2005: 20). Today it is broadly accepted that the current consumption model is not sustainable. The UN Agenda 21, Chap. 4 (1992) declares that "the major cause of the continued deterioration of the global environment is the unsustainable pattern of consumption and production" (UNCED: 1992: 2). The EEA (2010: 4) affirms: "The consumption of goods and services in EEA member countries is a major driver of global resources use – and associated environmental impacts".

In 2011, world population surpassed seven billion. OECD's model of consumption is a reference to imitate for the rest of the world, but it means that "we will need around five planets to provide the resources for these lifestyles by 2050"

R. Bermejo, *Handbook for a Sustainable Economy*,
DOI 10.1007/978-94-017-8981-3_19,

(Fredigo and Tukker 2009: 7). There is "a Global Consumer Class who shares certain elements of a common lifestyle (. . .) regardless of who they are and where they live" (UNEP 2002c: 21).

The development of economic globalisation determines that a growing percentage of environmental impacts linked to consumption in OECD countries come from imports. 65 % of air emissions from Norwegian consumption come from imports. 52 % of Slovakian CO_2 emissions come from production in other countries. EEA studies show that resources extracted in Europe are decreasing, while there are increasing imports of resources (most of them metals and fossil fuels, but also cultivated goods) (EEA 2005: 15). On average 40 % of total CO_2 emissions caused by consumption are embodied in goods imported from other member countries and from the rest of the world. This ratio rises to over 50 % in small countries, like Austria, Belgium, the Netherlands, Denmark and Sweden. It is estimated that half of the CO_2 associated to imports is coming from outside Europe. So "Europe's footprints continue to rise, while the bio-capacity per person, both in Europe and globally, is shrinking". And "due to increasing international trade the shares of non-domestic emissions in the carbon footprints of the EU Member States are likely to be on the increase" (EEA 2010: 10–12).

Due to the unsustainable consumption and production model (badly designed products multiply the impacts of consumption) the population is concerned about the future of humanity and the Earth. In the last two decades numerous studies and warnings from the scientific community have shown a growing concern. One of the latest warnings is the *Stockholm Memorandum*, signed by 20 Nobel Prize winners, which states: "Consumerism, inefficient resource use and inappropriate technologies are the primary drivers of humanity's growing impact on the planet (. . .) Science indicates that we are transgressing planetary boundaries that have kept civilization safe for the past 10,000 years. Evidence is growing that human pressures are starting to overwhelm the Earth's buffering capacity", and for this reason "humans are now the most significant driver of global change, propelling the planet into a new geological epoch, the Anthropocene" (Agre et al. 2011). So "reducing consumption is not an option, but is going to come anyway" (Lorek 2010).

2 Systems That Determine Consumption Patterns

Consumption is motivated by intrinsic and extrinsic factors. The former determine preferences and "comprise cognitive capacities, psychological factors, individual interests and philosophic or ethical norms, whereas the latter includes socioeconomic aspects like disposable income and time availability". They show possibilities and obstacles originated by the economic, social and legal context, which determines which preferences can be materialised (Spangenber and Lorek 2002: 136). The orthodox economy claims that preferences obey intrinsic factors. Consumers would be sovereign and manifest their preferences in the market, and companies would be forced to satisfy consumer demands. Summing up individual consumer preferences we would obtain the current model of consumption.

On the other hand, we find the theory that defends the premise that individual preferences are basically determined by extrinsic factors, like the dominant culture and the provision structures of goods and services. Several approaches have been proposed for study of the current pattern of consumption. The UNEP suggests an alternative approach, based on the SC literature: "In this model, the system is described from three perspectives: *provision* (the way that goods and services are produced and their systems of delivery and function); *motivation* (the incentives and disincentives which shape the market for goods and services) and *access* (factors which include or exclude consumers from participating in the market)" (2002c: 41).

2.1 Motivation Systems

The motivation system is determined by individual, social and cultural factors: "Actions and behaviours of people (. . .) can be attributed to motivations that arise from needs and desires (individual, social and cultural)" (UNEP 2002c: 42). The level of personal development determines to a large extent the objectives that each individual wants to reach in their life. It is obvious that everybody seeks happiness, but many people identify it with reaching materialistic objectives: "Financial success", "social recognition", and an "appealing appearance". Numerous studies, carried out all over the world, have shown, as Tim Kasser's book *The High Price of Materialism* manifests, that:

- The three aspirations form a unique entity: "We have found that people who value one of these values, such as fame, also tend to value money and image" (2002: 10).
- The quality of life of materialistic people is lower than that of the people who value less or not at all materialistic aspirations: "Adults who focused on money, image, and fame reported less self-actualization, and more depression than those less concerned with these values (. . .) they also reported significantly more experiences of physical unhealthy symptoms" (2002: 11).
- People who are focused on fulfilling materialistic aspirations are not able to have fulfilling social relations, which are indispensable for solving personality problems: Those adults "highly oriented towards materialistic goals reported fewer experiences of positive emotions and less overall satisfaction with their lives than did those with less materialistic goals" (2002: 13).
- On the contrary, "well-being and quality of life increase when these four sets of needs (safety, security, and sustenance; competence, efficacy, and self-esteem; connectedness; and autonomy and authenticity) are satisfied" (2002: 24, 25). (Figure 19.1 reflects these two types of evolution.).

Conclusions reached by psychology have been backed by sociology. Numerous sociological studies state that the levels of wealth and happiness are not correlated,

Fig. 19.1 Changes in well-being as a function of progress towards materialistic and non-materialistic goals (Source: Kasser 2002: 47)

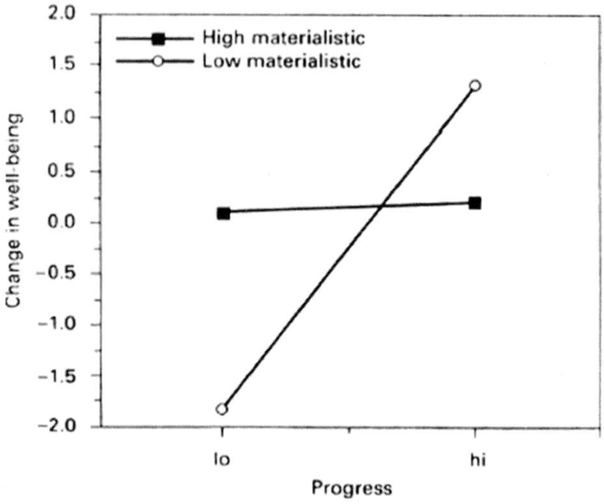

and that in the consumerist class happiness does not increase, once basic needs are satisfied: "It is only after an income level of about $15,000 per capita, that the life satisfaction score barely responds at all even to quite large increases in GDP" (Jackson 2009a: 33). During recent decades, in rich countries the happiness index remains stagnant or is dropping. In the UK the percentage of people who declare a high degree of happiness has fallen from 52 to 37 % since 1957. The same phenomenon has happened in China, where the population which claims to be satisfied has dropped by 15 % in the period from 1994 to 2005. In North America, every decade the number of people with depression and suicide attempts doubles. Suicides are the third cause of death. People's confidence in their community in North America and the UK has been sharply reduced in the last 50 years (Gardner and Prugh 2009: 50).

However, materialistic aspirations have been consolidated in recent decades. In the USA the percentage of students who believe that it is very important or essential to "develop a meaningful philosophy of life" dropped to about 40 % in the late 1990s, a decrease from over 80 % in the late 1960s (Kasser 2002: 104). This tendency can be a consequence of the weakening of the traditional social structures (like Churches, trade unions, political parties, etc.) which gave values, a sense of relevance and identity: "In a secular world, having something to hope for is particularly important when things are going badly. Retail therapy works for a reason" (Jackson 2009b: 99). Although other structures that promote values are emerging, they have not reached an adequate level to be significant for society.

Once basic physical needs are fulfilled consumption can be understood by its symbolic utility: reaching higher levels in the social strata; achieving a sense of leisure and freedom; buffering pains produced by failures; strengthening a sense of personal identity; buffering social conflicts: "The consumer products (. . .) are designed to motivate consumers, to create a desire for ownership and possession, to communicate value, identity, status, enjoyment and fulfilment" (UNEP 2002c: 11). For this reason, "once we accept a normal role in society (. . .) a great deal of *ordinary* consumption becomes almost inevitable" (Michaelis and Lorek 2004: 66).

The theory that justifies consumption as a means of expression and of personal identity building is based on the idea that work is losing ground to out-of-work activities as a main factor of personal identity. And for many people consumption is a key factor of their identity: "Consumer culture encourages many to establish a place in society or status through the purchase of material possessions and other lifestyle choices that we make". In this way, "we use consumption to help us construct our personal and collective identity" (EEA 2010: 21). But such behaviour means building a dependent identity, which is intrinsically pathological: "It is tempting to dismiss such a system as pathological. And in some senses it clearly is" (Jackson 2009b: 100).

Consumption is also a means for conflict solving in the family (conflicts between domains; conflicts originated in the domestic work sharing; conflicts between autonomy and cohesion, etc.). Tensions appear in the time distribution between several activities: remunerated work, household work, and external activities. There is a "fundamental tension between cohesion and separation. The individual needs the family as a secure base, yet at the same time searches individual development, which could put the cohesion of the family to the test" (Ropke 2002: 185). These individuals often try to solve this tension by several methods which mean an increase in consumption: proliferation of mobile phones as a tool for maintaining cohesion ("the mobile offers help to keep up a virtual community"); proliferation of gifts, especially to children, as a means of compensating for the lack of care; etc. (Ropke 2002: 185).

Also there is product hoarding with the aim of strengthening self-esteem and identity: "It allows more and more people to go around inventing and reinventing their social identities in the search for a credible place in society" (Jackson 2009b: 99). Another dimension is the search for compensation for the frustrations that emerge in social relations through consumption, but this does not solve anything either: "It is quite unlikely that buying new clothes will be effective if one is lovesick, or if they decide to buy a sports car through fear of being impotent" (Fischer-Kowalski and Haberl 2000). And although this behaviour does not solve anything, at least, it functions as a palliative: "And it is precisely because material goods are flawed, but somehow plausible, proxies for our dreams and aspirations, that consumer culture seems on the surface to work so well" (Jackson 2009b: 100).

In a nutshell, as psychologist Philip Cushman argues, the consumer "self is ultimately an *empty self* which stands in continual need of *being filled up* with" new products and services (Jackson 2009b: 66). So people lose wellbeing when they seek materialistic objectives, and it shows that "modern consumer society as being locked into a kind of *social pathology* -driven to consume by a mixture of greed, social norms, and the persuasive power of unscrupulous producers" (Jackson 2005: 21).

Last of all, the socio-economical factors which raise the level of consumption are: changes in the number of family members; market liberalisation; changes in marketing structures; etc. Mono-parental families are becoming more frequent, and young people abandon the family home in developed societies sooner than in the past. The average number of people per household in the EU-15 has fallen from 2.8 in 1980 to 2.4 in 2005 (EEA 2005: 20). It is estimated that the requirements for a person living alone are two-thirds greater than for two people. But these tendencies "can only explain a small part of the increased consumption (. . .) on the order of 10 %"(Sanne 2002: 9).

2.2 Systems of Provision

Until now, consumers have appeared as sovereign persons, although frequently they are able to choose an adequate consumption pattern for the fulfillment of their real needs. However, their behaviour is determined to a certain extent by provision and access systems. The provision system "is used to describe the combination of established industry processes and business practices, the accumulated physical production and delivery infrastructure and the corresponding social and cultural practises, which together define ways in which lifestyles and particular sets of products and services become mutually supporting structures" (UNEP 2002c: 41). That is, the analysis of the provision system offers knowledge about the extent to which consumption patterns "are determined by structures of creation, delivery, utility, disposal and information" (UNEP 2002b: 41). In the following part, we analyse the aspects which determine the structures of production and *marketing*.

2.2.1 Production Systems

The capital concentration and centralisation process allows us to understand that increasingly fewer corporations control the global market of different product groups and create global networks for distribution and sales. As a result, goods are designed to be sold all over the world, without taking into account cultural and environmental diversity. These structures greatly lengthen product chains, which also adds complexity to the chains and hinders the repair of products. For example, the Swedish Institute for Food and Biotechnology has analysed the production process of ketchup made in Sweden. The tomato cultivation and conversion into paste was made in Italy. The bags used to package the paste were produced in the Netherlands and transported to Italy to be filled. The five-layered, red bottles were either produced in the UK or Sweden with materials from Japan, Italy, Belgium, the USA and Denmark. The screw-cap of the bottle was produced in Denmark. Other parts of the commercial product (corrugated cardboard, label, glue, ink, etc.) were not included in the analysis (Andersson and Ohlson 1999).

Until the mid-twentieth century, "consumer durables were generally viewed as investments and (. . .) were designed to last as long as possible" (Cooper 2005: 57).

But economic globalisation determines a short life for product designs. This is so-called planned obsolescence. Which is "a business strategy in which the obsolescence (the process of becoming obsolete – that is, unfashionable or no longer usable) of a product is planned and built into it from its conception" (www.theeconomist.com/node/13354332). But in addition to technological obsolescence (produced by the design) there are two more. Functional obsolescence means that similar products, but with more functions, are put on the market. Psychological obsolescence is produced by changing trends. These types of obsolescence exert pressure on the consumer to replace *old* products with new ones. However, the new functions do not usually produce significant improvements. But the life span of products is diminishing. In Europe the mean life span of a computer is 3 years, 25 months for mobile phones, and in the case of young people only 20 months. It is possible to resist the pressure exerted by corporations, but it is very difficult when, for example, your *old* computer is not able to process programs designed for new computers (EEA 2008b).

Another consequence of this phenomenon is the increasing difficulty to repair products. Frequently there is a lack of spare parts. On the other hand, repair costs in OECD countries are rising at a greater pace than the prices of new goods, making repairs non-cost effective. In Great Britain, during the 1980s and the early 1990s the prices of televisions and washing machines increased by 20 % and 40 % respectively, while repair costs increased by 150 %. So repair work is declining in OECD countries "because labor costs are high, whereas manufacturing is increasingly relocated to countries with low costs" (Cooper 2005: 60).

The dominant approach concerning personal responsibility in consumption is based on the study of consumers' behaviour (water and electricity consumption, purchases with eco-labels, electrical appliances ownership, etc.). This approach systematically underestimates upstream and downstream environmental impacts, and consequently overestimates consumer responsibility. No other actors are taken into account when evaluating private consumption choices. However, many authors focus the blame on the product's design, and on the types of treatment at the end of the product's life: "Although 80–90 % of the impacts of a product occur in the use phase, they are largely (more than 80 %) determined in the design phase" (Spangenberg and Lorek 2002: 131). But also the distribution and sale structures must be taken into account. They are controlled by a small oligopoly formed by huge corporations (WalMart, Carrefour, Tesco-Lotus, etc.), which have created dense networks of big commercial areas located outside cities. This factor makes the car ownership culture stronger: "This upward spiralling, promoting car ownership, is a well-known fact in urban planning" (Sanne 2002: 10).

International trade makes it more and more difficult for the consumer to evaluate the environmental impacts generated by product chains, because they are increasingly longer and the goods are more complex. But even in the case of having a fair knowledge of them, their repercussion on consumer behaviour is weak, because impacts occurred abroad have a feeble influence on consumers. This phenomenon is known in economics as the 'discount of distance'.

2.2.2 Advertising and Marketing

The EEA states that the "advertising industry is increasingly creating new needs to ensure that we buy new products" (2010: 21). Sanne considers that "marketing is steadily becoming more intrusive as it turns to life-styling, taking a firm grip over the media and entering public institutions" (2002: 12). Publicity uses sophisticated artistic and dramatic techniques to create associations between certain brands and the values that are most deeply ingrained in our collective conscience. A car brand would symbolize reliability, a coffee brand sophistication, a clothes brand popularity and status, etc. Health and nutrition claims are used as a major marketing tool by the food industry in order to entice consumers into buying products: "Due to the huge number of exaggerated or unsubstantiated claims currently on the market, it is very difficult for consumers to know which ones to trust and ultimately make an informed choice. Too often claims stress only one positive aspect of a product, claiming for example a low level of sugar, but not mentioning the high levels of salt or saturated fat"(BEUC 2012; Sach and Finkelpearl 2010).

There is a huge lack of information on harmful chemical substances, on the impacts on health caused by the nano-elements incorporated to a growing number of goods, on the quality of goods imported from emerging countries, etc. Another source of concern is the precarious information about the quality of medicines. The BEUC has been denouncing the low level of control over the quality of medicines. In the case of the Commission's consultation about the legislative proposal on Pharmacovigilance, the BEUC considers that there is much to be improved, and it asks for: "Consumers to be enabled to report adverse reactions directly to national authorities (...) Clear and transparent safety information (...) The therapeutic and added value to be considered essential criteria for the marketing authorization" (BEUC 2008a). However, "the findings (coming from a Commission inquiry on the sector) confirm our concerns about the pharmaceutical sector especially with regard to the lack of innovation and the delay of entry of generics into the market" (BEUC 2008c).

There are two fields of advertising that are of the greatest concern: that of medicines directed at people, and that targeted at children with all kinds of goods. It is evident that the former has led to over-medication, as denounced by the medical establishment. Publicity aimed at children is pure manipulation, due to their inability to evaluate the messages received. It is estimated that on average a child receives over 100,000 radio and television advertisements before becoming an adult: "However, it can be difficult, also for children, to judge which things will actually be good for playing, so they tend to wish for nearly everything" (Ropke 2002: 185). Another consequence is "preventing the development of children's natural capacity for play" (Linn 2010: 62). However, the Sustainable Consumption and Production Action Plan does not include advertising (EEB 2008b).

As a result of the analysis carried out, we can coincide with the European Topic Centre on Sustainable Consumption and Production (ETC/SCP): "Although progress is made, the economic and regulatory framework can be criticised for failing to promote actions at a strong enough level" (2011: 7).

2.3 Systems of Access

Despite the fact that many products are being improved (more reusable, recyclable, energy efficiency, etc.), "environmental gains through technology improvements are being offset by continuous overall growth in consumption" (ETC/SCP 2011: 6). The UNEP states that the "patterns of consumption reflect systems and structures that control access to the existing market and to processes of influence in shaping the market" (2002c: 43). The critical factors determining access are: income, time, availability of infrastructures, availability of products and services, information and education (UNEP 2010a, b: 43). But the most important factor is a continuous increase of income.

Consumption is affected by income levels and by the fiscal system. This system is based on taxing income and added value, not on natural resource consumption, and usually subsidises the consumption of natural resources (fossil fuels, nuclear energy, unsustainable fishing, etc.). On the contrary, there are no tax rebates for ecological goods. Besides, environmental taxes are very low and scarce, and are falling, despite the fact that "experience indicates that economic signals are particularly effective at moderating consumption behaviour of both individuals and organisations. Thus, it is unfortunate that European countries have reduced the use of green taxation over the last decade" (ETC/SCP 2011: 7). Taxes on transport and on pollution have been maintained stable but are low. But energy taxes (by far the most relevant chapter of environmental taxes) have been sharply reduced. Accordingly, environmental tax revenue as a share of total tax revenue and as a share of GDP is steadily falling (ETC/SCP 2011: 92).

Historically, productivity growth was accompanied by reductions in working hours in Europe until the sixteenth century. From this point on the tendency was inverted, and working hours grew steadily until the end of the nineteenth century in developed countries. After this period the tendency changed. The big improvement in productivity continued to bring a reduction in working hours and salaries continued to rise. Also, the age of retirement was reduced, while life expectancy was increasing. But the primary aim of trade unions was a salary increase, instead of promoting shorter working hours. But, "had Americans opted to put the bounty of productivity growth into shorter hours, the average work year today would only be 860 hours" (Schor 2010a: 92; www.wikipedia.org).

However, in the 1980s the tendency to reduce working time was stopped, especially in countries in which liberalising ideas were stronger (the USA, Great Britain, Ireland, and Canada). After that it began to rise in many OECD countries. The USA, Denmark, Sweden, etc., has increased their working time. This tendency has been generalised in most OCDE countries. This tendency is a factor that makes consumption less sustainable: "People who are *time-poor* (that is, they work long hours) tend to make lifestyle choices that are more resource-intensive" (Schor 2010a: 93–95).

3 Elements of a Sustainable Consumption Strategy

3.1 General Aspects

The analysis performed in Sect. 2 shows that consumerism is the consequence basically of two factors. One is limited personal development, which leads to consuming to obtain some satisfaction which reduces the unhappiness produced by personality problems (but this means an incapability of facing these problems). The other factor is the outcome of economic structures which promote ever increasing consumption. Both factors bring about limited sovereignty, with a broad spectrum which goes from severe limitation (children, frequently older people, people who suffer from psychological problems) to little limitation, in the case of developed people who have significant knowledge about the causes of current unsustainability. We know that once basic needs are satisfied, the path to personal maturity is open, and an ample and growing minority is moving along this path. Kasser identifies it with the progress towards non-materialistic values. T. Jackson uses the term of prosperity, which means, apart from the fulfilment of materials needs, "the ability to give and receive love, to enjoy the respect of your peers, to contribute useful work, and to have a sense of belonging and trust in the community" (Jackson 2009a: 30).

Over the last two decades there has been a very important theoretical improvement regarding the concept of sustainable consumption, and regarding the policies to advance towards this objective. There is a broad agreement on the consumption groups which exert the most impacts. However, governments have not put into practise most of these policies. The Oslo Declaration states that "the emphasis to date has primarily been on questions of sustainable *production*. There has been considerably less effort devoted to charting the transitions that will be necessary to facilitate sustainable consumption" (Tukker et al. 2006). Governments have showed little interest in sustainable consumption, due to its potential of reducing consumption: "Sustainable consumption is a difficult area for governments (. . .) Governments still struggle to communicate sustainable consumption messages to significant or long-lasting positive effect" (EEB 2008b). So it is evident that "there is a huge, largely unused potential for encouraging environmentally less intensive consumption patterns". But it will not be easy to change them, because they "are institutionalized, economically, technically and socially" (EEA 2010: 7, 8).

Despite these factors, governments are the only entity capable of leading sustainability strategies. Chapter 4 of Earth Summit Agenda 21 (1992) was devoted to *Changing Consumption Patterns* by the leadership of OECD countries: "Achieving sustainable development will require both efficiency in production processes as well as changes in consumption patterns (. . .) which have predominantly emerged from developed countries". But this endeavour is impossible in the structural scenario determined by increasingly liberalised economies competing in a globalized market. This means relentless growth, and the creation of huge monopolies, which promote uncontrolled technological development. This dynamic determines that governments become moderators of the socioeconomic agents: "As long as national governments understand their roles in the governance of Sustainable Consumption as one of moderator providing opportunities for the exchange of opinions and voluntary commitments (. . .) any significant progress towards Strong Sustainable Consumption will fail to materialize" (Lorek and Spangenberg 2012: 10).

But taking into account that consumers are weak when faced with monopolies, governments have to adopt another role: "to empower consumers", as *A European Consumer Agenda - Boosting confidence and growth* (COM(2012) 225 final) declares. And "empowering consumers means providing a robust framework of principles and tools that enable them to drive a smart, sustainable and inclusive economy. Empowered consumers who can rely on a robust framework ensuring their safety, information, education, rights, means of redress and enforcement, can actively participate in the market and make it work for them by exercising their power of choice and by having their rights properly enforced." The WBCSD agreed with this vision: consumers "will reward brands for addressing their aspirations, including the need to *tread lightly*. They will increasingly avoid brands that feel unsustainable or irresponsible" (2012: 3). But reality goes in the opposite direction.

The sustainable consumption concept is being downgraded by power instances. On the one hand, they use many terms that in their opinion are equivalent to sustainability: "efficient consumption", "different consumption", "appropriate consumption", "responsible consumption", etc., (UNEP 2002b: 9). It brings a lack of clarity. On the other hand, they reject that sustainable consumption means less consumption for the consumer class. The OECD states: "Sustainable consumption is not about consuming less, it is about consuming differently" (OECD 1998). The WBCSD (2002) goes a little further: "Sustainable consumption is not about consuming less, it is about consuming differently, consuming efficiently". In fact, for them the concept of efficiency is the cornerstone of sustainability. For the UNEP (2008: 20) sustainable consumption is an alternative consumption "that results in reduced material and energy intensity per unit of functional utility". The WBCSD report *A vision for sustainable consumption* states that "*living well* is decoupled from consumption of physical products, and the materials that they contain" (2012: 5). But the decoupling theory means technological improvement, as we saw in Chap. 3. However, this approach does not take into account the rebound effect. That is, by using efficient products consumers reduce their consumption of resources. But in doing so, they save money which generates new consumption.

The EU is the political power that is adopting the most initiatives for environmental protection. But, at the same time, the EU is the leader of market liberalisation. This contradiction explains why sustainability has not been improved. In 2008, the European Commission approved the first *Sustainable Consumption and Production Action Plan*. The European Environmental Bureau (EEB) states that despite the positive aspects, "overall it considers it to **lack vision, clarity and ambition** (. . .) the proposals generally represent a **watering down of the Integrated Product Policy** (IPP)" (2008b). The Communication on IPP (COM(2003) 302 final) declares the purpose of establishing "the framework conditions for the continuous environmental improvement of all products throughout the production, use and disposal phases of their life-cycle; and developing a focus on products with the greatest potential for environmental improvement".

And, for the same reason, the *European Consumer Agenda* is not an adequate response to the challenges either. The European Parliament Report (2011/2149(INI)) about the Agenda states that "the Consumer Agenda needs to be visionary and holistic; it needs to reflect what is happening in the world. Currently we are facing a potential disastrous development with climate change, and therefore the consumption society has to change profoundly". The EEB (2008b) sees many gaps: "The further development of a sustainable consumption policy area represents a serious gap"; it "does not establish a framework for either product policy or sustainable industrial policy"; there is a lack of "product information" or "advertising".

3.2 Actions on Motivation Systems

We have seen that the strength of materialistic aspirations explains the dominant consumerism and the diminishing quality of life. So in order to advance towards responsible consumption, aspirations have to change. And this change will only be possible when the majority of the population understands that materialistic aspirations are hurdles for reaching real wellbeing. When people abandon their materialistic aspirations, they report an improvement in psychological health, richer social relations, and greater contributions to societal wellbeing and sustainability. On the basis of psychological research and theorising it can be concluded that are four sets of basic needs: "Needs for safety, security, and sustenance; for competence, efficiency, and self-esteem; for connectedness; and for autonomy and authenticity". The first set comprises food, shelter and clothing. The second set "involves a feeling that we are capable of doing what we set out to do and of obtaining the things we value". The third set is related to "intimacy and closeness with others". Finally, "we need to feel autonomous and authentically engaged in our behaviour" (Kasser 2002: 24, 25).

Numerous studies show that a significant number of people are emerging who have a holistic behaviour (as it is defined by some authors) or who are culturally creative, in the opinion of others. They are comprised mostly by middle-class people (more women than men), who prefer organic food, holistic medicine, renewable energy, ecologically produced goods, an equilibrium between work and leisure, socially responsible investment and inner development. They are cooperative and practice social activism at all levels (community, nation and planet). Several studies carried out in the USA, at the end of the last Century, reached the conclusion that 24.5–28.0 % of adults is culturally creative, and that this percentage is growing rapidly. A Budapest Club report concludes that 35 % of Italian people are culturally creative. Similar findings are reported in other European countries, in Japan, Australia and Brazil (Laszlo 2010: 56–59). However, these percentages seem too high, because if they were real, significant transformations would be visible, despite not acting as a collective.

On the other hand, sociologists reach the conclusion that most of the population shows a growing awareness about the environmental problems caused by consumption. According to UNEP's global consumer survey, "93 % are aware, on average, of the impact of their consumption patterns on the environment" (Fuchs and Lorek 2005: 278). The 2008 Eurobarometer shows that 95 % of the population considers that environmental protection is important; 75 % is inclined to buy ecological products, even if they have to pay a little more; but only 17 % state that they bought some ecological products during the month before the survey. The Special Eurobarometer (2011), dedicated to climate change, shows that "89 % see climate change as a serious problem, with 68 % considering it a very serious problem" (http://ec.eu/public_opinion/index). Although these surveys always show optimistic biases (due to the fact that people always want to give a positive image), it is evident that people's sensitivity towards environmental issues is rising.

3.3 *Actions on Provision Systems*

The UNEP states that "systems of provision means examining the various ways in which the consumption of products and services (. . .) are determined by structures of creation, delivery, disposal and information" (2002c: 41). And there is growing evidence of the huge impact of current patterns of consumption due to faulty product designs, and the lack of infrastructures for recycling and reusing waste.

The life-cycle perspective of the environmental impacts of a product "*captures* the whole production-consumption chain" (UNEP 2002c: 45). But "the inward responsibilities of households" are systematically overestimated. And "although 80–90 % of the impacts of a product occur in the use phase, they are largely (more than 80 %) determined in a design phase" (Spangemberg and Lorek 2002: 131). The UNEP states that "60–80 % of the (life-cycle) environmental impacts from products are determined at the design stage" (2002c: 45).

Taking into account that the previous chapters deal with sustainable production, here we analyse general policies to improve the product quality, the information, and selective policies. However, we realise that growing economic globalisation and the strength of oligopolies limit the positive impact of the policies.

3.3.1 Overall Policies

A Policy for the Overall Improvement of Product Quality

The consumption of eco-labelled products can be an important step towards responsive consumption, if most people consume them habitually, and if the eco-labelling systems are of high quality. However, the European Union is advancing slowly and with contradictions, due to several causes: the opposition of corporate groups to a high level quality of the EU system; only a small quantity of groups of

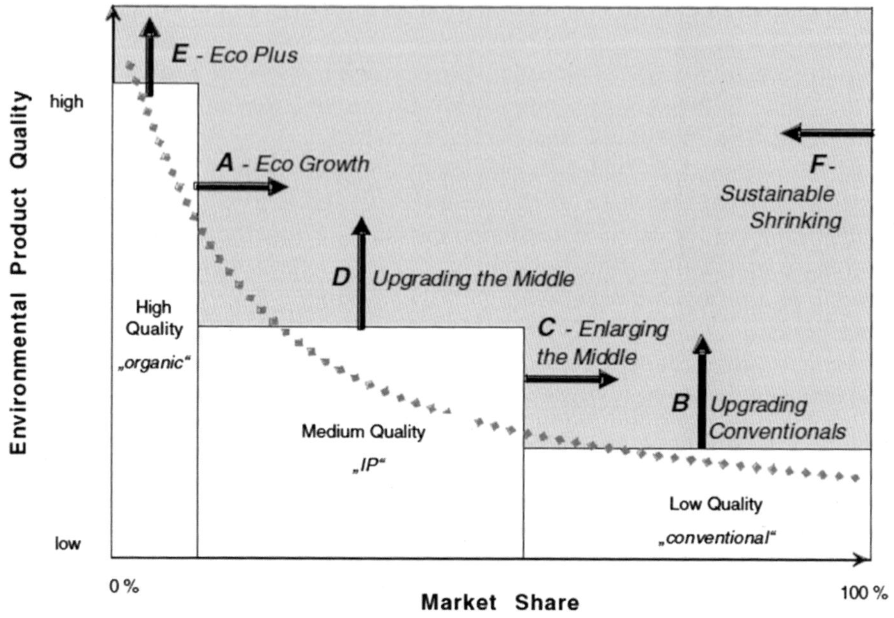

Fig. 19.2 Roadmap of the ecological mass market (Source: Lorek and Lucas 2003: 16)

goods have been certified; prices are relatively higher than non-certified products; the system is known only by a minority of the population. This outcome is caused by a low level of political will. The proposal of a Directive on eco-design states that there is a great potential for the improvement of products: "In the EU new products with deficient environmental behaviour, which could be highly improved with minor or zero costs, continue to be sold" (COM (2003) 453 final). For this reason a policy which seeks an improvement in quality of all products and services must be implemented. At the same time, the market quota of ecological goods and services must be improved, and those of low quality must be retired from the market, as Fig. 19.2 shows.

OECD countries, such as EU member States, the USA, Japan, etc., are promoting measures aimed at improving sustainability, like waste recycling, banning hazardous materials and substances, energy efficiency, and so on, although they are not doing so in an integrated manner. On the contrary, there are frequent contradictions. For instance, in the EU the treatment of hazardous products intended to minimize their environmental impacts is an obstacle for the agents that seek to recycle them (Directive 2009/28/CE).

Another negative tendency is the diminishing life span of products, although there are many cases of goods designed to last. They have "intrinsic qualities such as resistance to abrasion and wear, reliability, reparability, and upgradeability" (Cooper 2002: 61). It is estimated that it is technically possible to produce goods able to last twice as much as current ones, and with easy reparability, without incurring extra costs. Although a review of the many reasons why most goods have short lifetimes is beyond the scope of this chapter, we describe a pair of them. On the demand side, consumers "may prefer to purchase a large number of new goods and services frequently than buy a small number of high quality items occasionally". On the other, "in the saturated markets typical of industrialised countries there is an inherent commercial logic behind the production of goods short lived" (Cooper 2012).

But some governments are beginning to change their policies because of the growing scarcity of resources. The European Union's *Road Map to a Resource Efficient Europe* commits the Commission to assess the introduction of durability and reusability criteria for key products (COM (2010) 2020 final). The Directive on energy-using products (2009/125/EC) refers to the possibility of "extension of lifetime, minimum time availability of spare parts, modality, upgradeability, and reparability". However, the Consumer Agenda (COM (2012)) still hesitates on the need for taking measures in this direction: "The Commission will consider taking measures to make consumer goods more durable, including support for repair and maintenance services". The UK government appears to be more proactive. In the *Public Understanding of Product Lifetimes and Durability* Report it wants "people to see the benefits of extending the life of things they own and use", and it is looking for businesses to "design and manufacture goods that are more efficient, durable, reparable and recyclable" (Cooper 2012).

But, once more, governments must devise a systemic approach to the subject. They must select some types of goods in which it is easier to promote durability, such as kitchen appliances or furniture. There are various means (use standards, length of guarantee and life time labelling) by which durability may be communicated to consumers. On the supply side: "Another possibility is that companies adapt their business model and derive more revenue through the provision of services and less from selling goods" (Cooper 2012).

Information

Industrial societies are very complex, and this complexity is growing. The flow of information is increasing, but at the same time there is a lack of reliable information, and as far as some aspects are concerned, information is scarce. Besides, the information offered is frequently not comprehensible for the majority. In addition, the information is directed at specific collectives who are not able to comprehend it (like children or old people). It is evident that information has to be understood by the people concerned. This means that information directed at children and old people must be prohibited, that the marketing of medicines should be directed exclusively at health professionals.

The European Parliament Report (2011) "calls on the Commission to guarantee special protection for groups of consumers who are particularly vulnerable because of their mental, physical or psychological infirmity, age or credulity". Sweden and Greece have banned advertising directed at children, and Norway has limited it.

In addition, this country has banned car advertisements that claim they are ecological, green, etc. We have seen that health and nutrition claims are used as a major marketing tool by the food industry. In response to the problem, an EU regulation was adopted in 2006 laying down harmonized rules for the use of claims, which only partially solves the problem: "Unsafe consumer products, including products bearing the CE mark, are often found on the EU market". There is a lack of information on nano-materials, endocrine-disrupting chemicals (EDCs), cocktail effects, etc. (BEUC 2012a: 29–31). The European Parliament (2011) "stresses the urgent need to increase the general level of safety of consumer products in the EU".

The improvement of information not only involves eliminating the detected deficiencies, the flow must be adequate to the consumer demand (we will see that consumers prefer graphic symbols which allow them to rapidly detect the quality of goods). The reduction of the volume of information can be obtained by eliminating many certification systems, which do not have a minimal level of quality and generate confusion. Some authors estimate that there are about 800 systems in the world. There is a huge need for the improvement of the visibility of certificated products (Jackson 2009a: 129); (COM(2008) 400/2).

General Systems of Certification

The most trustworthy certification system is one with an independent organism as guarantor. Public certification systems are vulnerable to corporate pressure. General systems are normally public, while specific systems can be public and private.

Ecolabel Systems exist in the most important OECD countries. The "Nordic Swan" (Nordic Council) and the German "Blue Angel" are the systems that offer the highest level of quality and success. The former was created in 1989, and about 90 % of Nordic consumers know the system. It contains more than 42 product groups and 1,200 certified products by 350 firms. Some products have reached a dominant market quota. The latter was created in 1977, and contains 80 groups of products. In 2004 it was known by 83 % of consumers, and 49 % stated that they regularly bought these products (EEA 2008a, b: 276).

The EU began to apply its certification system in 1993. The system did not work, and the EU decided in 2000 to build the system on new foundations. The number of certified products began to grow rapidly. In 2011, there were 1,236 licenses. The most numerous groups were: tourist accommodation services (444), all-purpose & sanitary (174), indoor and outdoor paints and varnishes (101), and textile products (93), etc. (SCP/SIP Action Plan Newsletter, Issue No 2, February, 2011).

Despite the positive trends, the system suffers from shortcomings: its level of quality is lower than that of national systems; many reports show that some certified products produce huge environmental impacts (one of the last known cases is the Pindo Deli, which is referred to as the Indonesian paper certificate); the system has few certified groups; its market penetration is rather low; consumer knowledge of this certificate is very scarce; in the EU there are several certification systems, and "concern has been expressed on the growing number of national eco-labels, which may create confusion for consumers and constitute a barrier to intra-EU trade" (SEC(2009) 1707), so it is necessary to create a unique system and with the highest quality; etc.

Besides, the level of information about the system is very poor:

- "It seems that there is no common understanding on how much information a Competent Body is allowed to or should disclose upon request of consumers or citizens organizations.
- There is no formal complaint procedure in place.
- It is not possible to identify on what basis the eco-label license has been awarded to the product and how the on-going compliance is checked by the Competent Body" (Hammer 2011).

The Commission began in 2008 (too late, because the mandate was to do it by 2005) a new review of the eco-label, with a "Proposal for a Regulation", due to its poor development: "The experience (. . .) has shown the need to amend the Community eco-label scheme in order to increase its effectiveness and streamline its operation". There were only 26 product groups established and around 500 companies. Some measures were: "Introduce measures to encourage harmonization with other eco-labelling schemes"; "More product groups/quicker criteria development"; "Abolition of the annual fees and simplification of the assessment procedures"; "Boost marketing". And also proposed to establish "a European Union Eco-labelling Board (EUEB)", which "observes a balanced participation of all relevant interested parties" (COM (2008) 401 final). In 2010 the EU approved a Regulation on the functioning of the EUEB. The EUEB "aims at revising the EU Eco-label criteria and improving the overall framework for the implementation of the scheme, through a working plan established with the Commission". It also plans to assist the Commission to give recommendations on minimum environmental performance requirements (The flower, issue 01, 2011; EEB, 2008/30/10; Regulation (EC) No 66/2010).

Specific Certification Systems

Many States and the EU are developing certification systems for organic food. Due to a rapidly growing demand for it, in some States organic food is reaching remarkable market shares and percentages of land dedicated to it. Frequently their growth is greater than that obtained by other certified products. However, many experts state that requirements must be stricter in order to consider them as sustainable production. But we do not analyze this subject (nor many others) here because of space constraints. Global sales of organic products grew by 8.8 % in 2010. In Europe, sales are growing by 9 % a year. In China, organic production has quadrupled in the last 5 years. Market analysts predict that organic sales will grow by 20 % a year over the next 3 years. In 2010, 37 million hectares were dedicated to this type of agriculture. A large part of the land is dedicated to pasture, and for this reason Australia, New Zealand or Ethiopia are the world leaders in organic land (EEA 2008a, b: 270 and ss.; Soil Association 2012). The resource consumption of organic agriculture is less than the consumption of industrial agriculture. Compared to conventional production systems, "the energy use in organic agriculture is (. . .) about half" (Ziesemer 2007: 10).

By far the most important wood certification system is the Forest Stewardship Council (FSC), which is an independent and non-profit organization. FSC comprises a great number of wood sector enterprises and environmental organizations from 80 countries and five continents. Although it receives financial support from several States and the EU, they cannot be members. The FSC certifies wood following ten principles and 56 criteria types. It issues two kinds of certifications: for forest management and for chain of custody (CoC). CoC certification verifies company systems for tracking and handling materials used in FSC-certified forest products. This system comprised a forest area of 94 million hectares (Mh) in 2008 and 147 Mh in 2012. This represents about 11 % of the surface of all productive forests (www.sfc.org).

The Marine Stewardship Council (MSC) is the most important organization for the certification of sustainable fishing. It was created in 1997 by Unilever and the WWF, but in 1999 it became independent from them and currently it comprises a wide variety of organisations. In 2005, the FAO adopted its sustainability requirements. In 2012, there were 10,000 certified products and the number of products has duplicated annually over the last 4 years (www.msc.org).

Product Gradation Systems

These systems offer consumers a very simple eco-labelling system, in the form of a gradation scale of quality (mostly energy efficiency). The EU has approved a regulation which regularly strengthens the efficiency requirements of electrical appliances and prohibits non-efficient products. For instance, the EU is banning type-B refrigerators from the second half of 2010, 2 years later type-A's will be out-of-market, and from 2014 onwards only A + refrigerators or those with higher efficiency can be sold(2009/28/CE).

3.3.2 Selective Policies

General public policies have to be complemented with others aimed at specific product groups, consumer types, and kinds of companies, in order to obtain rapid improvements.

On Consumers

Some authors have classified consumers into several groups. Between consumers already behaving in an ecological way (innovators; they are called "positive greens" by the EEA) and the uninterested, ignorant and never-accessible consumers (laggards), there is a heterogeneous group. It is formed by three subgroups: early adopters, early majority and late majority. The first subgroup is interested in buying ecological products as soon as their price is adequate. The late majority will buy ecological products when the great majority is buying them (they are called "cautious participants" by the EEA). The early majority is the key subgroup because "their acceptance will trigger the take-off for ecological mass marketing" (Lorek and Lucas 2003: 18; EEA 2010: 23). The last subgroup has to be a key public target.

Green Public Procurement

Chapter 4 of Agenda 21 has a part titled "Leadership through public procurement" (UNCED, 1992). Green Public Procurement (GPP) is "a process whereby public authorities seek to procure goods, services and works with reduced environmental impact". The Commission acknowledges "that there is considerable scope for cost-effective green public procurement" (COM(2008) 400/2). This has a significant potential for transformation, because the public expenditure of OECD countries ranges between 7 and 19 % of GDP: "Public authorities, which are responsible for 16 % of all European consumption, can play a leading role through environmentally sustainable procurement" (EEA 2010: 47). Buildings are by far the biggest expenditure item (it can reach up to 60 %. Besides, it accounts for 6–7 % of European electricity consumption.

Besides direct savings and a reduction of the ecological footprint of the public sector, GPP "is a powerful instrument for stimulating innovation and encouraging companies to develop new products with enhanced environmental performance". And the criteria for GPP "can equally inform private procurement practices" (COM(2008) 400/2). But the EU lacks ambition in this field too. There are no mandatory targets to be achieved, only recommendations. Little by little the EC is defining minimal quality criteria for various groups of GPP consumption. And these criteria are only of the *core* type, which allows an "*easy* application of GPP". For example, "the *core* GPP criteria would be set at the level of the energy efficiency requirements of the Energy Star Regulation, whereas the *comprehensive* criteria would be set on the basis of eco-label criteria" (COM(2008) 400/2).

Three Focus Areas of Consumption

Many reports, some of them done by European institutions (EEA (2010), IPTS (2006) or ETC/SCP (2011)) have found that the majority identify three main kinds of consumption: "Private consumption (. . .) is dominated by housing, transport and food and drink, with the share of food and drink reducing as incomes increase" (EEA 2010: 9). The total requirement of the three clusters "makes up for nearly 70 % of material extraction and energy consumption and more than 90 % of land use (. . .) Each of these three clusters represents more than 15 % of the total energy and material consumption" (Spangemberg and Lorek 2002: 135). The percentage of income dedicated to these groups of consumption are, respectively, 27 %, 14 % and 16 % in the EU-15 (COM(2008) 397 final).

Transport consumes 30 % of the total energy in the EU. Oil provides 96 % (95 % in the world) of the total energy consumed by transport. And 74 % of oil consumption is produced by road transport. In the period from 1995 to 2006 domestic freight transport in the EU-26 grew (measured in tonnes-kilometre) at an average annual rate of 2.8 %. In the case of passenger transport (measured in passengers-kilometre) the increase was of 1.7 %. As the population grew less than mobility, there has been a per capita mobility increase in the period mentioned. Intra-EU passenger transport during the same period has been dominated by road (it accounts for over 83 % of the total). But air travel has seen the highest growth rate. The latest data shows a growth of 5.3 % in the period from 1993 to 2008. On the other hand, railway passenger transport has decreased. Overall rail demand was reduced by 1 % in the period from 1990 to 2006. In the period 1990–2006, road mobility increased by 61 % (EEA 2005: 38; Focus Group 2009: 8). A Flash Eurobarometer (2011) on the *Future of Transport* concluded that around 70 % of car users "felt that **public transport was not as convenient as a car"**, due to a "**lack of connections", "low frequency of services", and a "lack of reliability". However,** escalating oil prices and the economic crisis are producing a reversal of the trends mentioned, as we saw in Chap. 12.

The biggest environmental impacts of the food chain come from production, transport, distribution and industrial manufacturing. The impacts of buying and preparing food are minor. The food chain is formed by production, transport, preservation and distribution. The footprint of transport depends on the distance (the food-miles impact index is widely used) and mode used. Also, the environmental impact varies in relation to the kind of food. Those with the greatest impact include "in particular dairy farming, cattle farming and grain crops production, and fisheries" (EEA 2010: 28). Growing incomes are leading to a higher consumption of meat, dairy products, and drinks. The annual meat consumption in the world has been multiplied by four between 1961 and 2006, being 43 Kg/per capita the average meat consumption (in the EU-15 it is 91 Kg/per capita). This means a higher consumption of grains for feeding animals: "At least 35–40 % of all cereals produced worldwide are fed to livestock (. . .) Using cereals crops to feed animals is a highly inefficient use of calories and is very resource-intensive" (Tomlison 2012). But producing meat consumes a wide range of resources, depending on the type of meat. Raising cattle requires more land than for raising pigs, while chickens need less land than pigs. A surface equal to 10 % of European land is dedicated to producing beer, wine, tea and coffee. Taking into account also the land dedicated to satisfy fruit juice consumption, it is estimated that the land requirement for producing those drinks is similar to the land needed for cultivating food. Fish consumption has been multiplied by four between 1950 and 2005. This increase has destroyed most of the fishing grounds and, as a consequence, today fish-factories provide 40 % of the available fish (Halweil and Nierenberg 2009: 131). On the other hand, organic agriculture uses 40–50 % less energy than conventional agriculture, depending on the type of food cultivated (Ziesemer 2007: 10).

A drastic improvement of the sustainability of food consumption requires changing the type of diet (mostly reducing animal protein consumption), consuming eco-labelled, local and seasonal food, and little or no manufactured food, with minimal packaging. A policy in favour of sustainable nutrition depends on information, availability and prices. The population must know the nutritional qualities of ecological food, and its contribution to health and to improving environmental quality. Eco-labelled food prices have to be at least the same as those of conventional food. Governments ought to help with the improvement of distribution networks. But farmers with certificates do not have incentives to improve their practices, and there is ample scope for improvement. For this reason "the EU should develop a labelling system which would encourage farmers to incrementally improve their sustainability performance in a way that is visible for consumers. The sustainability criteria for such a scheme should be subject to regular review in order to allow for innovation to set new thresholds" (Buitenkamp 2008).

Key Enterprises

There are two company types which are very relevant for improving the quality of goods. On the one hand, there are small-niche enterprises (called Davids) with a strictly ecological philosophy. But they are not capable of reaching mass markets. On the other hand, there are large, well situated players in the market (called Goliaths) who have discovered that improving the ecological quality of their product may benefit them, although their willingness to do so is low. In order to transform the mass market into an eco-mass, governments must focus their industrial policies on helping to increase the number of eco-pioneers and on incentivising Goliaths to produce ecological goods (Lorek and Lucas 2003: 16, 17). The Action Plan has created a Retail Forum. Its "stated objective is 'that individual large retailers commit to a series of ambitious and concrete actions with clear objectives, timelines, deliverables, and monitoring indicators', to be set with stakeholders including consumer and environmental organisations" (EEB 2008b). But the EBB has decided "to suspend active participation" due to a lack of progress (Metamorphosis, February 2012).

On the other hand, many enterprises which distribute only ecological certified products are emerging. They are working in all fields. In the USA Ecofish and Cleanfish are specialized in products from sustainable fishing; Tyson Food sells chicks which have not been treated with antibiotics; Ikea sells furniture with the SFC certificate (Halweil and Nierenberg 2009: 146).

3.4 Actions on the Access Systems

The proposed changes in the provision system would allow people to access highly ecological quality goods and services. But the capability of accessing them depends on income and relative prices. In OECD countries most people have enough income to buy some ecological products and services. However, massive ecological consumption is not possible due to the low level of awareness among people. And the only short-term way to reach it is a fiscal system capable of balancing out the prices of both types of products. There are two methods: charging more on non-ecological products or reducing charges on ecological products.

We have seen that the consumers known as early adopters and the early majority are very important for spreading sustainable consumption. The first group will buy ecological goods as soon as their prices become a little lower. And after that this group can attract the second one. Thus, an ecological tax reform is needed to initiate the process. But there is not a single country which has approved such a policy. At an EU level, this change has not been possible, because (among other factors) unanimity is needed to change the fiscal policy. However, the EU defends the idea that a fiscal policy is the most effective tool to change conducts (www.consilium.europa.eu/Newsroom). It could be less difficult if the EU decides to reduce VAT for certified products. This measure was proposed by the Green Book on Product Integrated Policy (COM(2008)428final). The EEA defends "economic incentives at the point of sale as tax rebates for less-pressure-intensive products or services" (2010: 19). Historically, an increase in productivity has frequently produced a reduction of working hours and higher salaries. Opinion polls and surveys carried out in OECD countries discovered that the majority stated that they prefer shorter working hours, despite a lower salary. However, there is empirical evidence that most people do not want a salary reduction. This position shows the well-known behaviour by psychology of attaching more value to what we possess than to what we can achieve in the future. But a growing minority shows a solid preference for more free time, despite a salary reduction. In Sweden a sixth of workers prefer a reduction of working time sacrificing income. In the United States, "surveys done before the crash indicate that between 30 and 50 % of Americans say they would prefer to work fewer hours, even for less pay" (Schor 2010a, b).

However, this minority's demands frequently become impossible, due to the existing rigid norms. In the American case, "the structure of the labour market - including the need to work full-time to receive benefits- has made that difficult" (Schor 2010b). But there are exceptions: "Policies to influence the working patterns of individuals are by no means novel (...) in many high income countries (...) [they] have been in place for many years" (Pullinger 2010). The Netherlands has the highest rate of part-time work, as a result of facilitating laws. And from 2000, the law compels employers to accept workers' requests for part-time work. Besides, the law incentivises the reduction of working hours in the case of parents with children (each parent works 3/4 of the normal working hours). Belgium, Finland, Sweden and Denmark have also developed frameworks which regulate training leave, childcare and sabbatical periods. Sweden offers 15 months leave for each child, paying 80 % of the salary. In GB parents with children under 8 have the right to reduced working hours (de Graff 2010; Pullinger 2010; Schor 2010b).

And lastly, there exist laws that incentivize agreements between workers and employers to reduce working hours in the event of crises. This policy avoids the dismissal of workers. The state covers part of the salary reductions. This policy is particularly effective at the start of economic recessions, because employees do not know the possible development of the crisis. Such practices are applied in Germany, South Korea, Norway, Italy, Finland, Japan, etc., and in many states in the USA. In this country 166,000 jobs were saved in 2009 by implementing this policy. Germany is an outstanding example of applying the work-sharing scheme: "a federal scheme to replace lost wages (*Kurzarbeit*) accounted for about 20 % of the reduction in hours; private bargains between employers and unions, cancelled overtime, and flexible use of vacation and other time-off was responsible for the remainder". In 2009 unemployment grew very little, despite the fact that GDP was reduced by 4.7 %. So "work-share programs are probably the best way to respond to a short-term reduction in economic activity. But they also form a key pathway to a saner economy" (Schor 2010b).

Part IV
Evaluation of Transformability

Chapter 20
Overall Evaluation of Transformability and Its Trend

Keywords Resilience of socioeconomic systems • Transformability of socioeconomic systems • Transformability factors • System potential • System connectivity • System panarchy

In order to analyse the transformational capability of our civilisation, we retake the adaptive cycle theory explained in Chap. 7. We have seen that resilience is the ability of a complex system to create, test and maintain its adaptive capacity, that is, the capability of preserving its vital functions when faced with environmental changes. We have defined the concept of transformability in ecology and the factors which determine its level: potential, connectivity and panarchy. In ecosystems a high resilience is determined by a relatively low connectivity and a high potential, due to its huge biodiversity. These factors have to be adapted to human societies, but without losing their broad meaning, taking into account two main features: today's societies are highly unsustainable and human beings have a potential for foreseeing and averting risks.

This final chapter has the task of analysing the capability of our societies to transform themselves to reach sustainability. To do so we broadly evaluate each factor, as a prior step to carrying out the overall evaluation. However, we have to emphasise that we accomplish this task only based on the information contained in this book, as a first approach. In order to carry out a broad and in-depth analysis, a multidisciplinary group and a long period of time are necessary, due to the enormous amount and diversity of the information to be processed and evaluated.

R. Bermejo, *Handbook for a Sustainable Economy*,
DOI 10.1007/978-94-017-8981-3_20,

1 Potential

The potential determines the amount of elements that conform the system. In nature potential grows when its biomass, biotic diversity, redundancy, etc. increases. However, in socio-economic systems (SES) there are two kinds of potential: one whose growth undermines sustainability, because it reinforces the current trend of growing unsustainability and which we call perpetuation potential (PP); and another one that constitutes a basic element to build a sustainable economy; the transformational potential (TP). There are two basic kinds of potential: naturally and socially originated. This is composed of the following potentials: social, technological, built-up, institutional, financial and human.

1.1 Natural Potential

We have seen that many abiotic resources are on a course towards the rapid depletion and destruction of planetary ecosystems. The United Nations Millennium Ecosystem Evaluation (MEA) concludes: "Human activities have taken the planet to the edge of a massive wave of species extinctions" (MEA-SB 2006: 3). And "land use changes are perhaps the most critical aspect of anthropogenic global change in influencing the future of ecosystems and their services" (MEA-V2 2006: 450). The depletion process of fossil fuels and of many materials, besides being huge problems, leads to the reinforcement of the current process of destroying ecosystems.

The depletion of fossil fuels leads to the search of new fields which are frequently located in difficult to access areas and/or in important ecosystems (tropical forests), which are frequently very fragile (Arctic Pole or deep sea). Impacts can be huge and frequently irreversible. Beyond a depth of 1.5 km, oil spills cannot be stopped. The extraction of non-conventional oil causes enormous environmental impacts, because the fuel is found in diffuse quantities. Canadian bituminous sands are located in an area covered by the boreal forest. Its exploitation produces the forest's destruction, consumes 400,000 cubic metres of water per year and a huge amount of natural gas, generating very significant toxic water and air pollution (http://en.wikipedia.org). On the other hand, the diminishing ore richness determines the need for removing increasingly greater amounts of rocks to obtain the same quantity of mineral.

On the other hand, the political hierarchies are promoting the so-called bio-fuels. Apart from having low energy intensity, their whole production chain produces huge environmental and social impacts, especially in the case of biodiesel. Malaysia and Indonesia are by far the biggest producers of oil palm (the raw material) and export it mainly to the EU where it is refined. In order to obtain the enormous quantities of palm oil demanded, landowners rob land from small peasants and burn tropical forests to obtain new lands to be dedicated to oil palm cultivation. Many reports (many of them made by EU institutions) conclude that CO_2 emitted by burning and cultivating oil palms is greater than by burning oil in vehicles themselves.

1.2 *Potential from Social Sources*

1.2.1 Social Potential

Social potential is a very significant factor of societal transformability. It is formed by networks and norms which facilitate the social group's coordination for mutual benefit.

They are composed of three dimensions: structural (networks), relational (acknowledged norms which form communication codes and promote trust) and cognitive (common identity of the networks). We must distinguish between formal and informal social potential. The first type is formed by relationships established within organisations and which, at the same time, conform them. The second type refers to non-formal networks based on implicit norms, mutual trust and reciprocity.

The formal social potential refers to structured and hierarchical organisations, where relations are produced between individuals and organisations. Despite that, a high formal potential can only be obtained if relations are based on trust and reciprocity. These characteristics determine the institutional level of legitimacy and social cohesion of the organisations.

There are trends which destroy social potential and others which reinforce it. Economic globalisation promotes individualism and competition, causing social marginalisation and destroying social cohesion. It also degrades democracy, because the important decisions about the economy (which can produce strong social impacts: unemployment, reduction of social services, restrictions of labour rights, etc.) are taken by instances without democratic legitimacy. These trends downgrade de facto the democratic institutions, like parliaments, generating a growing mistrust among the population towards politics.

In parallel to the weakening process of the social potential, many initiatives are emerging (mainly at a local level) aimed at strengthening communities: the promotion of decentralised and sustainable economies; implementation of measures to protect the most vulnerable groups; promotion of people's participation in defining social policies and priorities; etc. These initiatives are being promoted by the so-called societies in energy emergency, eco-villages, municipalities which promote Local Agenda 21, etc. Besides, NGOs are being deployed and gaining strength very rapidly. Many of them have reached an international dimension and are taking part in United Nations environmental and social fora.

The negative process is far stronger than the positive one. The former is produced by the biggest power networks, which are working to strengthen the free market, and have developed today's international economic structure. On the other hand, the second trend is promoted by many grassroots groups. However, despite the huge imbalance of strength, it has to be taken into account that the first process is becoming weaker because of the increasing number of problems it is creating: an inability of world leaders to reach agreements upon how to continue the globalization trend; the diminishing role of the market in distribution of scarce raw materials; and a growing social opposition to economic globalisation. The second process is growing strongly (fuelled by the worsening problems) and is reaching a global dimension.

Technological and constructed potentials: Most of these potentials (especially the constructed potential) tend to perpetuate the dominant model, so they strengthen the potential for perpetuation (PP). They are capital-intensive, centralised and designed to be applied in any circumstance or location; they create little employment per unit of investment, use non-renewable energies, and produce enormous environmental impacts. Among the main technological trajectories we have: nuclear energy, technologies which use fossil fuels, most genetic engineering, some nanotechnologies, centralised computational applications, those which burn fossil fuels, and so on.

The constructed potential is formed by buildings, infrastructures, machines, durable consumption goods, etc. The constructed potential for perpetuation consumes fossil and nuclear energy, has a lower level of efficiency and is made with materials whose rate of recyclability is low. This potential is enormous because it has been accumulating since the start of the industrial revolution.

On the contrary, the transformational technological potential is decentralised (it is adapted to the specificities of each region or area), efficient, work-intensive, consumes renewable energy, and is formed by many technologies capable of being applied to local characteristics, suffering minor changes, and it contains infrastructures that can be used by many technologies at the same time. For example, railways designed for transport, passengers, freight, vehicles, etc. They are formed by technologies to capture solar and wind energies, to produce renewable hydrogen, for clean production, waste cycles waste, mass transport which produces the least environmental impacts, etc. But we also include new forms of organisation, such as industrial ecosystems, eco-efficient enterprises, organisations, etc.

Once more, contrasting the PP with the TP, the second is minor. But this is growing strongly. The constructed potential of renewable energy has been increasing at a greater pace than conventional energy in some OECD countries and regions. Also, metropolitan and long distance rail lines are steadily expanding mainly in emerging countries. We do not consider high speed railways as transformational technology.

1.2.2 Financial Potential

Financial potential is the investment capacity of a society. By far the greatest capacity is controlled by governments, banks and transnational corporations. The investment capability of governments comes from their money-collecting strength and their level of risk-free debt. Traditionally, States were able to borrow money from central banks and to print money. This action had to be carried out moderately in order not to cause a devaluation. But States have lost their control over central banks and with it they have lost the practice of borrowing and have to resort to the markets to obtain new funds. Being speculative markets, States with a high level of indebtedness can only borrow money with a very high interest rate when they decide that devaluating their currencies is not a good option. But countries which are members of the euro-zone have lost the capability of devaluating currencies, so they have no other alternative than to borrow money, and at the same time they are obliged to rapidly reduce their rate of indebtedness, by means of strong budgetary cuts. In this situation the capability for investment is nil.

On the other hand, the current financial system determines that banks create nearly all the monetary mass by borrowing money. That means that the economic system can only grow by increasing the level of indebtedness of societies. But during the crisis they strongly restrict its level, because without growth they do not want to take risks. Thus, when societies most need to invest to overcome the crisis, they lose the capacity for investment. Lastly, the financial system takes local money and normally invests it in the international markets, impoverishing the communities.

However, in some countries community development financial institutions (CDFI), which provide credit and financial services to populations are proliferating. This approach takes many forms: community development bank, community development credit union (CDCU), community development loan fund (CDLF), microenterprise development loan fund, community development corporation, and many more. These alternatives are particularly strong in the United States and Great Britain, and are frequently backed by law, like the USA's *Housing and Economic Recovery Act of 2008* (HERA). (www.neweconomics.org).

OECD countries have begun to take measures in order to reduce the vulnerability of the financial system: banks are obliged to create a reserve fund to help the bank to overcome crises; taxes on financial benefits obtained by rapid transactions; the banning of rapid transactions; splitting commercial banks and investment banks, etc. Although these measures reduce the system's vulnerability, the system is basically maintained. Thus, this system is part of the perpetuation potential.

1.2.3 Institutional Potential

The institutional PP is formed by institutions dedicated to preserving the current overall system. There are many types of institutions: for the development of fossil energy (primarily oil) (International Energy Agency IEA, Organisation of the Petroleum Exporting Countries (OPEC)); of nuclear energy (The International Atomic Energy Agency (IAEA), Nuclear Energy Agency (NEA), EURATOM, etc.); of the free market (WTO, EU, NAFTA, MERCOSUR, etc.); international financial organisations (International Monetary Fund (IMF), World Bank (WB)).

On the contrary, there is an acute scarcity of international institutions designed to promote transformations leading towards sustainability. Over the last decades there have been many proposals for the creation of a body to promote sustainability with an equal status to the WTO. At the moment, the only proposal that has at least a minimal possibility of being fulfilled is the integration of all United Nations environmental responsibilities in the UNEP. However, in recent years some positive changes have been produced. In 2009, the International Renewable Energy Agency (IRENA) was created. Also, some governments have created Ministries for

Sustainable Development or Sustainability Ministries. In the case of Queensland this Ministry has the status of Vice Premier Ministry, and it has the task of coordinating all ministries in order to apply the approved strategy for sustainable development.

However, the WTO has been losing strength during recent years. Besides, the WM has begun to take into account the environmental dimension in order to decide about financing projects. Another step in the right direction is the G-20 proposition for strengthening and democratising the IMF. Also, the increasing significance of the G-20 in detriment of the G-8. But, in spite of the above mentioned improvements, the institutional potential for transformation is one of the weakest ones, and we do not foresee any significant advance in this field in the near future. This result is a consequence of the governmental lack of will to assume the challenge of working hard and urgently to avert the threat of the mounting problems, especially the process of collision with the Earth.

1.2.4 Human Potential

This potential refers to the capacity level of workers. Among other elements, it comprises the knowledge for developing new technologies and for using the constructed potential. But, as we have seen in other potentials, here there are also potentials for the perpetuation and transformation of our societies. And taking into account that the technological potential for perpetuation is by far the strongest one, the human potential for perpetuation must be dominant. Besides, there is another element for strengthening this potential: educational institutions and above all universities usually have a great inertia. So their curricula are permanently lagging behind society's needs. Perhaps the primary factor of inertia could be the established evaluation system of researcher merits. Frequently, the merit evaluation systems show a bias in favour of conventional approaches and technologies, and they also receive most funds for research. On the other hand, groundbreaking theoretical and technological approaches suffer from a lack of funding and merits are not recognised.

However, the mounting evidence of the problems which humanity faces and the advances in some significant fields (ecology, sustainable economics, architecture, medicine, psychology and so on) are beginning to change curricula, merit ratings and funding policies.

2 Connectivity

Connectivity means the level of flexibility of a system to cope with changes. When connectivity is high a system is incapable of adapting itself to new circumstances. On the other hand, when a system has a low connectivity it is not able to control itself and to avert external threats. In both cases a system's collapse is the most likely outcome. For this reason the system needs a medium connectivity, one with which it is capable of adapting to external changes and of maintaining its self-control. The range of situations in which a system shows adaptability is called *window of viability*. The level of connectivity of a human system is determined basically by three factors: hierarchy, fluxes and paradigm.

2.1 Hierarchy

The hierarchy of socio-economic systems is qualitatively different from that of nature. The natural hierarchy is scalar, embedded and non- exclusive, and is determined by the organisational structure of the network of life. On the contrary, in current societies the hierarchy is determined by privileges, by control. It is not embedded because the system's structure does not determine it as the hierarchy is based on the decision-making structure (the existing level of democracy, social participation in the decision-making process, level of informative transparency, and so on). The current societal hierarchy is exclusive, because power is highly centralised, so a small group of people hold the power. This situation is worsening because economic centres are being strengthened and as a result of this their ability to influence government decisions also.

Being a hierarchy mostly based on obtaining privileges, agents of these economic centres have a huge interest in maintaining the current situation and work strongly towards multiplying connections with the primary power networks. In short, power centres are working towards furthering the globalisation process, and national power networks are trying to strengthen the international status of their States and national corporations.

The main power networks are controlled by the symbiotic power of the big political powers, primary financial entities and the largest corporations. These corporations control the biggest and most significant part of the industry, international trade, and the main communication holdings. These symbiotic relationships are reinforced through exchanges of people between main political parties and corporations. It is logical, because governments promote the creation of national monopolies to be able to compete in the international markets, and defend them when conflicts appear between corporations or between national corporations and other States. For instance, USA governments reject the attempts of strategic national company takeovers by Chinese enterprises. In parallel, corporations are the largest source of political party financing.

But in a competitive world it is difficult for the main powers to reach agreements to face the main general risks. And when they make decisions they are the result of compromises between conflictive interests, so they lack depth and, frequently, coherence. Corporations seek to control sectors and frequently decide upon conflicting strategies, as in the case of new technological trajectories. For instance, while some corporations decide to maintain the use of conventional technologies (for instance, conventional energies), others decide to develop alternative technologies because they believe that they will dominate in the near future (renewable energies).

Consequently, they compete for public funds. Frequently, new developments in some sectors negatively affect companies located in other sectors. Cargill controls the international grain market, and it is denouncing the strong development of bio fuels because they reduce the amount of cereals dedicated to feeding humans. Dow Chemical is pressing the USA government to make car makers to adopt high standards of vehicle efficiency, a means of reducing oil consumption and consequently prices, because high oil prices reduce Dow's profits (Vaidya 2008).

At a geostrategic level changes are occurring that reduce the overall connectivity of the system. When the Soviet Union collapsed the previous bipolar world became a mono-polar one. However, this world soon began to become weaker, leading to an increasingly multi-polar situation, mainly due to the emergence of China, though other emerging countries have also become stronger. At the same time, the United States is losing strength as a superpower. On the other hand, the European Union's status is steadily falling, because of its inability to solve serious economic and institutional problems. In general, the fact that the economic crisis is primarily affecting OECD countries explains, at least partially, the rapid weakening of the status of the traditional powers. The strengthening of a multi-polar world to a large extent explains that the superpowers have been incapable of culminating the liberalising trade Round at the WTO.

The WTO's status is also becoming weaker. Superpowers are adopting measures that the WTO has decided are barriers for the free market, though they do not accept the WTO's resolutions and continue with their behaviour, even though they receive sanctions. For instance, the EU does not accept the resolution on beef hormones. The USA continues to subsidise cereals and ethanol exports, and prohibits foreign companies from buying strategic companies. On the other hand, there is a global dynamic of nationalising many of the most important raw materials, especially gas and oil. Exporting countries control over 80 % of oil and natural gas reserves. And emerging countries are obtaining secure access to raw materials by signing long-term bilateral agreements with exporting countries.

2.2 *Fluxes*

Goods: Economic globalization increases international fluxes of goods (raw materials and finished products) and passengers, because it leads to the specialization of countries and areas. It also promotes the development of huge international monopolies, and as they specialize their plants, which are located in many countries, they produce a huge intra-firm trade. It represents 40 % of international trade. There are three main types of fluxes: raw materials, semi-manufactured products and finished goods. Two thirds of physical exports of materials are fossil fuels (60 % in weight) and abiotic materials and semi-manufactured products account for 20 % in weight. They are exported by Non-OECD countries to OECD countries, which use them to produce high-value goods and services.

Although exchanges between countries are very numerous, some studies show that most exchanges are concentrated in less than ten countries (this phenomenon is frequent in nature too), but as globalisation develops the number of countries with which they will have the main exchanges will grow. Besides, some of them determine a big dependency (that is, extensive connectivity). There are three which are particularly significant in relation to dependency: energy (mainly fossil fuels), strategic materials, and food (especially cereals). Frequently, a small group of countries control most exports. The United States, Canada, Argentina, Australia, France and Russia monopolise most cereal exports. On the other hand, the number of cereal importing countries is very high and it is growing steadily. In particular the trade of some strategic metals (aluminium, steel, nickel, copper, etc.) and of fossil fuels has grown much in recent decades. Exports of each of these materials are controlled mainly by three countries. Thus, the growing dependency of most countries on exports from an increasingly limited number of countries determines an extraordinary level of connectivity.

However, there are other factors which reduce the system's general connectivity. We have seen that in the period from 2005 to 2008 trade in weight was reduced due to rising oil prices, which determined rising transport costs. The reduction of trade was particularly pronounced in very long distance exports. The new process of escalating oil prices is beginning to produce the same effect and also its main consequence: a strengthening of the level of self-sufficiency of most economies, and reduced connectivity.

Financial fluxes: In 2008, the intensity and speed of financial exchanges reached the highest level in history, due to the extraordinary liberalisation produced in the sector. This phenomenon contributed to a strengthening of connectivity, but also generated highly vulnerability, which led to the sector's collapse. This has also become the main factor of the economic slowdown. The G-20 is trying to define a new framework for the international finance system. Its member countries have reached a broad agreement on two main issues: the IMF will control the sector at an international level, and a tax on financial transactions will be imposed. This agreement could bring down the system's connectivity a little, because would reduce exchanges. But it has not been put into practice yet, because there are many details to be fixed, which determine a new power balance in the control of the institution.

Information fluxes: In nature information is created by genetic changes as a result of adaptive processes, and it is stored in genetic code. In socio-economic systems information generates cultural changes, and produces socio-political changes. The information level has three elements: the wealth of knowledge, the level of transparency of government institutions and the capacity of knowledge-generating institutions to extend it to society as a whole.

The first element determines the social potential for improvement, and the other two the degree of application of the existing potential. Transparent management determines that the population knows about decisions and their repercussions. On the other hand, there is a growing gap between the large amount of knowledge generated and the population's ability to comprehend it. This means that the population is increasingly less able to understand and evaluate the benefits and risks of, for instance, new technologies. In essence, the main issue is that the population can evaluate what parts of the knowledge generated bring about a strengthening of the transformational or perpetuation potential.

The mass media seek to legitimise the main elements of the existing socio-economical and political system, by propagating the idea that it is the only way that has the capacity for problem-solving and the creation of increasing well-being (identifying this as growing consumption and a longer life expectancy). On the other hand, the mass media filter (by not informing of, or more frequently by giving less importance to) information that damages the positive image. But this image is becoming weaker, because it is less credible in a world that shows the many degrading elements of well-being. Populations show a growing concern about the polarisation of wealth, the drop in the well-being of most of the population, worsening environmental impacts, the scarcity of resources, and these factors and many others show a growing societal vulnerability to problems generated at an international level. Inevitably these subjects are being dealt with by the mass media with growing intensity. For example the case of the Fukushima nuclear reactor meltdown or the destruction caused by hurricane Sandy on the East coast of the United States.

However, the growing awareness of the population is not only caused by the evidence of mounting problems, but is also due to new sources of information, especially the Internet and mobile telephony. Through them the population can rapidly obtain information (which the mass media frequently do not publicise or downplay its relevance) and analyse the causes and importance of problems and their future outcomes. These new sources of information increase the connectivity of the transformational movement, which until now has been weak, limiting its full potential for effecting change. This growing connectivity between transformative actors is detrimental to the connectivity of the general system, contributing to the process of weakening the dominant paradigm.

2.3 Paradigms

The information available determines the strength of existing paradigms. We have commented that humanity is developing a confrontation process with nature, due to the dominant paradigm, which is based on the following premises: the human species is the most important (the Latin word primus means the first) and lives apart from nature; nature is chaotic and risky, so it must be dominated (humanised) in order to extract its wealth; happiness is obtained primarily through a process of increasing income, meaning that the economy must grow endlessly; there is no scarcity of resources; technological development will give us access to new resources; and the best model for producing the greatest rate of growth is based on a free market. This paradigm is the main factor of strength (great connectivity) of the general system.

Despite this, there is a great amount of data that shows that, although today's level of erosion of the dominant paradigm is relatively low, its weakening dynamics are rapidly becoming greater: numerous scientists have denounced its non-scientific nature; there is growing evidence of the destruction process of nature, the scarcity of resources, a massive loss of species, climate change, etc.; the degradation process of democracy; the growing disparity between incomes. As a result of these and many other problems the general system is becoming more vulnerable, and its crises more frequent, more profound and longer-lasting. All premises are suffering an erosion process, though not at the same rate. We arrange them here in an approximate decreasing order of erosion: separation between the human species and the rest of nature; nature is chaotic and risky; there is no scarcity of resources; the free market as the best tool to produce general well-being; and the correlation between income and happiness. The weakening process of the dominant paradigm runs counter to economic orthodoxy. In the last decade most of the economists who received the Nobel Prize are critical of the orthodox economy (Stiglitz, Akerlof, Hahneman, Aumann, Schelling, Sen, Krugman, Ostrom, etc.). Besides, orthodox economists have been unable to predict the economic crisis.

However, despite the degradation process that the dominant paradigm is suffering and growing dissatisfaction with the situation, most people do not have a clear vision of the mounting problems and the challenges that humanity must urgently face. This is the reason why they do not stand in favour of making profound changes, due to their inability to comprehend the situation and dynamics of today's world, and to envisage other possible alternatives. However, there is mounting evidence of the proliferation of initiatives to strengthen communities and decentralised and sustainable economies. This means that there is an ample transformative minority.

3 Panarchy

In nature each ecosystem is subjected to two forces. One exerted by the ecosystem of which the referential ecosystem is part and its force tends to stabilise it or when it collapses, it helps it to recover. The other force is exerted by a smaller ecosystem integrated in the referential ecosystem or simply by a part of this ecosystem. It promotes change because it is the first in reacting the changes that take place in its environment, tending to adapt to them, and in doing so they cause changes in the referential ecosystem. Thus, the convergence of the two forces gives stability and adaptability, producing a perfect equilibrium of forces tending to maintain basic functions, coherence and identity in an ever-changing environment.

Socio-economic systems function differently. The general system is acting not to stabilise societies which constitute parts of it, but to promote changes in them in order to fulfil its preferences, in essence, promoting economic liberalisation. However, this behaviour produces negative effects, especially in the weakest societies, although the strongest networks obtain benefits. Then, weak societies (mostly municipalities and regions) react trying to avert the impacts or at least to minimise them. These processes basically imply a reversed panarchy. When this happens in nature the referential ecosystem collapses. So the socio-economic systems with reversed panarchy are greatly unstable, and if they do not change rapidly, they will become more and more unstable, leading to their collapse.

However, we have seen that there are forces that tend to weaken this reversed panarchy. Briefly, the world is passing from the mono-polar hierarchy exerted by the United States to a multi-polar one, due to the growing strength of emerging countries (China, India, Brazil, Russia, etc.). So these main actors are incapable of reaching an agreed strategy, meaning that the liberalisation process is nearly stagnant. On the contrary, a rapidly increasing number of communities are working to create sustainable and self-reliant economies. So here the function of revolt appears, which some species exercise in ecosystems. Another aspect which is weakening this reversed panarchy is the changes produced by the increasing scarcity of resources. The main political powers, and especially emerging countries, are promoting bilateral agreements with exporting countries to secure long range access to resources. So the market is withdrawing in the strategic case of scarce resources. Although it is difficult to evaluate the state of reversed panarchy, it is obvious that the transformative movement is becoming stronger and the conflicting interests of superpowers are stagnating the liberalisation process. But it is regretful that superpowers are not able to promote positive changes, like a reform of the financial system or policies to protect the global environment.

4 An Attempt to Evaluate Global Transformability and Its Trend

The analysis of the potential shows the disparate situation and dynamics of the constituent elements. The natural potential is being reduced at an increasing pace. And this potential greatly determines the evolution of the other elements. In Chap. 1 we saw that several seminal reports warn us that there is no time to lose in order to avert the threat of destroying a large part of the natural potential. If it happens, essential services for preserving life could be highly degraded; arable land, water and critical materials would be severely scarce.

The different elements of the socially-originated transformative potential show rather low levels, although most of them are quickly becoming stronger. The stored technological potential for transformation has a significant dimension in some leading European countries (Germany, Denmark, Switzerland, etc.), especially in the case of renewable energy or conventional rail systems. On the other hand, we have the institutional and financial potential. The human and social potential reach an intermediate position, although they are very weak. As a whole, the socially-originated potential is growing rapidly, but coming from a very feeble position. This dynamic determines that in the mid-term this potential could be enormous.

The evaluation of the general system's connectivity shows that it is very significant. There are two main causes: economic globalisation and the dominant paradigm. Economic globalisation increases international trade, which strengthens the dependency on imports of most countries. In particular, three items increase connectivity and dependency: energy, critical materials and food. But connectivity is weakened because escalating oil prices increase transport costs and this deters trade, especially in the case of long distances.

The dominant paradigm is a major force for connectivity, because nothing significant can be changed if a large majority of people believe that there is no alternative to the current economic and social model and that a *good* government and technological development can solve the problems. But there can be a worse perspective, it is the case of the majority of people believing that the current situation is bad, but have no hope that it will improve because they do not envisage any possible alternative.

Another element which is detrimental to connectivity is the evolution which is taking place in information. The amount and quality which people receive determines whether a paradigm can change or not. We have seen that the dominant paradigm is becoming weaker due to the ever-growing evidence of the mounting problems, and to the increasing flow of non-biased information that people have access to via the Internet. This information is not only focused on problems, but also on solutions which are being improved by the contribution of many people. Besides, there is information about a rapidly increasing amount of alternative experiences.

On the other hand, superpowers struggle to improve their international status and privileges. But this priority makes it very difficult to reach agreements, which leads to a stagnant situation. This outcome has the positive dimension that superpowers are unable to strengthen the free market. The negative dimension is their inability to solve problems. The first dimension produces a loss of connectivity. The second increases the system's vulnerability. If the stagnant situation is consolidated, the general systems will continue forward without control until their collapse.

Panarchy is partially reversed because superpowers have been developing a free market economic system. This means that they are not trying to stabilize the global socioeconomic system, but are working to change the system to increase their power. This attempt is losing strength, due to the current process of going from a monopolar world to a strong multi-polar one. Thus, the power system is less and less able to control the world, and a loose system is incapable of survive in nature. On the contrary transformative movements are emerging at a strong pace. Thus, panarchy is partially reversed, but the reversion is losing strength.

We can conclude that transformability is very weak (due to the feeble transformational potential, to the high connectivity and to the panarchy that has been partially reversed), but it is growing fast. However, if transformability does not grow fast enough, the process of destroying the natural capital could lead to an enormous lack of natural potential. In this case, humanity would not be able to build a sustainable world, capable of satisfying our basic needs. Thus, increasing the rate of change is the main challenge of the current generation.

Bibliography

Abalofia, M., & Biggart, N. (1992). Competitive systems: A sociological view. In P. Ekins & M. Max-Neef (Eds.), *Real-life economics*. London: Routledge.

Abel, T., & Stepp, J. R. (2003). A new ecosystems ecology for anthropology. *Conservation Ecology, 7*(3), 12.

Abel, N., et al. (2006). Collapse and reorganization in social-ecological systems: Questions, some ideas, and policy implications. *Ecology and Society, 11*(1), 17.

Ad-Hoc Working Group. (2010). *Critical raw materials for the EU*. Brussels: European Commission.

Adamides, E. D., & Mouzakitis, Y. (2008). Industrial ecosystems as technological niches. *Journal of Cleaner Production, 17*, 172–180.

Adamson, K. -A. (2009). *Small stationary survey*. Fuel Cell Today.

Adamson, K.-A., & Callaghan, L. (2009). *2009 Niche Transport Survey*.www.fuelcelltoday.com

Adriaanse, A., et al. (1997). *Resource flows. The material basis of industrial economies*. Washington, DC: World Resource Institute.

Agre, P., et al. (2011). *The Stockholm memorandum*. 3rd Nobel Laureate symposium on global sustainability, Stockholm.

Aguilera Klink, F. (1992). *Sobre la irrelevancia conceptual de la economía ambiental*. Barcelona: III Jornadas de Economía Crítica.

Aleklett, K. (2010). *Peak coal in China*. www.energybulletin.net/print/55011

Allenby, B. R. (1999). *Industrial ecology: Policy framework and implementation*. Upper Sadle River: Prentice Hall.

Allenby, B. R. (2009). The industrial ecology of emerging technologies. *Journal of Industrial Ecology, 13*(2), 168–183.

Allenby, B. R., & Rejeski, D. (2008). The industrial ecology of emerging technologies. *Journal of Industrial Ecology, 12*(3).

Alterra, et al. (2007). *Review of existing information on the interrelations between soil and climate change*. Wageringen: Alterra, Wageringen UR.

Altmann, M., et al. (2004). *Potential for hydrogen as a fuel for transport in the long term (2020–2030)*. Brussels: European Commission.

Altvater, E. (1993). *The future of the market*. London: Verso.

Ambah, F. S. (2008). *Saudies prepare for post-petroleum era*. www.Philly.Com

ANC (The National Academies). (2007). *Coal: Research and development to support national energy policy*. Washington, DC: The National Academies.

Anderies, J. M., et al. (2004). A framework to analyze the robustness of social-ecological systems from institutional perspective. *Ecology and Society, 9*(1), 18.

Anderson, R. (1998). *Mid-course correction*. Atlanta: The Perregrinzilla Press.

R. Bermejo, *Handbook for a Sustainable Economy*,
DOI 10.1007/978-94-017-8981-3,

Anderson, R. (2003). *Ants, Galileo, and Gandhi*. Sheffield: Greenleaf Publishing.
Anderson, K., & Ohlson, T. (1999). *Including environmental aspects in production development: A case study of tomato ketchup*. Academy Press.
Anderson, et al. (1995). *Trading with the environment, ecological economics, institution and policy*. London: Earthscan.
Anderson, R, et al. (2010). Changing business cultures from within. *State of the World 2010*. www.worldwatch.org
Anton, P., et al. (2001). *The global technological revolution*. Santa Monica: National Defence Research Institute.
Appleyard, D. (2010). *PV global outlook: A bright future shines on PV*. Renewable Energy World.
Arrow, K. et al. (1995, April 28). Economic growth, carrying capacity and the environment. *Science*.
Ashton, W. S. (2009). The structure, function, and evolution of a regional industrial ecosystem. *Journal of Cleaner Production, 13*(2).
Aunos. (1991). *Actualidad forestal del País Vasco*. Fundación BBV: Bilbao.
Ayres, R. U. (1989). Industrial metabolism. In J. Ausubel & H. Sladovich (Eds.), *Technology and environment*. Washington, DC: National Academy Press.
Ayres, R. U. (2002). On industrial ecosystems. In R. U. Ayres & L. W. Ayres (Eds.), *Handbook of industrial ecology*. London: Edward Elgar.
Ayres, R. U. (2008). Sustainability economics: Where do we stand. *Ecological Economics, 67*, 281–310.
Ayres, R. U., et al. (1992). Industry and waste. In J. I. C. Dooge et al. (Eds.), *An agenda of science for environment and development into 21st century*. Cambridge: Cambridge University Press.
Baas, L. W. (2001). Developing an industrial ecosystem in Rotterdam. *Journal of Clean Production, 4*(2), 4–6.
Baas, L. W., & Boons, F. A. (2004). An industrial ecology project in practice exploring the boundaries of decision-making levels in regional industrial ecosystems. *Journal of Cleaner Production, 12*, 1073–1085
Baines, T. S., et al. (2007). *State-of-art in product-service systems*. Proceedings of the IMechE (Vol. 221).
Bakhtiari, A. M. S. (2006). *After 'Peak Oil', 'Peak Gas' too*. www.sfu.ca, March.
Bardi, H., & Pagani, M. (2007). *Peak minerals*. http://europe.theoildrum.com/node/3086
Bardi, U., et al. (2011). Modelling EROEI and net energy in the explotation of non renewable resources. *Ecological Modelling, 233*(1), 54–58.
Bartelmus, P. (2003). Accounting for sustainable development? *Ecological Economics, 61*, 613–616.
Bartolomeo, M., et al. (2003). Eco-efficient producer services. What are they, how do they benefit customary and the environment and how likely are they to develop and be intensively utilised? *Journal of Clean Production, 11*, 829–837.
Batson, A. (2009, March 24). China takes aim at dollar. *The Wall Street Journal*.
Beder, S. (1996). Charging the earth. The promotion of price-based measures for pollution control. *Ecological Economics, 16*, 51–63.
Behrendt, D., et al. (2003). *Eco-service development*. Sheffield: Greenleaf.
Behrendt, D., et al. (2004). *Environmental policy instruments for dematerialization of European Union*. Vienne: SERI.
Beherens, A. (2004). *Environmental policy instruments for dematerialisation of the European union*. Vienne: SERI.
Benyus, J. (1997). *Biomimicry. Innovation inspired by nature*. New York: Quill/William Morrow.
Benyus, J. (2006). *Biomimicry. Innovation inspired by nature*. Aspen Ideas Festival.
Benyus, J. (2009). *What do you mean by the term biomimicry?*. www.biomimicryinstitute.org
Benyus, J. (2010). *The Biomimicry Institute. Inspiring, educating, and connecting biomimics throughout the world*. Biomimicry Institute. www.biomimicryinstitute.org
Bergin, P. R., & Glick, R. (2006). Global price dispersion: Are prices converging or diverging? (Working Paper 50). Federal Reserve Bank of San Francisco.

Berman, A. (2010). *ExxonMobil's acquisition of XTO energy: The fallacy of the manufacturing model in shale plays*. www.theoildrum.com/node/6229

Berman, A., & Pittinger, L. (2011). *US shale gas: Less abundance, higher costs*. www.theoildrum.com/node/8212

Bermejo, R. (2005). *Análisis socioeconómico del Plan Estratégico de Infraestructuras y Transporte 2005–2020*. Cuaderno Bakeaz 69.

Bermejo, R. (2008). *Un futuro son petróleo. Colapsos y transformaciones socioeconómicas*. Madrid: CIP y Catarata.

Bermejo, R. (2009). La política de transporte española ante el fin de la era de los combustibles fósiles. In S. Álvarez Cantalapiedra & O. Carpintero (Eds.), *Economía ecológica: Reflexiones y perspectivas*. Madrid: Ediciones Ciencias Sociales.

Bermejo, R. (2011). *Manual para una economía sostenible*. Madrid: Catarata.

Bermejo, R., Arto, I., & Hoyos, D. (2010). The concept of sustainable development and its manipulation. In K. Unceta & A. Arrinda (Eds.), *Development cooperation. Facing the challenges of global change, current research* (Vol. 3). University of Nevada, Reno.

Berthoud, G. (1993). Market. In W. Sachs (Ed.), *The development dictionary*. London: Zed Books.

BEUC (European Consumer's Organization). (2002). *Labels and green claims*. Bruxelles: BEUC.

BEUC (European Consumer's Organization). (2003). *BEUC memorandum for the Irish presidency*. Bruxelles: BEUC.

BEUC (European Consumer's Organization). (2008a). *A strong, transparent and proactive pharmacovigilance system*. Bruxelles: BEUC.

BEUC (European Consumer's Organization). (2008b). *Input to the commission consultation document on the revision of the EU energy labelling scheme*. Brussels: BEUC.

BEUC (European Consumer's Organization). (2008c). *Pharmaceutical sector inquiry*. Brussels: BEUC.

BEUC (European Consumer's Organization). (2012, February 21). *European chemicals safety net flawed Commission study reveals*.

Bina, C. (2009). *The geopolitics of energy in the epoch of globalization*. Workshop International Competition, Rivalry and Conflict.

BIS (Bank for International Settlements). (2007). *77º Annual report*. Basle: BIS.

Blau, J. (2010). *Stakeholders team up to expand Europe's super-grid*. Renewable Energy World.

Bleischwitz, R., & Bringezu, S. (2007). Global resources management. *Policy Paper, 27*.

BNEF (Bloomberg New Energy Finance). (2010). *Fossil fuel subsidies outpace renewables*. BNEF.

Bloomberg New Energy Finance (BNEF). (2011). *Onshore wind energy to reach grid parity with fossil fuels by 2016*. BNEF.

Boons, F., & Roome, N. (2001). Industrial ecology as a cultural phenomenon. *Journal of Industrial Ecology, 4*(2).

Boons, F., & Spekkink, W. (2012). Levels of institutional capacity and actor expectations about industrial symbiosis. *Journal of Industrial Ecology, 16*(1), 61–69.

Boscá, J. E., et al. (2004). Efectos macroeconómicos de las inversiones en infraestructuras públicas. Jornadas sobre el PEIT. Madrid.

Boston University Student Research. (2011). *Car sharing industry*. CFA Institute Research Challenge.

Braclay, C. (2012). Food miles, *Science and Environment Section*. House of Commons Librery.

Bradley, R. L. (2007). Resourceship: An Austrian theory of minerals resources. *Review of Austrian Economy, 20*, 63–90.

Brand, F. S., & Jax, K. (2007). Focusing the meaning(s) of resilience: Resilience as a descriptive concept and a boundary object. *Ecology and Society, 12*(1).

Brangwyn, B., & Hopkins, R. (2008). *Transition initiatives primer. Becoming a transition town, City, District, Village, Community or even Island*. Transition Network.

Braungart, M., & Bollinger, A. (2004). *Cradle to cradle: Redesigning the relationship between industry and nature*. 9th European Round Table on Sustainable consumption and production, Bilbao.

Braungart, M., et al. (2003). Eco-effective design of products and production systems: Eight theses on methodological and institutional prerequisites. In J. D. Seiler-Hausman et al. (Eds.), *Eco-efficiency and beyond*. Sheffield: Greenleaf.
Braungart, M., et al. (2006). Cradle-to-cradle design: Creating healthy emissions – A strategy for eco-effective product and system design. *Journal of Cleaner Production, 15*, 1337–1348.
Braungart, M., et al. (2007). Cradle to cradle design: Creating a healthy emissions- a strategy for eco-effective products and systems design. *Journal of Cleaner Production, 15*(13–14), 1337–1348.
Breese, J., & Room, D. (2005, October 21). Powering down America: Local government's role in the transition to a post-petroleum world. *Energy Bulleting*, 49–58.
Bright, C. (2003): "La historia de nuestro futuro", en Bright, C. et al., *La situación del mundo 2003,* Barcelona, Icaria/WWI.
Bringezu, S. (2006). *Materializing policies for sustainable use and economy-wide management of resources*. Wuppertal: Wuppertal Institute.
Bringezu, S., & Schutz, H. (2001). *Total material requirement of the European Union*. European Environmental Agency (Technical Report nº 55).
Bringezu, S., et al. (2007). *Towards a sustainable biomass strategy* (Wuppertal Papers Nº 163). Wuppertal Institute.
Bringezu, S., & Schutz, H., et al. (1998). *Material flows account*. Eurostat.
Bringezu, S., & Schutz, H., et al. (2001, January). *Towards sustainable resource management in European Union* (Wuppertal Papers Nº 121). Wuppertal.
Bringezu, S., & Schutz, H., et al. (2004, June). *Towards a sustainable biomass strategy* (Wuppertal Papers No. 163). Wuppertal.
Brisbane City Council. (2007). *Final report. A call for action*. Brisbane City Council.
Bromley, W. D. (1995, August). Property rights and natural resource damage assessment. *Ecological Economics*, 129–135.
Bromley, W. D. (1998). Searching for sustainability. The poverty of spontaneous order. *Ecological Economics, 24*, 231–240.
Brook, D. (2010). *The other next big thing: Peer to peer carsharing*. Carsharing USA.
Brown, L. R. (2001). *Eco-economy. Building an economy for the earth*. Washington, DC: Earth Policy Institute; New York: Norton.
Brown, L. R. (2008). *Plan B 3.0 mobilizing to save the civilization*. Washington, DC: Earth Policy Institute.
Brown, L. R. (2009). *Plan B 4.0 mobilizing to save the civilization*. Washington, DC: Earth Policy Institute.
Brown, L. R. (2010). Rising temperatures raise food prices – Heat, drought and a failed harvest in Russia. *Plan B Updates*. Washington, DC: Earth Policy Institute.
BTC (Bundeswehr Transformation Centre). (2010). *Peak oil. Implications of resource scarcity on (National) security*. BTC.
Buitenkamp, M. (2008). *EEB's vision to European agriculture 2008–2020*. EEB.
Bukold, S. (2010). *Offshore oil drilling: Public costs and risks are too high* (Global Oil Briefing Nº 50). Hamburg: Energy Comment.
Burnaby City Council. (2006). *Global peak in oil production: The municipal context*. Burnaby.
Butler, J. (2009). *Portable fuel cell survey 2009*. www.fuelcelltoday.com
CAGT (Center for German Army Transformation). (2010). *Implications of resource scarcity on (National) security*. CAGT.
Cairncross. (1996). *Ecología S. A.* Madrid: Ecoespaña.
Callaghan, L. (2008). *2008 bus survey*. Fuel Cell Today. www.fuelcelltoday.com
Callaghan, L., et al. (2010). *2010 transport fuel cell analysis*. Fuel Cell Today. www.fuelcelltoday.com
Callahan, J. (2011). *2011 ASPO-USA: Day 2*. www.theoildrum.com/node/8573
Callahan, J. (2012). *Gas boom goes bust*, www.theoildrum.com/node/9744
Calleja, I., et al. (1999). *Natural resources and the environment panel report*. Sevilla: European Commission (Joint Research Centre).

Callicott, J. B. (1999). *Beyond the land ethic: More essays in environmental philosophy* (CID Working Paper No. 120). Albany: State University of New York Press.
Callicott, J. B. (2002). The pragmatic and promise of theoretical environmental ethics: Forging a new discourse. *Environmental values, 11.*
Cameron, A. (2011). *Tracking the CPV global market: Ready to fulfil its potential?* www.renewableenergyworld.com
Campbell, C., & Sivertsson, A. (2003, May). *Updating the depletion model.* Proceedings of the second international workshop, Paris, France.
CAN Europe et al. (2009). *Comments on ERGES smart grids position paper.* World Future Council. www.ceer.eu
Cantlon, J. E., & Koening, H. E. (1999). Sustainable ecological economics. *Ecological Economics, 31*, 107–121.
Capra, F. (1996). *The web of life*. London: HarperCollins.
Capra, F. (2002). *The hidden connections. Integrating the biological cognitive, and social dimensions of life into a science of sustainability*. New York: Doubleday.
Carpenter, A. (2007). *Making decisions for the future: Climate change. The premier's climate change action statement*. Perth: Western Australia Government.
Carpenter, S. R., & Forlke, C. (2006). Ecology for transformation. In *TRENDS in ecology and evolution*. Amsterdam: Elsevier.
Carpintero, O. (2006). *La bioeconomía de Georgescu-Roegen*. Madrid: Montesinos.
Casey, V. (2012). Biomimicry 3.8: What would you ask nature? *Core 77*. www.core77.com
CE-DGXI. (1998). *Integrated product policy*. Brussels: European Commission.
CERA (Cambridge Energy Research Associates). (2007). *Engineering talent squeeze –"people deficit" – likely to cause further delays in some oil & gas projects through 2010*. CERA. www.cera.com, www.ihs.com
CES. (1992). *Dictamen sobre resolución del Consejo de las Comunidades Europeas sobre un programa comunitario de política y actuación en materia de medio ambiente y desarrollo sostenible*. Comisión Europea: Bruselas.
CGAT (Centre for German Army Transformation). (2010). *Implications of resource scarcity on national security*. CGAT.
Chaykovskaya, E. (2010, October 29). Trillions necessary to maintain Russia's oil production. *The Moscow News*.
Chen, G. et al. (2007). *Peak oil and its implications for Hong Kong*. HKSAR Government.
Chertow, M. R. (2002). Industrial symbiosis: The legacy of Kalundborg. In R. Ayres & L. Ayres (Eds.), *A handbook of industrial ecology*. London: Edward Elgar.
Chertow, M. R. (2007). 'Uncovering' industrial symbiosis. *Journal of Industrial Ecology, 11*(1), 11–30.
Chertow, M. R. (2008). *Industrial symbiosis*. www.eoearth.org/article/Industrial_symbiosis
Chertow, M. R. (2009). Dynamics of geographically base industrial ecosystems. In M. Ruth & B. Davidsdottir (Eds.), *The dynamics of regional and networks in industrial ecosystems*. Cheltenham: Edward Elgar.
Chertow, M. R., & Ehrenfeld, J. (2012). Organizing self-organizing systems. Toward a theory of industrial symbiosis. *Journal of Industrial Ecology, 16*(1), 13–27.
Chew, A. (2010). *Germany no longer critical to PV market growth*. Renewable Energy World. www.renwwableenergyworld.com
Christensen, P. P. (2001). Early links between sciences of nature and economics. In P. Cleveland et al. (Eds.), *The economics of nature and the nature of economics*. Cheltenham: E. Elgar.
Christensen, E., & Geffin, S. (1991). GATT sets its net on environmental regulation: The GATT panel ruling on Mexican yellowfin tuna imports and the need for reform of international trade system. *Inter-American Law Review, 23*, 569–612.
Christou, M. (2010). *Crops2Industry "Non-food Crops-to-Industry schemes in EU27*. 2nd Thematic Workshop, Winschoten, Netherlands.
Clark, W. C., et al. (2005). *Science for global sustainability: Toward a new paradigm* (CID Working Paper N. 120), Harvard University.

Coase, R. (1960). The problem of social cost. *Journal of Law and Economics, 3*.
Coase, R. H. (1981). El problema del coste social. *Hacienda Pública Española* nº 68.
Cohen, D. (2009a). *The next oil shock*. www.energybulletin.net/node/49857
Cohen, D. (2009b). *A year we will live to regret*. www.energybulletin.net/node/51013
Cohen, D. (2010). *The shale game shell game*. www.energybulltin.net/print/55004
Cohen-Rosenthal, E. (2004). Making sense out of industrial ecology: A framework for analysis and action. *Journal of Cleaner Production, 12*, 1111–1123.
COI (Commission on Oil Independence). (2006). *Making Sweden OIL-FREE society*, G. Persson, Sweden.
Colen, P., & Lambrecht, M. (2010). *Product service systems as a vehicle for sustainability: Exploring service operations strategies*, Casablanca.
Cole, K., Cameron, J., & Edwards, C. (1983). *Why ecomomists disagree: The political economy of economics*. London: Longman.
Coleman, J. (2008). *Energy chief: Flat production behind oil prices*. Associated Press.
Collins, T. (2001, January 5). Toward a sustainable chemistry. *Science, 291*, 48–49.
Comisión Brandt. (1981). *Diálogo Norte-Sur (Informe Brandt)*. México City: Nueva Imagen.
Coob, K. (2011). *Not so much: Shale gas shows its limitations*. www.energybulletin.net/print/59541
Cook, M. B., et al. (2006). The transfer and application of product service systems: From academia to UK manufacturing firms. *Journal of Cleaner Production, 14*(17), 1455–1465.
Cooper, T. (2002). Durable consumption: Reflections on product life cycles and the throwaway society. In E. Hertwich (Ed.), *Life-cycle approaches to sustainable consumption*. Luxembourg: International Institute for Applied Systems Analysis.
Cooper, T. (2005). Slower consumption. *Journal of Industrial Ecology, 9*(1–2), 51–68.
Cooper, T. (2012). *The value of longevity: Product quality and sustainable consumption*. Global Research Forum on Sustainable Consumption and Production, Workshop, Río de Janeiro.
Costanza, R. (1993). Developing ecological research that is relevant for achieving sustainability. *Ecological Applications, 3*(4), 579–581.
Coté, R. P. (2001). *The evolution of an industrial park: The case of burnside*. Halifax, Canada. www.uneptie.org/pc/estate
Coté, R. P., & Crawford, P. (2003). A case study in eco-industrial development: The transformation of Burnside industrial park. In E. Cohen-Rosenthal (Ed.), *Eco-industrial strategies*. Sheffield: Greenleaf.
Cottrell, J. (2009). *The double benefit of EFR: Good practices and interesting ideas in some EU countries*. Seminar: Environmental Fiscal Reform and its Contribution to a Green Economy, Madrid.
Cour des comptes. (2006). *Le reseau Ferroviaire. Une reforme inachevée, une stratégie incertaine*. Paris: Gobierno Francés.
Crosby, A. W. (1988). *El Imperialismo Ecológico de Europa*. Barcelona: Crítica.
Crowell, R. (2011). *With wind power smaller is bigger*. www.renewableenergyworld.com. Accessed 15 Apr 2011.
Curtis, F. (2003, March). Eco-localism and sustainability. en *Ecological Economics* 46.
Daily, G. C., et al. (1996). Managing earth's life support systems: The game, the players, and getting everyone to play. *Ecological Applications, 6*, 19–20.
Dajian, Z. (2006). *Circular economy and China's development under the scarcity of natural capital*. International Conference on Industrial Ecology, Lausanne, Tonji University. www.pmpp.cn
Daly, H. E. (1974). The economics of the steady state. *American Economic Review, 64*, 15–21.
Daly, H. E. (1989). *Steady-state and growth concepts for the next century, economy and ecology: Towards sustainable development*. London: Kluwer Academic.
Daly, H. E. (1992). Allocation, distribution, and scale: towards an economics that is efficient, just, and sustainable. *Ecological Economics, 6*(3), 185–193.
Daly, H. E., & Goodland, R. (1994). An ecological-economic assessment of reregulation of international commerce under GATT. *Ecological Economics, 9*, 73–92.

Danish Board of Technology. (2004). *Oil-based technology and economy. Prospects for the future*. Copenhagen: Teknologiradet.
Danish Ecological Council. (2002). *Environmental budget reform in Denmark. Policy paper*. www.ecocouncil.dk
Davignon, E. (2008). *Informe anual del coordinador europeo. Proyecto prioritario nº 3. Eje ferroviario de alta velocidad del sudoeste de Europa*. Bruselas: Comisión Europea.
de Belleuve, E. B., et al. (1994). The North American free trade agreement. An ecological-economic synthesis for the United States and Mexico. *Ecological Economics, 9*, 53–72.
de Graff, J. (2010). Reducing work time as a path to sustainability. In WWI (Ed.), *La situación del mundo 2009*. Barcelona: Icaria.
de Groot, R. S. (1992). *Functions of nature. Evaluation of nature in environment, planning management and decision making*. Amsterdam: Wolter-Noordhoff.
De Groot, R. S. (1994). Environmental functions and the economic value of natural ecosystems. en A. Jansson, et al. (Coed.), *Investing in natural capital: The ecological economics approach to sustainability*. Washington, DC: Island Press.
de Jonn, P. (1999). *Our common journey*. London: Zed Books.
de Sousa, L. (2009). *Andris Piebalgs: It may have peaked*. http://europe.theoildrum.com/node/5397
de Sousa, L. (2010). *Copper peak*. http://europe.theoildrum.com/node/6307
de Vries, E. (2009). Speaking of wind: Discussions from Germany. *Renewable Energy World*. www.renwableenergyworld.com.
DEFRA (Department for Environment, Food and Rural Affairs). (2007). *UK biomass strategy*. Defra. www.defra.gov.uk
Diario de Sesiones del Congreso de los Diputados. (2008). *Discurso de don José Luis Rodriguez Zapatero en la sesión de investidura como Presidente del Gobierno,* Sesión plenaria nº2, celebrada el 8 de Abril de 2008.
Dick-Madsen, S. (2009). *The double benefit of EFR: Good practices and interesting ideas in some EU countries*. Seminar: Environmental Fiscal Reform and its Contribution to a Green Economy, Madrid.
Diederen, A. (2009). *Metal minerals scarcity: A call for managed austerity and the elements of hope*. http://europe.theoildrum.com/node/5239
Diederen, A. (2010). *Global resource depletion*. Delft: Eburon.
Dietz, T., et al. (2003). The struggle to govern the commons. *Science, 302*, 1907–1912.
Dinda, S. (2004). Environmental Kuznets curve hypothesis: A survey. *Ecological Economics, 49*(4), 431–455.
DiPaola, A. (2013). OPEC nations seek cash for .5 billion solar power plan, Bloomberg. www.renewableenrgyworld.com
Dobers, P. (2009). Corporate social responsibility: Management and methods. *Corporate Social Responsibility and Environmental Management, 16*(4), 63–69.
Dodge, J. C. I., et al. (1992). *An agenda of science environment and development into 21st century*. New York: Cambridge University Press.
DOE (US Department of Energy). (2004a). *Report for the Congress. A hydrogen economy and fuel cells: An overview*. Washington, DC: DoE.
DOE (US Department of Energy). (2004b). *Hydrogen posture plan*. Washington, DC: DoE.
DOE (US Department of Energy). (2006). *Hydrogen posture plan. An integrated research, development and demonstration plan*. Washington, DC: DoE.
Dubyne, D. (2007b, October 21). The new silk road: Rail links from China. *Energy Bulletin*.
Durning, A. T. (1994). *Cuanto es bastante. La sociedad de consumo y el futuro de la Tierra*. Barcelona: Apóstrofe.
Dyson, B., et al. (2010). *Towards a twenty-first century banking and monetary system,* Positive Money. NEF y University of Southampton.
EC-DGXI. (1998). *Integrated product policy*. Brussels: European Commission.
ECORYS. (2008). *Progress on EU sustainable development strategy*. Brussels: European Commission.

ECOS (Environemental Council of the States). (2007). *Regional collaboration among states to improve joint planning and priority setting: A new England case study*. ECOS.

Eddington, R. (2006). *The Eddington transport study*. www.dft.gov.uk

Eder, P., & Sotoudeh, M. (2001). *Innovation and cleaner technologies as a key to sustainable development: The case of the chemical industry*. Seville: Institute for Prospective Technological Studies, European Commission.

Edwards, A. (2007). Property in Totnes: Wizards of wacky west. www.telegraph.co.uk

Edwards, A., et al. (2008). *Biofuels in the European context: Facts and uncertainties*.

EEA (European Environment Agency). (1996). *El tributo ambiental*. Copenhagen: EEA.

EEA (European Environment Agency). (1997). *Towards sustainable development for local authorities: Approaches, experiences and sources*. Copenhagen: EEA.

EEA (European Environment Agency). (1998, December 17). Making sustainability accountable. *Newsletter*.

EEA. (2000). *Environmental taxes: Recent developments in tool for integration*. Copenhagen: European Commission.

EEA (European Environment Agency). (2004). *Energy subsidies in the European Union: A brief overview*. Copenhagen: EEA.

EEA (European Environment Agency). (2005). *Market-based instruments for environmental policy in Europe*. Copenhagen: EEA.

EEA (European Environment Agency). (2006). *Using the market for cost-effective environmental policy*. Copenhagen: EEA.

EEA (European Environment Agency). (2007). *Size, structure and distribution of transport subsidies in Europe*. Copenhagen: EEA.

EEA. (2008). *Effectiveness of environmental taxes and charges for managing sand, gravel and rock extraction in selected EU countries*. Copenhagen: EEA.

EEA (European Environment Agency). (2008a). *Effectiveness of environmental taxes and charges for managing sand, gravel and rock extraction in selected EU countries*. Copenhagen: EEA.

EEA (European Environment Agency). (2008b). *Time for action – Towards sustainable consumption and production in Europe*. Copenhagen: EEA.

EEA (European Environment Agency). (2009). *Europe's onshore and offshore wind energy potential*. Copenhagen: EEA.

EEA (European Environment Agency). (2010). *The European environment. State and outlook 2010. Consumption and the environment*. Copenhagen: EEA.

EEA (European Environment Agency). (2011). *Laying the foundations for greener transport*. Copenhagen: EEA.

EEAC (European Environment and Sustainable Development Advisory Councils). (2008). *Statement: Sustaining Europe for a long way ahead*. EEAC. www.eeac-net.org

EEB. (2002). Environmental fiscal reform. Making prices work for the environment. *Campaign Newsletter*, 4, December.

EEB (European Environmental Bureau). (2004a). *European chemicals policy reform. From emotions to facts*. www.eeb.org

EEB. (2004b). *EEB workshop on the thematic strategy for soil protection*. www.eeb.org

EEB (European Environmental Bureau). (2008a): *EEB response to the sustainable consumption and production and sustainable industrial policy action plan*. www.eeb.org

EEB (European Environmental Bureau). (2008b). *EEB 2008 annual conference report*. www.eeb.org

EGFTF (European Expert Group on Future Transport Fuels). (2011). *Future transport fuels*, EGFTF.

EHA (European Hydrogen Association). (2010). *EHA contribution to the consultation on the energy strategy for Europe EU 2011–2020*. EHA.

Ehrenfeld, J. R. (2004a). Industrial ecology: A new field or only a metaphor? *Journal of Cleaner Production, 12*, 825–831.

Ehrenfeld, J. R. (2004b). Can industrial ecology be the "science of sustainability". *Journal of Cleaner Production, 8*, 1–3.

Ehrenfeld, J. R. (2007). Would industrial ecology exist without sustainability in the background? *Journal of Industrial Ecology, 11*(1), 73–84.

Ehrenfeld, J. R. (2009). Understanding complexity expands the reach of industrial ecology. *Journal of Industrial Ecology, 13*(2), 165–167.

Ekins, P. (1992). *Real-life economics. Understanding wealth creation*. London: Routledge.

Engdahl, W. (2006). *The emerging Russian giant play its cards strategically*. Centre for Research on Globalization. www.globalresearch.ca

EPA (US Enviromental Protection Agency). (2002). *Wastewise update. Building for the future*. Washington, DC: EPA.

EPA (US Enviromental Protection Agency). (2009). *"Green servicizing" for a more sustainable US economy*. Washington, DC: EPA.

EPIA. (2010a). *Global market outlook for photovoltaic until 2014*. EPIA. www.epia.org

EPIA. (2010b). *Unlocking the sunbelt. Potential of photovoltaic*. EPIA. www.epia.org

EPIA. (2010c). *Solar generation 6*. EPIA. www.epia.org

EPIA. (2010d). *Thin film go mainstream*. EPIA. www.epia.org

EPIA. (2010e). *Concentrator photovoltaics technology has reached commercialization*. EPIA. www.epia.org

EPIA. (2011). *Global market outlook for photovoltaics until 2015*. EPIA. www.epia.org

EPIA. (2012). *Global market outlook for photovoltaic until 2016*. EPIA.

EPIA (European Photovoltaic Energy Association)/Greenpeace. (2007). *Solar generation IV-2007*. EPIA y Greenpeace. www.epia.org/www.greeneace.org

EPIA (European Photovoltaic Industry Association). (2012). Market report 2011.

Erdmenger, J. R. (2005). The financial power end environmental benefits of green purchasing. In C. Erdmenger (Ed.), *Buying into the environment*. Sheffield: Greenleaf.

EREC (European Renewable Energy Council). (2010a). *Re-thinking 2050. A 100% renewable energy vision for the European Union*. Brussels: EREC.

EREC (European Renewable Energy Council). (2010b). *Energy infrastructure*. Brussels: EREC.

EREC (European Renewable Energy Council), & Greenpeace. (2010). *Energy (R)evolution. Towards a fully renewable energy supply in the EU 27*. Brussels: EREC y Greenpeace.

ERGEG (European Regulators Group for Electricity & Gas). (2009). *Position paper on smart grids. An ERGEG public consultation paper*. Brussels: ERGEG. E09-EQS-30-40.

Eriksen, T. (2009). *World oil forecast-update November 2009*. www.energybulletin.net/print/50793

Erkman, S. (1998). *Vers une écologie industrialle*. Paris: Editions Charles Léopold Mayer.

Erkman, S. (2002). The recent history of industrial ecology. In R. Ayres & L. Ayres (Eds.), *A handbook of industrial ecology*. London: Edward Elgar.

ESC (Economic and Social Committee). (1992). *Proposal for a resolution of the council of the European communities on a community programme of policy and action in relation to the environment and sustainability* (COM(1992) 23 final). European Commission.

ESC (Economic and Social Committee). (1996). *Opinion on international trade and the environment* (ESC 1264/96 N/GW/ss). European Commission.

ESC (Economic and Social Committee). (2000). *Opinion of the ESC on the communication of the commission Europe's environment: What directions to Future?* (CES 593/2000). European Commission.

ESTELA (European Solar Thermal Electricity Association). (2012). *The first five years of ESTELA*. Brussels: ESTELA.

ESTIA, et al. (2005). *Concentrating solar power – Now*. Brussels: ESTIA.

ETC/SCP (Euopean Topic Center for Sustainable Consumption and Production). (2011). *Progress in sustainable consumption and production in Europe*. ETC/SCP.

Etzioni, A. (1992). The I and we paradigm. In P. Ekins & M. Max-Neef (Eds.), *Real-life economics*. Londres: Routledge.

European Commission. (1996). *Communication on trade and environment*. Brussels: European Commission.

European Commission. (1997). *Communication on taxes and charges in the single market*. Brussels: European Commission.

European Commission. (1999a). *From Cardiff to Helsinki and beyond. Report to European Council on integrating environmental concerns and sustainable development into communities policies* (SEC(1999) 1941). Brussels: European Commission.
European Commission. (1999b). *The EU approach to the millennium round* (COM(1999) 331 final). Brussels: European Commission.
European Commission. (2001a). *WHITE PAPER. European transport policy for 2010: Time to decide* (COM(2001) 370). Brussels: European Commission.
European Commission. (2001b). *Libro Verde sobre la política de productos integrada* (COM (2001) 68 final). Brussels: European Commission.
European Commission. (2001c). *Comunicación sobre el Sexto Programa de Acción de la Comunidad Europea en materia de Medio Ambiente* (COM(2001) 31 final). Brussels: European Commission.
European Commission. (2001d). *Sustainable production. Challenges & objetives for EU research policy*. Brussels: European Commission.
European Commission. (2001e). *Propuesta de estrategia de desarrollo sostenible* (COM(2001)264 final). Brussels: European Commission.
European Commission. (2002a). *Towards a global partnership for sustainable development* (COM(2002) 82 final). Brussels: European Commission.
European Commission. (2002b). *La tecnología medioambiental en pro del desarrollo sostenible* (COM(2002) 122 final). Brussels: European Commission.
European Commission. (2003a). *Hydrogen energy fuel cells*. Commission European High Level Group. Brussels
European Commission. (2003b). *Report on the implementation of community waste legislation* (COM(2003) 250 final/3). Brussels: European Commission.
European Commission. (2003c). *Política de Producto Integrada* (COM(2003) 302 final). Brussels: European Commission.
European Commission. (2003d). *Communication on VAT rates other than standard VAT rates* (COM(2003) 380 final). Brussels: European Commission.
European Commission. (2003e). *2003 environmental policy review (Communication)* (COM(2003) 745 final). Brussels: European Commission.
European Commission. (2004a). *Nanotechnologies: A preliminary risk analysis on the basis of a workshop organized in Brussels on 1–2 March 2004*. Brussels: DG Health and Consumer Protection.
European Commission. (2004b). *National sustainable development strategies in the European Union. A first analysis by the European Union*. Brussels: European Commission.
European Commission. (2004c). *Stimulating technologies for sustainable development: An environmental technologies action plan for the European Union* (COM (2004) 38 final). Brussels: European Commission.
European Commission. (2007a). *Handbook for trade sustainability impact assessment*. Brussels: European Commission.
European Commission. (2007b). *Sustainable urban transport plans* (Technical Report 2007/018). Brussels: European Commission.
European Commission. (2007c). *Comunicación interpretativa sobre residuos y subproductos* (COM(2007) 59 final). Brussels: European Commission.
European Commission. (2007d). *Green paper on market-based instruments for environment and related policy purposes* (COM(2007) 140 final). Brussels: European Commission.
European Commission. (2007e). *Report of environmental technologies (2005–2006)* (COM(2007) 162 final). Brussels: European Commission.
European Commission. (2008a). *Proposal for regulation of the European parliament and the council of a community Ecolabel scheme* (COM(2008) 0401 final). Brussels: European Commission.
European Commission. (2008b). *Proposal for amending directive 2003/87/EC so as to improve and extend the EU greenhouse gas emissions trading in the single market* (COM(2008) 16 final). Brussels: European Commission.

European Commission. (2008c). *Communication on the sustainable consumption and production and sustainable industrial policy* (COM(2008) 397 final). Brussels: European Commission.
European Commission. (2008d). *Directive of the European parliament and of the council establishing a framework for the setting of eco-design requirement as for energy related products* (COM(2008) 399). Brussels: European Commission.
European Commission. (2008e). *Public procurement for a better environment* (COM(2008) 400/2). Brussels: European Commission.
European Commission. (2008f). *Proposal for a amending Directive 2006112/EC as regards reduced rates of value added tax* (COM(2008) 428 final). Brussels: European Commission.
European Commission. (2008g). *Second strategy energy review. An EU energy security and solidarity action plan* (SEC(2008) 2794). Brussels: European Commission.
European Commission. (2008h). *The raw materials initiative – Meeting our critical needs for growth and jobs in Europe* (COM(2008) 699). Brussels: European Commission.
European Commission. (2008i). *Communication on the directive on the energy performance of building* (COM(2008) 780 final). Brussels: European Commission.
European Commission. (2008j). *Guide to COST-BENEFIT ANALYSIS of invest projects*. Brussels: European Commission.
European Commission. (2009a). *Green paper. RTE-T: A policy review* (COM(2009) 44final). Brussels: European Commission.
European Commission. (2009b). *A sustainable future for transport: Towards an integrated, technology-led an user friendly system* (COM(2009) 279 (final). Brussels: European Commission.
European Commission. (2009c). *EU energy and transport in figures*. Brussels: European Commission.
European Commission. (2009d). *Incorporación del Desarrollo Sostenible en las políticas de la Unión Europea: Informe de 2009 sobre la estrategia de la Unión Europea para el Desarrollo Sostenible* (COM(2009) 400 final). Brussels: European Commission.
European Commission. (2009e). *Report on the state of implementation of integrated product policy* (COM(2009)693 final). Brussels: European Commission.
European Commission. (2010a). *EU energy trends to 2030 – UPDATE 2009*. Brussels: European Commission.
European Commission. (2010b). *Seventh annual report from the Commission to the European parliament: Overview of thirst country trade defence actions against the EU* (COM(2010) 334 final). Brussels: European Commission.
European Commission. (2010c). *Communication concerning the development of a single European rail network* (COM(2010) 474 final). Brussels: European Commission.
European Commission. (2010d). *A strategy for competitive, sustainable and secure energy* (COM(2010) 639 final). Brussels: European Commission.
European Commission. (2010e). *Europa 2020. Una estrategia para un crecimiento inteligente, sostenible e integrador* (COM(2010) 2020 final). Brussels: European Commission.
European Commission. (2011a). *WHITE PAPER. Roadmap to a single European transport area*. Brussels: European Commission.
European Commission. (2011b). *Roadmap to a resource efficient Europe* (COM(2011) 571 final). Brussels: European Commission.
European Commission. (2011c). *Working document: Accompanying the white paper* (SEC(2011) 391 final). Brussels: European Commission.
European Commission. (2012). *EU transport in figures. Statistical pocketbook 2012*. Brussels: EC.
European Council. (2001). *Conclusiones de la Presidencia* (SN 200/01). Brussels: European Council.
European Council. (2005). *Conclusiones de la Presidencia*. Consejo Europeo de Bruselas 10255/1/05. Brussels: European Council.
European Council. (2006). *Renewed EU sustainable development strategy*. Council of the European Union 10117/06. Brussels: European Council.

European Parliament. (1993). *Dictamen sobre el V Programa de Medio Ambiente*. European Parliament.
European Parliament. (1996). *Informe sobre la Organización Mundial de Comercio*. Strasbourg: Comisión de Relaciones Económicas Exteriores del Parlamento Europeo.
European Parliament. (2001). *Informe sobre la propuesta de decisión del Parlamento Europeo y del Consejo por el que se establece el Programa Comunitario de Acción en materia de Medio Ambiente para 2001–2010* (final AS-0175/2001). Strasbourg: European Parliament.
European Parliament. (2010a). *Decentralised energy systems*. Strasbourg: European Parliament.
European Parliament. (2010b). *Europa 2020. Una estrategia para un crecimiento inteligente, sostenible e integrador* (COM(2010) 2020 final). Strasbourg: European Parliament.
European Parliament. (2011a). *WHITE PAPER. Roadmap to a single European transport area*. Brussels: EC.
European Parliament. (2011b). *Roadmap to a resource efficient Europe* (COM(2011) 571 final). Strasbourg: European Parliament.
European Parliament. (2011c). *Working document: Accompanying the white paper* (SEC(2011) 391 final). Strasbourg: European Parliament.
European Parliament. (2012). *EU transport in figures. Statistical pocketbook 2012*. Brussels: EC.
European Parliament and Council. (2002). *Decisión relativa al sexto programa marco de la Comunidad Europea para acciones de investigación, desarrollo tecnológico y demostración, destinado a contribuir a la creación del Espacio Europeo de Investigación e Innovación (2002–2006)* (nº 1513/2002/CE). European Commission.
European Parliament and Council. (2003). *Directiva por la que se instaura un marco para el establecimiento de requisitos de diseño ecológico aplicables a los productos que utilizan energía* (COM (2003) 453 final). European Commission.
European Parliament and Council. (2008a). *Propuesta de Reglamento sobre la red ferroviaria europea para un transporte de mercancías competitivo* (COM(2008) 852final). European Commission.
European Parliament and Council. (2008b). Framework waste *Directive* (Directive 2008/98/CE). European Commission.
EUROSTAT. (2008a). *Energy, transport and environmental indicators*. Luxembourg: Eurostat.
EUROSTAT. (2008b). *Statistics in focus. Agriculture and fisheries*. Luxembourg: Eurostat.
EWEA (European Wind Energy Association). (2006). *NOFUEL. Wind power without fuel*. www.ewea.org
EWEA (European Wind Energy Association). (2009a). *Oceans of opportunity. Harnessing Europe's largest domestic energy resource*. www.ewea.org
EWEA (European Wind Energy Association). (2009b). *Pure power. Wind energy targets for 2020 and 2030*. www.ewea.org
EWEA (European Wind Energy Association). (2010). *Powering Europe*. www.ewea.org
EWEA (European Wind Energy Association). (2011). *Wind in our sails*. www.ewea.org
EWEA (European Wind Energy Association). (2012). *EWEA annual report 2011*. www.ewea.org
EWG (Energy Watch Group). 2007). *Coal: Resources and future production*. EWG-Paper Nº 1.
Fabe, D., & O'Connors, J. (1990). La lucha por la naturaleza: la crisis ambiental y la crisis del ambientalismo en los Estados Unidos. *Ecología Política, Ecología Política,* nº1, Madrid.
Faber, M., et al. (1996). *Ecological economics: Concepts and methods*. Cheltenham: Edward Elgar.
FAESTA (The Foundation for the Economics of Sustainability). (2007). *Envisioning a sustainable Ireland from an energy availability perspective*. Dublin: Environmental Protection Agency.
Farley, J., & Constanza, R. (2002). Envisioning shared goals for humanity: A detailed, shared vision of a sustainable USA in 2100. *Ecological Economics, 43*, 245–259.
FCH JU (Fuel Cells and Hydrogen Joint Undertaking). (2009). *Multi-annual implementation plan 2008 – 2013*. FCH JU.
Federal Environment Ministry. (2002). *German strategy of sustainable development*. Bonn: Federal Environment Ministry.
Fedrigo, D., & Tukker, A. (2009). *Blueprint for European sustainable consumption and production*. Brussels: EEB.

FEM (Foro Económico Mundial). (2001). *2001 environmental index*. IMF.
Ferguson, N. (2010). *Complexity and collapse. Empires on the edge of chaos*. www.informationclearinghouse.info/article24874.htm
Ferrer, G., & Guide, D. R. (2002). Remanufacturing cases and state of the art. In R. Ayres & L. Ayres (Eds.), *A handbook of industrial ecology*. Cheltenham/London: Edward Elgar.
Figenschou, A., & Simmons, M. (2008). *A peak-oiler, but still in the closet? IEA's 2008 report*. www.energybulletin.net/node/47234
Fisher-Kowalski, M., & Haberl, H. (2000). El metabolismo socioeconómico. *Ecología Política, 19*.
Fitch, A. (2008). *CGC urged to consider dollar policy*. www.thenational.ae/article
Flavin, C., et al. (2003). Prólogo. In C. Bright et al. (Eds.), *La situación del mundo 2003*. Barcelona: Icaria/WWI.
Focus Groups. (2009). *Report: The future of transport*. Brussels: Directorate-General for Energy and Transport, European Commission.
Folke, C., et al. (2005). Adaptive governance of social-ecological systems. *Environmental Resources, 30*, 441–473.
Folke, C., et al. (2006). Resilience: The emergence of a perspective for social-ecological systems analysis. *Global Environmental Change, 16*.
Foucher, S. (2011). *Crude oil and liquids capacity. Additions: 2011–2015*. www.theoildrum.com/node/7785
Foucher, S., & Border, J. M. (2011). *Crude oil and liquids capacity additions: 2010–2015*. www.theoildrum.com/node/7785
French, H. (2000). *Vanishing borders: Protecting the planet in the age of globalization*. New York: W.W. Norton & Company.
French, H. (2002). Remodelando el gobierno mundial. en C. Flavin y otros (Eds.), *La situación del mundo 2002*. Barcelona: Icaria.
French, H., & Mastny, L. (2001). El control de los delitos internacionales en contra el medio ambiente. In L. R. Brown et al. (Eds.), *La situación del mundo 2001*. Barcelona: Icaria.
Friedemann, A. (2007, April 13). Peak soil: Why cellulosic ethanol, biofuels are unsustainable and a threat to America. *Energy Bulletin*.
Frosch, R. (1992). *Two tramps in mud times*, Workshop on industrial ecology, snow, Colorado.
Frosch, R., & Gallopoulos, N. (1989). Strategies for manufacturing. *American Scientific, 261*(3), 144–152.
Frost & Sullivan. (2010). *Sustainable and innovative personal transport solutions – Strategic analysis of Carsharing market in Europe*. Frost & Sullivan.
Fuchs, D., & Lorek, S. (2005). Sustainable consumption governance: A history of promises and failures. *Journal of Consumer Policy, 28*, 261–288.
Fuel Cell Today. (2010). *2010 portable fuel cell analysis*. www.fuelcelltoday.com
Fuel Cell Today. (2011). *Industry review 2011*. www.fuelcelltoday.com
Fuel Cell Today. (2012). *Industry review 2012*. www.fuelcelltoday.com
Funtowics, S. O., & Ravetz, J. R. (1993). Science for the post-normal age. *Futures, 25*(7), 739–755.
Funtowicz, S. O., & Ravetz, J. (1994). Emergent complex systems. *Futures, 26*, 568–582.
Funtowicz, et al. (1999). *Information tool for environmental policy under conditions of complexity*. EEA.
Gallopin, G. C. (1991). Human dimensions of global change: Linking the global and the local processes. *International Social Science Journal, 130*, 707–718.
Gallopin, G. C. (2002). Planning for resiliency: Scenarios, surprises and branch points. In L. H. Gunderson & C. S. Holling (Eds.), *Panarchy. Understanding transformations in human and natural systems*. London: Island Press.
Gallopin, G. C. (2006). Linkages between vulnerability, resilience, and adaptive capacity. *Global Environmental Change, 16*, 293–303.
GAO (Government Accountability Office). (2007). *Crude oil. Uncertainty about future oil supply makes it important to develop a strategy for addressing a peak and decline in oil production*, GAO-07-283.

García, E. (2002). *Consumo y medio ambiente en el País Valenciano*. Barcelona: Centre de Cultura Contemporánea.

Gardner, G., & Assadourian, E. (2004). "Reconsiderando la vida buena", en WWI, *La situación del mundo*, Barcelona, Icaria.

Gardner, G., & Prugh, T. (2009). "Sembrar una economía sostenible", *La situación del mundo 2009*, WWI, Barcelona, Icaria.

Gardner, G., & Sampat, P. (1999). Forging a Sustainable Materials Economy. In *La situación del mundo*. Barcelona, Icaria: WWI.

Garner, A., & Keoleian, G. A. (1995). *Industrial ecology: An introduction*. Ann Arbor: Michigan University.

Gay, C. (2010). *Accelerating solar: A look at the next decade*. Renewable Energy World.

Geiser, K. (2003). Química verde: diseño de materiales y procesos sostenibles. In E. Blount (Ed.), *Industria como naturaleza. Hacia la producción limpia*. Madrid: Catarata.

Geng, Y., & Coté, R. P. (2002). Scavengers and decomposers in an eco-industrial park. *International Journal of Sustainable Development and World Ecology, 9*, 333–340.

Geng, Y., & Coté, R. P., et al. (2006). Empirical analysis of eco-industrial development in China. *Sustainable Development*. www.interscience.wiley.com

Geng, Y., Coté, R. P., et al. (2009). Assessment of the national eco-industrial park standard for promoting industrial symbiosis in China. *Journal of Industrial Ecology, 13*(1), 15–26.

Georgescu-Roegen, N. (1971). *The entropy law and the economic process*. Cambridge: Harvard University Press.

Gertsakis, J., et al. (2002). Industrial ecology and extended producer responsibility. In R. Ayres & L. Ayres (Eds.), *A handbook of industrial ecology*. London: Edward Elgar.

Giampietro, M. (2008). The future of agriculture: GMOs and the agonizing paradigm of industrial agriculture. In A. Guimaraes Pereira & S. Funtowicz (Eds.), *Science for policy: Challenges and opportunities*. New Delhi: Oxford University Press.

GIFNFC (Government-Industry Forum on Non-Food Uses of Crops). (2002). *Annual report of the government-industry forum on non-food uses of crops*. Londres: DEFRA (Department for Environment, Food & Rural Affairs).

GIFNFC (Government-Industry Forum on Non-Food Uses of Crops). (2003). *Annual report of the government-industry forum on non-food uses of crops*. London: DEFRA.

Giljun, S. (2010). *European resource consumption and resource efficiency in global context*. www.seri.at

Glinka, et al. (2007). Estrategias de reciclaje y reutilización de residuos sólidos de construcción y demolición. *España, 5*.

Gobierno Federal de México. (2008). *Prospectiva de Mercado de gas natural*. Gobierno Federal/SENER.

Goedkoop, M., et al. (1999). *Product service systems, ecological and economic basics*. The Hague: Dutch Ministries of Environment and Economic Affairs.

Goldsmith, E. (1992). *The way: An ecological world view*. Devon: Themis Books.

Goldsmith, E. (1996). *The way: An ecological world view* (Rev. and enlarged ed.). Devon, England: Themis Books.

Goldsmith, E. (2003). *Towards a biosphere ethic, Centre of Human Ecology*. Edinburgh University.

Gómez Uranga, M. (1997). *Proyecto docente investigador*. Bilbao: Basque Country University.

Gorelick, S. (1998). *Small is beautiful, big is subsidised*. Devon: International Society for Ecology and Culture.

Gorlach, B., et al. (1999). *From for Helsinki. Environmental requirements in the process of integrating environmental issues into other policy areas of the European Union* (European Policy Paper Nº 4). Wuppertal: Wuppertal Institute.

Gosselin, P. G. (2008, May 20). Relentless rise in oil prices tests economy's resilience. *Los Angeles Times*.

Gossen, M., & Scholl, G. (2011). *Latest trends in car-sharing*. Institute for Ecological Economy Research (IÖW).

Gotts, N. M. (2007). Resilience, panarchy, and world-systems analysis. *Ecology and Society, 12*(1).

Gowdy, J. M. (1999). Hierarchies in human affairs: Microfoundations of environmental sustainability. In J. Kohn et al. (Eds.), *Sustainability in question*. London: Edward Elgar.
Gowdy, J. M. (2006). Evolutionary theory and economic policy with reference to sustainability. *Journal of Bioeconomics, 8*, 1–19.
Gowdy, J. M. (2007). *Behavioral economics and climate change policy* (Rensselaer Working Paper in Economics), Rensselaer Polytechnic Institute.
Grantham, J. (2011, April). Time to wake up: Days of abundant resources and falling prices are over forever. *GMO Quaterly Letter.*
Grantham, J. (2012, November). On the road to Zero Growth. *GMO Quarterly Letter*.
Gray, C., & Charter, M. (2007). *Remanufacturing and product design*. The Centre for Sustainable Design (South East England Development Agency). Kingston upon Thames.
Green Fiscal Commission. (2010). *Competitiveness and environmental reform* (Briefing Paper Seven March).
Green Tax Commission. (2009). *Lessons from two green tax shifts in the United Kingdom*. London: Policy Studies Institute. www.greenfiscalcommission.org.uk
Greer, J. M. (2009). *The ecotechnic future*. Gabriola Island: New Society Publishers.
Greño, P. (2004). *Comercio de derechos de emisión: Implicaciones para el País Vasco*. Bilbao: Unidad de Economía Ambiental, UPV-EHU.
Guide, D. R., & van Wassenhove, L. N. (2002). Closed-loop supply chains. In R. Ayres & L. Ayres (Eds.), *A handbook of industry ecology*. London: Edward Elgar.
Gunderson, L. (1999). Resilience, flexibility and adaptive management—Antidotes for spurious certitude? *Conservation Ecology, 3*(1), 7.
Gustafsson, B. (1998, February–March). Scope and limits of the market mechanism in environmental management. *Ecological Economics*.
GWEC (Global Wind Energy Council). (2010). *Global wind 2009 report*. GWEC. www.gwec.net/
GWEC (Global Wind Energy Council). (2011). *Global wind 2010 report*. GWEC.
GWEC (Global Wind Energy Council). (2012). *Global wind 2011 report*. GWEC.
GWEC (Global Wind Energy Council), & Greenpeace (2010). *Global wind energy. Outlook 2010*. GWEC y Greenpeace.
Hagens, N. (2010). Fractal adaptive cycles in natural and human systems. www.theolidrum.com/node/6099
Hagen, N. (2011). Will natural gas fuel America in 21st century? www.theoildrum.com/node/7912
Halada, K. (2007). *Ecomaterials go into the next stage*. In 8th international conference of ecomaterials – ICEMS.
Hall, J., & Kerr, R. (2003). Innovation dynamics and environmental technologies: The emergence of fuel cell technology. *Journal of Cleaner Production, 11*, 459–471.
Hall, C. A., & Klitgaard, K. A. (2006). The need for a new, biophysical-based paradigm in economics for the second half of the age of oil. *International Journal of Transdisciplinary Research, 1*(1), 4–22.
Halweil, B., & Nierenberg, D. (2009). Carne y pescado: los ingredientes más caros de la alimentación mundial. *La situación del mundo 2009*, WWI, Barcelona, Icaria.
Hamilton, J. (2009). *Causes and consequences of the oil shock of 2007–08* (Brooking Papers on Economic Activity). Washington, DC
Hamilton, J. (2011, online). Oil prices, exhaustible resources, and economic growth, prepared for *Handbook of Energy and Climate Change*.
Hammer, L. (2011). *Improving the EU ecolabel – Discussion paper for the Belgian presidency meeting 2010*. Brussels: BEUC.
Hansen, L., & Lovins, A. L. (2010). *Keeping the lights on while transforming electric utilities*. Snowmass: Rocky Mountain Institute.
Hara. K. (2011). *Research and development of fuel cells vehicles and hydrogen in Japan*, Ministry of Economy. Trade and Industry (METI).
Hart, P. (2010). *Oil demand to decline in the west, according to International Energy Agency*. www.energybulletin.net/print/54991
Hartzog, P. B. (2003). *21st century governance as a complex adaptive system*. www.panarchy.com

Hatch, C., & Price, M. (2008). *Canada's toxic sands*. Report by Environmental Defense Canada.
Hawken, P., et al. (1999). *Natural capitalism*. London: Little, Brown and Company.
Heinberg, R., & Fridley, D. (2010). The end of cheap coal. *Nature, 468*, 367–369.
Hemmingsen, E. (2010). At the base of Hubbert's Peak: Grounding the debate on petroleum scarcity. *Geoforum, 41*(4), 531–540.
Herman, R., et al. (1990). Dematerialization. *Technological Forecasting and Social Change, 38*(4), 333–347.
HFTP (Hydrogen Fuel Cell Technology Platform). (2005). *Deployment strategy*. Brussels: HFTP.
Hinton, S. (2006). *Why the Swedish oil commission fails to deliver route to oil independence*. www.energybulletin.net
Hirsch, F. (1984). *Los límites sociales al crecimiento*. Méjico: FCE.
Hirsch, R. L., et al. (2005). *Peaking of world oil production: Impacts, mitigation, & risk management*. Washington, DC: Department of Energy.
Hirsch, R. L., et al. (2010). *The impending world energy mess*. Amazon.
Hjorth, P., & Bagheri, A. (2005–6). Navigating towards sustainable development: A system dynamics approach. *Futures, 38*, 74–92.
Ho, M. W. (2005, July 11). Making the world sustainable. *Energy Bulletin*.
Hobson, J. (2010). EU commission discriminates hemp fibres. www.hanf-info.ch
Hodgson, G. (1992). The amazing growth of LETs in UK. In *New economics*. London: New Economics Foundation.
Hodgson, G. (1993). Rationality and the influence of institutions. In P. Ekins & M. Max-Neef (Eds.), *Real-Life Economics*.
Hoffmann, H. (1997). Comercio y medio ambiente: ¿luz roja o luz verde? *Revista del CEPAL*, agosto.
Holliday, C., & Pepper, J. (2000). *Sustainability through the market. Seven keys for success*. Conches-Geneva: WBCSD.
Holling, C. S. (1973). Resilience and stability of ecological systems. *Annual Review of Ecology and Systematics, 4*, 1–23.
Holling, C. S. (1993). New science and new investments for sustainable biosphere. In A. M. Janson et al. (Eds.), *Investing in natural capital*. Washington: Island Press.
Holling, C. S. (Ed.). (1978). *Adaptive environmental assessment and management*. New York: Wiley.
Holling, C. S. (Ed.). (2002). *Panarchy: Understanding transformations in human and natural systems*. Washington, DC: Island Press.
Holling, C. S. (2004). From complex regions to complex worlds. *Ecology and Society, 9*(1), 11.
Holling, C. S. (2009). *Collapse and renewal*. Resilience Alliance. www.resience...Internet
Holling, C. S. (2010). The resilience of terrestrial ecosystems. Local surprise and global change. In L. H. Gunderson et al. (Eds.), *Foundations of ecological resilience*. Washington, DC: Island Press.
Holling, C. S., & Gunderson, L. (2002). Resilience and adaptive cycles. In L. H. Gunderson & C. S. Holling (Eds.), *Panarchy. Understanding transformations in human and natural systems*. London/Washington, DC: Island Press.
Holling, C. S., et al. (2002). Discoveries for sustainable futures. In L. H. Gunderson & C. S. Holling (Eds.), *Panarchy. Understanding transformations in human and natural systems*. London: Island Press.
Hontelez, J. (2009). *The double benefit of EFR: Good practices and interesting ideas in some EU countries*. Seminar: Environmental Fiscal Reform and its Contribution to a Green Economy, Madrid.
Höök, M. et al. (2010). Validity of the fossil fuel production outlooks in the IPCC Emission Scenarios. *Natural Resources Research*.
Hopkins, R. (2008). *The transition handbook. From oil dependency to local resilience*. London: Green Books.
Hopkins, R. (2010). *Why community might not need organizing*. www.postcarbon.org

Hopkins, R. (2011). *Transition companion. Making your community more resilient in uncertain times*. www.amazon.com
Hopson, J. (2010). *EU commission discriminates hemp fibres*. www.hanf-info.ch
Hoyos, D. (2004). La deuda ecológica vasca. In I. Barcena (Ed.), *Euskal Herria nora zoaz? Retos sociales y ambientales de la sostenibilidad*. Bilbao: Ekologistak Martxan.
Hueting, R. (1991). The use of the discount rate in a cost-benefit analysis for different uses of a humid tropical forest. *Ecological Economics, 3*, 43–57.
Hughes, S. (2007, October 12). US energy security: Oil supply unable to keep up with demand. *Dow Jones Newswires*.
Hukkinen, J. (2003). From groundless universalism to grounded generalism: improving ecological economic indicators of human-environmental interaction. *Ecological Economics, 44*, 11–27.
Hummels, D. (2009). *Globalization and freight transport costs in maritime shipping and aviation*. Paris: OECD and International Transport Forum.
Hwang, A. (2002). Semiconductors have hidden costs. In *VitalSigns 2002/2003*. London: WWI/Earthscan.
HyWays (The European Hydrogen Roadmap). (2007a). *Action plan*. www.HyWays.de
HyWays (The European Hydrogen Roadmap). (2007b). *Roadmap*. www.HyWays.de
ICLEI (Conferencia Europea De Ciudades Sostenibles). (1994). *Carta de Aalborg*. New York: ICLEI.
ICSU (International Council For Science). (2005). *Science and society: Rights and responsibilities*. Paris: ICSU.
ICSU (International Council For Science)., & IGFA (International Group of Funding Agencies for Global Change Research). (2008). *IGSU-IGFA review of the Earth Systems Science Partnership (ESSP)*. Paris: ICSU.
IEA (International Energy Agency). (2008). *World energy outlook 2008*. Paris: IEA.
IEA. (2009). *World energy outlook 2009*. Paris: IEA.
IEA (International Energy Agency). (2010a). *Medium-term oil and gas 2010*. Paris: IEA.
IEA (International Energy Agency). (2010b). *World energy outlook 2010*. Paris: IEA.
IEA (International Energy Agency). (2010c). *Technology roadmap. Concentrating solar power*. Paris: IEA.
IEA (International Energy Agency). (2011). *Outlook 2011*. Paris: IEA.
IEA (International Energy Agency). (2012). *Outlook 2012*. Paris: IEA.
IEA, Solar Paces et al. (2009). *Concentrating solar power. Global outlook 09*, Paris: IEA.
IEEP (Institute European Environmental Policy). (2010). *Anticipated indirect land use change associated with expanded use of biofuels and bioliquids in the EU – An analysis of the national renewable energy action plans*. London: IEEP.
IEEP. (2011). *The land-use implications of EU bioenergy policy. Going beyond ILUC*. IEEP.
IENICA (Interactive European Network for Industrial Crops and their Applications). (2006). *Summary report for the European Union 2000–2005*. Brussels: IENICA.
IETC (International Environmental Technology Center). (2003). *Environmentally sound technologies for sustainable development*. Paris: PNUMA.
IISD (International Institute for Sustainable Development). (2010). *The global subsidies initiative. Relative subsidies initiative to energy sources: GSI estimates*. IISD. www.iisd.org
Illum, K. (2004). *IEA oil projections disputed*. Copenhagen: ECO Consultant.
IP (Implementation Panel). (2007). *Implementation plan*. European Hydrogen & Fuel Cell Technology Platform.
IPCC (Intergubernamental Panel of Climate Change). (2007). *Climate change 2007. Mitigation of climate change*. Cambridge: Cambridge University Press.
IPTOES (Industry Taskforce on Peak Oil & Energy Security). (2008). *Oil crunch, securing the UK's energy future*. IPTOES.
IPTOES (Industry Taskforce on Peak Oil & Energy Security). (2010a). *Oil crunch (Second report)*. London: IPTOES.
IPTOES (Industry Taskforce on Peak Oil & Energy Security). (2010b). *Peak oil. Implications of the Gulf of Mexico oil spill*. London: IPTOES.

IPTS (Institute for Prospective Technological Studies), & ESTO (European Science Technology Observatory). (2006). ENVIRONMENTAL IMPACT OF PRODUCTS (EIPRO). Analysis of the life cycle environmental impacts related to the final consumption of the EU25, EIPRO&ESTO.

Jackson, T. (2005). Live better by consuming less? *Journal of Industrial Ecology, 9*(1–2), 19–36.

Jackson, T. (2009a). El reto de un mundo sostenible. *La situación del mundo 2009*. London: WWI.

Jackson, T. (2009b). *Prosperity without growth?* London: Sustainable Development Commission.

Janicke, M., et al. (2001). *Governance for sustainable development in Germany: Institutions and policy making*. Paris: OECD.

Jansson, A. M., & Jansson, B. O. (1994). Ecosystems properties as a basis for sustainability. In A. M. Jansson (Ed.), *Investing in natural capital*. Washington, DC: Island Press.

Japanese Government. (2002). *Basic plan for establishing a recycling in Japan*. www.uneptie.org

Jespersen, J. (1999). Reconciling environment and emplyment by switching from goods to services? A review of Danish experience. *European Environment, 9*(1), 17–23.

JOE (Join Operating Environment). (2010). *The joint operating environment 2010*. Suffolk: United States Joint Forces Command.

Johnson, C. (2008, January 23). Going bankrupt. Why the debt crisis is now the threat to the American Republic. *Common Dreams News Center*.

Johnstone, S., et al. (2008). Integrating products and services through life: An aerospace experience. *International Journal of Operations & Production Management, 29*(5), 520–538.

Jones, M. (2005). The battles of titans. In A. Mckillop & S. Newman (Eds.), *The final energy crisis*. London: Pluto Press.

Jones, J. (2010). *MENA taps its green reserves*. Renewable Energy World.

Jorgensen, C. (2003). *Environmental fiscal reform: Perspectives for progress in the European Union*. Bruselas: European Environmental Bureau.

Joung, O. R., et al. (2006). The globalization of socio-ecological systems: An agenda for scientific research. *Global Environmental Change, 16*, 304–316.

Kaizuka, I., & Ikki, O. (2010). *Japanese PV power: New support framework boots the PV Market*. Renewable Energy World. www.renewableenergyworld.com

Kasser, T. (2002). *The high price of materialism* (A Bradford book). Cambridge, MA: MIT Press.

Kavalov, B., & Peteves, S. D. (2007). *The future of coal*. Brussels: Joint Research Centre, European Commission.

Kay, J. J., & Regier, H. (2000). Uncertainty, complexity, and ecological integrity: Insights from an ecosystem approach. In P. Crabbé, A. Holland, L. Ryszkowski, & L. Westra (Eds.), *Implementing ecological integrity: Restoring regional and global environmental and human health* (NATO science series, environmental security). Dordrecht: Kluwer.

Kearney, A. T., & ESTELA (European Solar Thermal Electricity Association). (2010). *Solar thermal electricity 2025*. AT Kearny y ESTELA.

Keijzers, G. (2002). The transition to the sustainable enterprise. *Journal of Cleaner Production*.

Kerr, W., & Ryan, C. (2001). Eco-efficiency gains from remanufacturing. A case study of photocopier remanufacturing at Fuji Xerox Australia. *Journal of Cleaner Production, 9*, 75–81.

Keynes, J. M. (1933). National self-sufficiency. *Yale Review, 22*(4), 755–769.

Keynes, J. M. (1964). *General theory of employment, interest and money*. Lóndres: Macmillan.

Kharecha, P. A., & Hansen, J. E. (2008). Implications of "peak oil" for atmospheric CO_2 and climate. *Global Geochemical Cycles, 22*(3). Stockholm

Kilian, L. (2010). Comment on 'causes and consequences of the oil shock of 2007-08' by J. D. Hamilton. *Brooking panel on economic activity*.

Kincaid, J. (2003). Region-wide eco-industrial networking in North Carolina. In E. Cohen-Rosenthal (Ed.), *Eco-industrial strategies*. Sheffield: Greenleaf.

Klare, M. T. (2007, June 14). The Pentagon v. peak oil. *Energy Bulletin*.

Klare, M. T. (2008). *How scarce energy resources can quickly lead to deadly wars*. www.alternet.org

Kobayashi, Y. (2010). *Natural gas situation and LNG supply/demand trends in Asia-Pacific and Atlantic markets*. Tokyo: Institute of Energy Economics.

Konyrova, K. (2008, May). Kashagan field oil production delayed. *New Europe*.

Kopits, S. (2011, March). Oil, the economy and policy. *Our Finite World*.

Koppelaar, R. (2008). *Oilwatch M, monthly – March 2008*. Amsterdam: ASPO Netherlands.

Koppelaar, R. (2009a). *Oilwatch monthly February*. www.theoildrum.com

Koppelaar, R. (2009b). *Costs and environmental impacts of electric cars*. http://theoildrum.com/node/5104

Koppelaar, R. (2010a). *Oilwatch monthly – April 2010*. Amsterdam: ASPO Netherlands.

Koppelaar, R. (2010b). *Oilwatch monthly – June 2010*. Amsterdam: ASPO Netherlands.

Korhonen, J. (2001). Some suggestions for regional industrial ecosystems. Extended industrial ecology. *Eco-Management and Auditing, 8*, 57–69.

Korhonen, J. (2004). Industrial ecology in the strategic sustainable development model: Strategic applications of industrial ecology. *Journal of Cleaner Production, 12*.

Korowicz, D. (2010). *Tipping point. Near-term systemic implications of the peak in global oil production. An outline review*. Dublin: Faesta.

Korten, D. (1996). *When corporations rule the world*. London: Earthscan.

Kortman, J., et al. (2006). *Chemical product services in the European Union*. Seville: IPTS, European Commission.

Krauss, C. (2008). *Gas prices send surge of riders to mass transit*. www.nytimes.com

Kuhndt, M., & Liedtke, C. (2000). *Setting a factor X into practice*. Wuppertal: Wuppertal Institute.

Lacey, S. (2010). *EPA lowers cellulosic ethanol targets: A sign of the times*. Renewable Energy World.

Lacey, S. (2013). Anatomy of a solar PV system: How to continue "Ferocious Cost Reductions" for solar electricity, *Climate Progress*. www.renewableenergyworld.com

Laherrere, J. (2001, June). *Stimates of oil reserves*. Paper presented at the EMF/IEA/IEW meeting, International Institute for Applied Systems Analysis (IIASA), Laxenburg, Austria

Laherrere, J. (2010). *Copper peak*. www.theoildrum.com/node/6307

Lambin, E. F. (2005). Conditions for sustainability of human-environmental systems: information, motivation, and capacity. *Global Environmental Change, 15*(3), 177–180.

Lang, T., & Hines, C. (1993). *The new protectionism. Protecting the future against free trade*. London: Earthscan.

Laszlo, E. (2007). *Science and akashic field: An integral theory of everything*. Rochester (VT): Inner Traditions.

Laszlo, E. (2010). *Chaos point, 2012 and beyond*. Charlottesville: Hampton Road.

Latouche, S. (2008). *La apuesta por el decrecimiento*. Barcelona: Icaria.

Lawrence, F. (2005). *Food study reveals hidden costs of transport*. www.gardian.co.uk. July 15, 2005.

Lawton, J. H. (2000). *Community ecology in a changing world*. Germany: Ecology Institute.

Layburn, P., & Lombardi, R. (2012). Industrial symbiosis in European policy. *Journal of Industrial Ecology, 16*(1), 11–12.

LBST (Ludwig-Bölkow-Systemtechnik). (2011a). *Assessment of potential and promotion of new generation of tenable technologies*. Ottobrunn: LBST.

LBST (Ludwig-Bölkow-Systemtechnik). (2011b). *Decentralised energy systems*. Strasbourg: European Parliament.

Leake, J., & Swindford, S. (2007, June 10). Top scientist says biofuels are scam, *The Sunday Times*.

Lele, S. (1991). Sustainable development. A critical review. *World Development, 19*(6), 601–621.

Lerch, D. (2007). *Post carbon cities: Planning for energy and global warming for local governments*. Sebastopol: Post Carbon Institute.

Lesley, V., & Bullock, H. (2008). *The future proofers*. www.forumforthefuture.org

Levesque, C. (2007). *What is the percentage of federal subsidies allotted for wind power*. www.worldrenewableenergy.com

Levin, S. A. (2000). *Fragile dominion: Complexity and the commons*. Cambridge: MM, Perseus.

Levine, S. (2009). *China changes calculus for petro-rulers*. A blog on Russia, Central Asia and the Caucasus.

Lewis, J. I. (2012). *Emerging conflicts in renewable energy policy and international trade law*. Washington, DC: Georgetown University.

Lewis, N. S., et al. (2005). *Basic research needs for solar energy utilization*. Washington, DC: U.S. Department of Energy.

Liedtke, C. (2003). Toward sustainable products and services. In J.-D. Seiler-Hausmann et al. (Eds.), *Eco-efficiency and beyond*. Sheffield: Greenleaf.

Lietaert, M. (2007). *The growth of cohousing in Europe*. www.cohousing.org

Lietaert, M. (2009). Cohousing's relevance to degrowth theories. *Journal of Cleaner Production, 18*, 576–580.

Likvern, R. (2010). *Global oil supplies as reported by EIA's International Petroleum Monthly for September 2010*. www.energybulletin.net/print/54211

Linch, J. P. (2010) *PV – A question of scale*. Renewable Energy World. www.renewableenergyworld.com

Linde, A. G., et al. (2009). *Initiative 'H2 mobility'*. www.pr-inside.com

Lindhal, M., et al. (2009). *Product-service systems experiences – An international survey of Swedish, Japanese, Italian and German Manufacturing Companies*. CIRP IPS2 Conference.

Lindhqvist, T. (2000). *Extended producer responsibility in cleaner production*. Doctoral dissertation, International Institute for Industrial Environmental Economics, University of Lund.

Lindhqvist, M. R. (2003). The role of public policy in advancement of product service systems. *Journal of Cleaner Production, 11*, 905–914.

Linn, S. (2010). Commercialism in children's lives. In *The state of the world 2010*. WWI, Barcelona: Icaria.

Liu, C. (2010). *Financial crisis paves the way for Chinese solar giants*. Renewable Energy World. www.renewableenergyworld.com

Loose, W. (2011). *The state of European car-sharing*, Bundesverband CarSharing e. V.

Lorek, S. (2010). *Towards strong sustainable consumption governance*. Saarbrücken: LAP Publishing.

Lorek, S., & Lucas, R. (2003). *Towards sustainable market strategies. A case study on eco-textiles and green power*. Wuppertal: Wuppertal Institute.

Lorek, S., & Spangenberg, J. (2012). *Sustainable consumption within a sustainable economy – Debunking buzzwords to develop the content*. Lunenburg: Sustainable European Research Institute.

Lovelace, R. (2009). *Transport and adaptive capacity: An integrated approach to UK policy evaluation*. http://europe.theoildrum.com/node/5365

Lovins, A. B., & Lovins, L. H. (1997). *Factor 4*. Barcelona: Galaxia Gutenberg.

Lowe, E. (2003). Eco-industrial development in Asia developing countries. In E. Cohen-Rosenthal (Ed.), *Eco-industrial strategies*. Sheffield: Greenleaf.

Macalister, T. (2009). *Key oil figures were distorted by US pressure, says whistleblower*. www.guardian.co.uk

Mackey, B. G., et al. (2008). *Green carbon. The role of natural forests in carbon storage*. Acton: Australian National University Press.

Mactiernan, H. A. (2004). *Sustainable transport coalition's, oil: Living with less (Conference)*. Perth, Western Australia: Ministry for Planning and Infrastructure.

Maier-Rigaud, G. (1991). *Ecological economics and global change*. Bonn: Institute fur Europerische Umweltpolitik.

Makower, J. (2001). *The clean revolution: Technologies from the leading edge*. Oakland: Clean Edge.

Makover, J., & Pernick, R. (2001). *Clean Tech: Profits and potential*. Clean Edge.

Makover, J., & Pernick, R. (2002). *Clean energy trends 2002*, Clean Edge.

Marglin, S. A. (1974). What the bosses do? The origins and functions of hierarchy in capitalist production. *The Review of Radical Political Economics, 6*(2), 60–112.

Margolis, D., & Entin, D. (2011). *Report on survey of cohousing communities 2011*. Cohousing Association of the United States (Coho/US).

Margulis, L. (1998). *The symbiotic planet*. London: Science Masters.

Maribyrnong City Council. (2008). *Maribyrnong, Australia's peak oil policy and action plan.* Maribyrnong: Maribyrnong City Council.

Maro, P. (2007). *EEB's response to the European Commission's Green Paper on market instruments for environment and energy related policy purposes.* Brussels: EEB.

Martin, E., & Shaheen, S. (2011). The impact of carsharing on public transit and non-motorized travel: An exploration of North American chasharing survey data. *Energies, 4*, 2094–2114.

Martinez Alier, J. (2004). *El ecologismo de los pobres. Conflictos ambientales y lenguajes de valoración.* Barcelona: Icaria.

Martinez Alier, J. (2008, April 18–19). Sustainable de-growth. In *Proceedings of the first international conference on economic de-growth for ecological sustainability and social equity.* Paris.

Martinez Alier, J., et al. (2010). Sustainable de-growth: Mapping the context, criticisms and future prospects of an emergent paradigm. *Ecological Economics, 69*(9), 1741–1747.

Martinot, E., & Junfeng, L. (2010, July 21). *Renewable energy policy update for China.* www.renewableenergyworld.com

Marx, C. (1990). *The capital: A critical analysis of capitalistic production.* Electronic Text Center, University of Virginia Library.

Marx, C., & Engels, F. (1975). *Obras Escogidas (Tomos I y II).* Madrid: Editorial Ayuso.

Mastny, L. (2003). La situación del mundo: un año de revisión. en C. Bright, y otros (Eds.), *La situación del mundo 2003.* Barcelona: Icaria.

Mathews, J., & Tan, H. (2011). Progress toward a circular economy in China. *Journal of Industrial Ecology, 15*(3).

Mathews, E., et al. (2000). *The weight of nations. Material outflows from industrial economies.* Washington, DC: World Resource Institute.

Matutinovic, I. (2001). The aspects and the role of diversity in socioeconomic systems: an evolutionary perspective. *Ecological Economics, 39*(2), 239–256.

Matutinovic, I. (2002). Organizational patterns of economies: an ecological perspective. *Ecological Economics, 40*(3), 421–440.

Matutinovic, I. (2006). Mass migrations, income inequality and ecosystems health in the second wave of a globalization. *Ecological Economics, 59*, 199–203.

Max-Neef, M. (2005). Foundations of transdisciplinarity. *Ecological Economics, 53*(1), 5–16.

Mayumi, K. (2009). *The biofuel delusion: The fallacy behind large-scale agro-biofuel production.* London: Earthscan Research Edition.

Mcconnell, J., & Harzok, A. (1997). Pennsylvania's success with local tax reform. In T. Schroyer (Ed.), *Building blocks for a just and sustainable society.* New York: The Bootstrap Press.

McDonough, W., & Braungart, M. (2002). *Cradle to cradle. Remaking the way we make things.* New York: North Point Press.

Mckinsey & Company. (2010). *A portfolio of power-trains for Europe: A fact-based analysis.* Enterprises Consortium.

McManus, J. (2011). *Comments on S. Fucher's article.* www.theoildrum.com/node/7785

Mcnamara, A. (2007) *Queensland's vulnerability to rising oil prices.* www.epa.gld.gov.au

Mcnamara, A. (2008, March 4). *Highway of diamonds.* Conference in the Brisbane Institute.

MEA-CF. (2006). MA conceptual framework. In *Millennium ecosystem assessment.* New York: United Nations.

Meadows, D. H., Meadows, D. L., Randers, J., & III Behrens, W. W. (1972). *The limits of growth.* New York: Universe Books.

Mearns, E. (2009a). *Oil price: Where next? – And thoughts for 2010.* www.energybulletin.net/print/51028

Mearns, E. (2009b). *Is Europe running low on natural gas?* http://europe.theoildrum.com/node/4933

Mearns, E. (2010a). *The Chinese coal monster – Running out of puff.* http://europe.theoildrum.com/node/7123

Mearns, E. (2010b). *Fractal adaptive cycles in natural and human systems.* www.theoildrum.com/node/6099

Mearns, E. (2011a). *Countdown to $100 oil- Deja Vu?* www.theoildrum.com/node/8273
Mearns, E. (2011b). *Countdown to $100 oil – No normal recession.* www.theoildrum.com/node/8483
MEA-SB (Millennium Ecosystem Assessment). (2006). Natural assets and human well-being (statement from the board). In *Millennium ecosystem assessment.* Nueva York: United Nations.
MEA-SDM. (2006). Summary for decision makers. In *Millennium ecosystem assessment.* New York: United Nations.
MEA-V2. (2006). Lessons learned for scenarios analysis. In *Millennium ecosystem assessment.* Nueva York: United Nations.
MED-EMIP (Euro-Mediterranean Energy Market Integration Project). (2010). *Medring update study.* Brussels: MED-EMIP.
Meeker-Lowry, S. (1995). *Invested in common good.* Philadelphia: New Society Publishers.
Mesarovic, M. D., & Pestel, E. (1974). *Mankind at the turning point: The second report to the Club of Rome.* E.P. New York: Dutton and Company.
Michelis, L., & Lorek, S. (2004). *Consumption and the environment in Europe.* Danish Environmental Protection Agency: Copenhagen.
Mill. (1996). *Principios de Economía Política.* Madrid: Fondo de Cultura Económica.
Milmo, D., & Adetunji, J. (2008). *Britain goes slow as trains, planes and ships cut fuel costs.* www.guardian.co.uk
Ministerio de Industria. (2008). *Plan de Ahorro y Eficiencia Energética 2008–2011.* Madrid: Ministerio de Industria.
Ministry of Economic Development. (2006). *Draft New Zealand energy strategy for 2050.* Wellington: Ministry of Economic Development.
Ministry of Public Works and Transport. (2005). *Plan Estratégico de Infraestructuras y Transporte.* Madrid: Ministerio de Fomento.
Ministry of Public Works and Transport. (2010). *Plan estratégico para el impulso del transporte de mercancías en España.* Madrid: Ministerio de Fomento.
Ministry of the Environment. (2002). *Sweden's national strategy for sustainable development.* Ministry of the Environment.
Mint, P. (2010). Five PV history lessons; EPIA takeaways. *Renewable Energy World.* www.renewableenergyworld.com
Mint, P. (2012). *PV still facing a bumpy ride: Working in a low-incentive world.* www.renewableenergyworld.com
Mint, P. (2013). The often un profitable, always challnaging road to success of the photovoltaic industry, *SPV Market Research.* www.renewableenergyworld.com
Modvar, C., & Gallopin, G. C. (2005). *Sustainable development: Epistemological challenges to science and technology.* Santiago: CEPAL.
Moll, S., et al. (2003). *Resource use in European countries.* Copenhagen: European Environmental Agency.
Moll, S., et al. (2005). *Resource use in European countries.* Wuppertal: Wuppertal Institute.
Mont, O. (2002). Clarifying the concept of product-service system. *Journal of Cleaner Production, 10*, 237–245.
Mont, O., & Lindhqvist, T. (2003). The role of public policy in the advancement of product service systems. *Journal of Cleaner Production, 11*, 905–914.
Montgomery, J. (2012a). *Germany's new solar containment policy.* www.renewableenergyworld.com
Montgomery, J. (2012b). *Europe's 2011–2012 PV installs: Two tales of growth.* www.renewableenergyworld.com
Morikawa, M. (2000). Eco-industrial development in Japan (Indigo Development Working Paper 11). Emeryville: Indigo Development Center.
Mueller, T. (2008, April). Biomimetics: Design by nature. *National Geographic.*
Mulder, K., et al. (2006). The contribution of built, human, social and natural capital to quality of life in International and unintentional communities. *Ecological Economics, 59*, 13.
Mumford, L. (1971). *Ciencia, técnica y civilización.* Madrid: Alianza Editorial.

Murphy, D. (2009). *Further evidence of the influence of energy on the U. S. economy*. http://netenergy.theoildrum.com/node/5304

Murphy, D., & Hall, C. A. S. (2011). Energy return on investment, peak oil, and the end of economic growth. *Annals of the New York Academy of Sciences, 1219*, 52–72.

Naredo, J. M. (1987). *La economía en evolución. Historia y perspectivas de las categorçias básicas del pensamiento económico*. Madrid: Siglo XXI.

Naredo, J. M. (1990). Comentario que suscita la reedición del libro de Polanyi. *Archipiélago, 5*.

Naredo, J. M. (2006). *Raíces económicas del deterioro ecológico y social*. Madrid: Siglo XXI.

NAS (National Academy of Sciences). (2007). *Coal: Research and development to support national energy policy*. Washington, DC: NAS.

Ndefo E. O., et al. (2007, February 15). Russia: A critical evaluation of its natural gas resources. *Energy Tribune*.

Nelder, C. (2008). *Notes on ASPO-USA peak oil conference*. ASPO-USA.

Neurath, O. (1973). *Empiricism and sociology*. Dordrecht: Reidel.

New Zealand Labour Party. (2005). *Energy policy. Labour's vision*. www.scoop.co.nz

NIC (National Energy Council). (2008). *Global trends 2025: A transformed world*. NIC 2008–003.

Nielsen, S. N. (2007). What has modern ecosystem theory to offer to cleaner production, industrial ecology and society? The views of an ecologist. *Journal of Cleaner Production, 15*, 1639–1653.

Nijdam, D. S., et al. (2005). Environmental load from Dutch private consumption. *Journal of Industrial Ecology, 9*(1–2), 147–168.

Nikiforuk, A. (2012). *Alberta plays catch-up on frack front*. www.energybulletin.net/print/61238

Norgaard, J. (1998). Tecnología eficiente en una economía ineficiente. In J. Riechmann (Ed.), *Necesitar, desear, vivir. Sobre necesidades, desarrollo humano, crecimiento económico y sustentabilidad*. Madrid: Catarata.

NPC (National Petroleum Council). (2007). *National Petroleum Council scenario for net regional oil imports and exports in 2005 y 2030*. Washington, DC: NPC.

Nygren, E. (2009). *Aviation fuels and peak oil*. Thesis. Uppsala University, Uppsala.

O'Connor, M., et al. (1996). Emergent complexity and procedural rationality: Post-normal science for sustainability. In R. Costanza, O. Segura, & J. Martinez-Alier (Eds.), *Getting down to earth—Practical applications of ecological economics*. Washington, DC/Covelo: Island Press.

Oakland City Council. (2008). *Oil independent Oakland action plan*. City of Oakland.

Oates, W. E., & Portney, P. R. (1996). Economic incentives and the containment of global warming. In W. E. Oates (Ed.), *The economic of environmental regulation*. Cheltenham: Edward Elgar.

Odum, E. P. (1992). *Ecología: Bases Científicas para un nuevo paradigma*. Barcelona: Vedrá.

Odum, E. P., & Sarmiento, F. O. (1998). *Ecología: el puente entre ciencia y sostenibilidad*. México D.F: McGraw-Hill Interamericana.

OECD (Organization for Economic Cooperation and Development). (1985). *Environment and economics*. Paris: OECD.

OECD (Organization for Economic Cooperation and Development). (1993). *Fiscalidad y medio ambiente*. Paris: OECD.

OECD (Organization for Economic Cooperation and Development). (1994). *The environmental effects of trade*. Paris: OECD.

OECD (Organization for Economic Cooperation and Development). (1997a). *Economic globalization and the environment*. Paris: OECD.

OECD (Organization for Economic Cooperation and Development). (1997b). *Environmental taxes and green tax reform*. Paris: OECD.

OECD (Organization for Economic Cooperation and Development). (1997c). *Reforming environmental regulation in OECD countries*. Paris: OECD.

OECD (Organization for Economic Cooperation and Development). (1997d). *Evaluating economic instruments for environmental policy*. Paris: OECD.

OECD (Organization for Economic Cooperation and Development). (1998). *Eco-efficiency*. Paris: OECD.

OECD. (2001). *Environmental strategy for the first decade of the twenty-first century*. Paris: OECD.

OECD (Organization for Economic Cooperation and Development). (2002a). *OECD environmental strategy for the first decade of the 21st century*. Paris: OECD.
OECD (Organization for Economic Cooperation and Development). (2002b). *OECD workshop on environmentally harmful subsidies*. Paris: OECD.
OECD (Organization for Economic Cooperation and Development). (2005). *Environmental fiscal reform for poverty reduction*. Paris: OECD.
OECD (Organization for Economic Cooperation and Development). (2006). *Good practices in the national sustainable development strategies of OECD countries*. Paris: OECD.
OECD (Organization for Economic Cooperation and Development). (2007). *Environmental taxes*. Paris: OECD.
OECD (Organization for Economic Cooperation and Development). (2008a). *OECD science, technology and industry outlook 2008*. Paris: OECD.
OECD (Organization for Economic Cooperation and Development). (2008b). *Environmental innovation and global markets*. Paris: OECD.
OECD (Organization for Economic Cooperation and Development). (2008c). *Promoting sustainable consumption*. Paris: OECD.
OECD (Organization for Economic Cooperation and Development). (2008d). *OECD guidelines for multinational enterprises*. Paris: OECD.
OECD (Organization for Economic Cooperation and Development). (2009a). *Key transport statistics 2008*. Paris: OECD.
OECD (Organization for Economic Cooperation and Development). (2009b). *The bioeconomy to 2030. Designing a policy agenda*. Paris: OECD.
OECD (Organization for Economic Cooperation and Development). (2009c). *Sustainable manufacturing and eco-innovation. Framework, practices and management*. Paris: OECD.
OECD (Organization for Economic Cooperation and Development). (2009d). *Statistics on international trade in services* (Vol. I). Paris: OECD.
Öko-Institut. (2009). *Critical metals for future sustainable technologies and their recycling potential*. Darmstadt: Öko-Institut.
Olsson, P., et al. (2006, online). Shooting the rapids: Navigating transitions to adaptive Governance of social-ecological systems. *Ecology and Society, 1*(1).
Omann, I. (2007). *A multicriteria tool for evaluating of impact of product service an sustainable development: An application in Austrian companies*. Vienna: Sustainable Europe Research Institute (SERI).
Omann, I., et al. (2007). *The environmental effects of food consumption for different household categories*. Vienna: Sustainable Europe Research Institute (SERI).
Orr, D. W. (2002). *The nature of design*. New York: Oxford University Press.
Oster, S., & Davis, A. (2008). *China spurs coal-price surge*. http://online.wsj.com
Paleocrassas, Y. (2002). *A radical fiscal reform*. www.eeb.org
Paris, A. J. (2009a). *I am a banker. Some of us did not f*ck up*. http://energybulletin.net/print/48463
Paris, A. J. (2009b). *The cost of wind, the price of wind, the value of wind*. http://europe.theoildrum.com/node/5354
Patzek, T. W. (2006, June 26). The real biofuels cycles. *Online Supporting for Science Letter, 312*.
Patzek, T. W., & Croft, G. D. (2010). A global coal production forecast with multi-Hubbert cycle analysis. *Energy, 35*, 3109–3122.
Patzek, T. W., et al. (2005). Ethanol from corn: Clean renewable fuel for the future, or drain on our resources and pockets? *Environment, Development and Sustainability, 7*, 319–336.
Pauli, G. (1998). *Upsizing*. Sheffield: Greenleaf Publishing.
Pauli, G. (2010). *The blue economy*. Taos: Paradigm Publications.
Pearce. (2002). *Environmentally harmful subsidies: Barriers to sustainable development. OECD workshop on environmentally harmful subsidies*. Paris: OECD Development.
Pearce, D., & Markandya, A. (1989). *Environmental policy benefits: Monetary valuation*. Paris: OECD.
Pearce, D., & Turner, R. K. (1990). *Economics of natural resources and the environment*. London: Harvester Wheatsleaf.

Pearce, D., et al. (1989). *Blueprint for a green economy*. London: Earthscan.
Pearce, D., et al. (1994). *Blueprint for a green economy*. London: Earthscan.
Pearce, D., et al. (2004). *Blueprint for a green economy*. London: Earthscan.
Permick, R. (2002). *Clean energy markets*. San Francisco: Clean Edge.
Permick, R., et al. (2009). *Clean tech job trends 2009*. San Francisco: Clean Edge.
Permick, R., et al. (2010). *Clean tech job trends 2010*. San Francisco: Clean Edge.
Petrov, K. (2006, January 18). The proposed Iranian oil bource. *Energy Bulletin*.
Phillips, L. (2009, May 13). Europe's overseas push into biofuels. *Business Week*.
Pimentel, & Patzek. (2005). Ethanol production using corn, switchgrass, and wood; Biodiesel production using soybean and sunflower. *Natural Resources Research, 14*(1).
Pimentel, D., & Patzek, T. W. (2006). Ethanol production using corn, switchgrass, and wood; biodiesel production using soybean and sunflower. *Natural Resources Research, 14*(1), 65–76.
Pimentel, D., & Pimentel, M. (2006). Global environmental resources versus world population. *Ecological Economics, 59*, 195–198.
Plastics Europe. (2003). *Plastics in Europe*. PlasticsEurope (Association of Plastics Manufacturers).
Plastics Europe. (2009). *The compelling facts about plastics for 2008*. Plastics Europe.
Plastic Europe, et al. (2008). *The compelling facts about plastics for 2007, EUPCE/EUPR/EPRO/PlasticsEurope/.*
Platt, A. (2000, November). Why poison ourselves? A precautionary approach to synthetic chemicals. *Worldwatch Paper, 153*.
Platt, A. (2002). Reduciendo nuestra carga tóxica. en C. Flavin, y otros (Eds.), *La situación del mundo*. Barcelona: Apóstrofe.
PNUD (Programa de Naciones Unidas Para El Desarrollo), et al. (2001). *Informe Mundial sobre Energía*. Madrid: IDAE.
Polanyi, K. (1989). *The great transformation*. Boston: Beacon.
Portland City Council. (2007). *Report of the City of Portland*. Portland City Council.
Post Carbon Institute. (2009). *Post Carbon Institute Manifesto. The time for change has come*. PCI.
President's Council of Advisers. (2006). *The energy imperative technology and the role of emerging companies*. Washington, DC: President's Council of Advisers.
Proctor, A., et al. (2002). *Eco-industrial park survey: Feedback report for respondents*. Yorkshire: Department of Geography, University of Hull.
Public Citizen Report. (2003). *The U.S. threats against Europe's GMO policy and the WTO SPS agreements*. Public Citizen.
Pullinger, M. (2010). *Workshop 8: Employment*. In The steady state economy conference.
Quang, T. T. (1996). *Buddhism and environmental protection*. www.budsas.org
Queenland's Oil Vulnerability Taskforce. (2007). *Queensland's vulnerability to rising oil prices. Taskforce report*, A. McNamara MP, Chair Oil Vulnerability Taskforce.
Rammel, C., & Staudinger, M. (2002). Evolution, variability y sustainable development. *International Journal of Sustainable Development. World Ecology, 9*, 3001–3313.
Rammel, C., & Staudinger, M. (2004). The bridge between diversity and adaptivity: Answering McIntosh and Jeffrey. *International Journal of Sustainable Development. World Ecology, 11*(1), 9–23.
Rammel, C., & Van der Bergh, J. (2003). Evolutionary policies for sustainable development: Adaptive flexibility and risk minimising. *Ecological Economics, 47*(2–3), 121–133.
Rammel, C., Stalgl, S., & Wilfing, H. (2007). Managing complex adaptive systems—A coevolutionary perspective on natural resource management. *Ecological Economics, 62*(1).
Ramonet, I. (1997). Pensamiento único y nuevos amos del mundo. In *Como nos venden la moto*. Barcelona: Icaria.
Ramos, D. (2010). Análisis del proceso de liberalización del mercado comunitario de transporte aéreo. en D. Hoyos, & A. Sanz Coord., *Nueva Política de Transporte para el Siglo XXI*, EKONOMIAZ 73.
Ramos Gorostiza, J. L. (2009). El medio físico en el pensamiento agrario español: del arbitrismo al regeneracionismo. *Historia Agraria, 49*.

Rapier, R. (2009). *Book review: Oil 101*. www.energybulletin.com/node/5340
Rapier, R. (2010). *Next feneration biofuels: Five challenges and five positive notes*. www.theoildrum.com/node/6677
Raskin, P., et al. (2002). *Great transition. The promise and lure of the times ahead*. Stockholm: Environmental Institute.
Ravaioli, C., & Ekins, P. (1995). *Economists and the environment. What the top economists say about the environment*. London: Zed Books.
RCEP (Royal Commission on Environmental Pollution). (2003). *Chemicals in products. Safeguarding the environment and human health*. Londres: Cámara de los Comunes.
Reguly, E. (2007, November 14). UE sleepwalking into gas crisis. ENI says. *The Global and Mai*
Ren, X. (2002). Biodegradable plastics: A solution or a challenge. *Journal of Cleaner Production, 11*.
REN21. (2010). *Renewables 2010. Global status report*. REN21, PNUMA.
Renton, A. (2008, January 20). Is this the end of cheap food? *The Observer*
Rickerson, W., & Jungjohann, A. (2011). *No nukes, no problem? Germany's race for a Renewable future*. www.rnewableenergyworld.com
Riele, H., et al. (2001). *Dematerialization*. Amsterdam: Dutch Ministry of Environment.
Rihani, S. (2002). *Complex systems theory and development practice*. London: Zed Books.
Ring, I. (1997). Evolutionary strategies in environmental policy. *Ecological Economics, 23*(3), 237–250.
Ritter, S. K. (2001). Green Chemistry. *CENEAR, 79*(29).
Rizhkov, N. (1986). *Sobre las orientaciones fundamentales del desarrollo económico y social de la URSS en 1986–1990 y hasta el año 2000*. Moscú: Novósti.
Roads2HyCom. (2009). *Fuel cells and hydrogen in sustainable energy economy*. www.roads2hy.com
Rodrigue, J. P. (2008). *International oil transportation*. Hofstra University. http://people.hofstra.edu
Roll, E. (1966). *A history of economic thought*. London: Faber.
Roodman, D. M. (1998). *The natural wealth of nations. Harnessing the market for the environment*. Nueva York: W.W. Norton & Company.
Rooney, L. (2006). *Kinsale adopts descent action plan*. http://localplanet.ie/content/view/191/49/
Rooney, L. (2007). Kinsale adopts energy descent action plan. *The Local Planet*. www.localplanet.ie
Ropke, I. (1999). The dynamics of willingness to consume. *Ecological Economics, 28*(3), 399–420.
Ropke, I. (2001, September). New technology in everyday life-social processes and environmental impact. *Ecological Economics*.
Ropke, I. (2002, November). Consumption dynamics and technological change – Exemplified by mobile phone and related technologies. *Ecological Economics*.
Ropke, I. (2004). Work-related consumption drivers and consumption at work. In L. A. Reisch & I. Ropke (Eds.), *The ecological economics of consumption*. Cheltenham: E. Elgard.
Ropke, I. (2009). Theories of practice – New inspirations for ecological economic studies on consumption. *Ecological Economics, 68*(10), 2490–2497.
Rother, L. (2008). *Shipping costs start to crimp globalization*. www.nytimes.com/2008/08/03
Rubin, J. (2009). *Why your world is about to get a whole lot smaller*. New York: Random House.
Rubin, J., & Buchanan, P. (2008a, January). Delays will tighten global oil markets. *StrategEcon*, New York: CIBC World Markets Inc.
Rubin, J., & Buchanan, P. (2008b). *What's the real cause of the global recession?* New York: CIBC World Markets Inc.
Rubin, J., & Tal, B. (2005). When will energy brake US consumers? *Monthly Indicators*, CIBC World Markets Inc.
Rubin, J., & Tal, B. (2008, May). Will soaring transport costs reverse globalization? *StrategEcon*, New York: CIBC World Markets Inc.
Rutledge, D. (2007, June 25). *The coal question and climate change*. www.theoildrum.com
Sach, I. (1981). *Initiation a l'ecodeveloppment*. Tolouse: Privat.

Sachs, W. (1992). The economist's prejudice. In P. Ekins & M. Max-Neef (Eds.), *Real – Life economics*. London: Routledge.
Sachs, W. (1999). *Planet dialectics: Explorations in environment and development*. London: Zed Books.
Sachs, J., & Finkelpearl, S. (2010). From selling soap to selling sustainability: Social marketing. *State of the World 2010*. WorldWatch Institute.
SACTRA. (2000). *SACTRA report*. London: Department for Transport.
San Francisco City Council. (2006). *Resolution acknowledging the challenge of Peak Oil*. San Francisco: San Francisco City Council.
Sanne, S. (2002). Willing consumers – or lock-in? Policies for sustainable consumption. *Ecological Economics, 42*, 273–287.
Santarius, T. et al. (2004, febrero). *Balancing trade and environment. An Ecologiacal Reform de la OMC como un Desafío a la Gobernabilidad Mundial Sostenible* (Wuppertal Papers, nº 133e).
Sbert, J. M. (1993). Progress. In W. Sach (Ed.), *The development dictionary*. London: Zed Books.
SCAG (Southern California Association of Governors). (2008). *Regional comprehensive plan*. SCAG.
Schettkat, R. (2009). *Analyzing rebound effects*. Wuppertal: Schumpeter School of Business and Economics.
Schindler, J. et al. (2008). *Crude oil – The supply outlook*. Energy Watch Group.
Schlumberger, C. E. (2008). *The result of speculation or an early indicator of a major and growing future challenge to the airline industry?* Washington, DC: The World Bank.
Schmidt, P. R., et al. (2009). *Sustainability of trnasport fuels*. In: LBST, Electric Vehicle Conference, Stavanger, Norway.
Schmidt, P. R. (2011). *Sustainability issues of 2nd generation biofuels*. Ottobrunn: Ludwig-Bolkow-Systemtechnik.
Schor, J. (2010a). Sustainable work schedules for all. In *State of the world 2010*. Washington, DC: WorldWatch Institute.
Schor, J. (2010b). *The work-sharing boom: Exit ramp to a new economy?* www.yesmagazine.org
Schumacher, E. F. (1978). *Lo pequeño es hermoso*. Madrid: Hermann Blume.
Schumacher, E. F. (1980). *El buen trabajo*. Madrid: Debate.
Schutz, J. (1999). The value of economic reasoning. *Ecological Economics, 31*, 23–29.
Schutz, J. (1999a). Organizing diversity. In J. Khon et al. (Eds.), *Sustainability in question*. London: Edward Elgar.
Schwartz, L. (2010). China policy: A defining decade for the would-be leader. *Renewable Energy Magazine*.
SCI/Verkher. (2008, September). *La Vie du Rail Internat*ional.
SCI/Verkher. (2009). *Worldwide rail transport markets 2009–2019*. Berlin: SCI/Verkher.
Scott, A. (1999). Trust, law, sustainability, and responsible action. *Ecological Economics, 31*, 139–154.
Seager, T. P., & Theis, T. L. (2002). An uniform definition and quantitative basis for industrial ecology. *Journal of Cleaner Production, 10*, 225–235.
SForward. (2008). *San Francisco's environmental plan 2008. Building a bright future*. San Francisco.
Shahan, Z. (2012). *Offshore wind could power the world*. http://cleantechnica.com
Shaheen, S., et al. (2003). *U.S. shared-use vehicle survey findings: Opportunities and obstacles for carsharing & station car growth*. Transportation Research Board.
Shaheen, S. A., & Cohen, A. P. (2007). *Woldwide carsharing growth: An international comparison*. Berkeley: PATH University of California.
Shalins, M. (1986). *La economía de la Edad de Piedra*. Madrid: Akal.
Sharp, R. (2008, May 01). They don't just shop in Totnes – They have their very own currency. *The Independent*.
Shi, H., et al. (2012a). China's quest for eco-industrial parks, Part 1. *Journal of Industrial Ecology, 16*(1), 8–10.

Shi, H., et al. (2012b). China's quest for eco-industrial parks, Part 2. *Journal of Industrial Ecology, 16*(1), 290–292.

Shiao, M. J. (2012). Thin film 2012–2016: Technologies, markets and strategies for survival. *GTM Research.*

Shimamoto, M. (2008). Forests sustainability and trade policies. *Ecological Economics, 66*(4), 605–614.

Shorgen, J. F., & Nowell, C. (1992, May). Economics and ecology. A comparison of experimental methodologies and philosophies. *Ecological Economics.*

Simmons, M. (2006, February, online). What a difference 20 years make in crude oil prices. *World Oil Magazine, 2.*

Skrebowski, C. (2010). Opinion A. In ITPOES (Ed.), *The oil crunch*. ITPOES.

Skrebowski, C. (2011). *ODAC Newsletter – Sept 16.* www.energybulletin.net/print/58911

Smil, V. (2002). Global biogeochemical cycles. In R. Ayres & L. Ayres (Eds.), *A handbook of industrial ecology*. London: Edward Elgar.

Smith, A. (2005). *The wealth of nations*. University Park: The Pensylvania State University.

Smith, J. J., & Nelson, K. (1998). Giving life to the Property Tax Shift (PTS). *Redefining Progress.* www.progress.org

Soderbaun. (1992, May). Neoclasical and institutional approaches to development and the environment. *Ecological Economics.*

Soil Association. (2012). *Organic market report 2012.* www.soilassociation.org

Spangenberg, J. (2003). Integrating economic, social and environmental policies: Who calls the tune? *SERI Background Papers* (3).

Spangenberg, J. (2004). Reconciling sustainability and growth: criteria, indicators, policy. *Sustainable Development, 12*(2), 74–86.

Spangenberg, J., & Lorek, S. (2002). Environmentally sustainable household consumption: From aggregate environmental pressures to priority fields of action. *Ecological Economics, 43*, 127–140.

Spangenberg, J., & O'Connor, M. (2003). *Sustainability science in the European research area: Descriptions, definitions, and challenges*. Bureau Cologne: Sustainable Europe Research Institute.

Spangenberg, J., & O'Connor, M. (2010). Sustainable science: A new mode of science. Another step in the evolution of science paradigm. *INES Global Responsibility, Newsletter, 61.*

Spangenberg, J., et al. (1998). *Material flow-based indicators in environmental reporting*. Coopenhague: EEA.

Speck, S. (2009, October 29). Ecological tax reforms in Europe. In *Ecological Tax Reforms I – A key to sustainability and competitiveness*. Vienne: Forum Finanz.

Staniford, S. (2012). *January oil supply*. www.theoildrum.com/node/8959

State of Connecticut. (2008). *An act concerning energy scarcity and security, renewable and lean energy and a state solar strategy*. Public Act No. 08–168.

Steer, T., & Ott, T. (2004). The industrial region as a promising unit for eco-industrial development-reflections, practical experience and establishment of innovative instruments to support industrial ecology. *Journal of Cleaner Production, 12*, 947–965.

Stiglitz, J. E. (2009, January). Capitalist fools. *Vanity Fair Magazine.*

Stimpson, C., & Kleekamp, C. (2007, July 12). *Offshore score since 2001: Europe 24, U.S. 0.* www.renewableenergyaccess.com

Stoughton, M., & Votta, T. (2003). Implementing service-based chemical procurement: Lessons and results. *Journal of Cleaner Production*, December.

Sumaila, U. R., et al. (2007). The World Trade Organization and global fisheries sustainability. *Fisheries Research, 88*, 1–4.

Sundin, E., et al. (2009). Using company-academia networks to improving product/service systems at large companies. In T. Fischer et al. (Eds.), *Introduction to product/service systems design.* London: Springer.

Sundin, E. (2010). Life-cycle perspectives of product/service-systems: In design theory. In T. Sakao & M. Ludehl (Eds.), *Introduction to product/service-system design*. Springer.

Swisher, J. N. (2002). *Cleaner energy, greener profits: Fuel cells as cost-effective distributed energy resources*. Snowmass: Rocky Mountain Institute.

Tainter, J. A. (2005). *The collapse of complex societies*. Cambridge: Cambridge University Press.

Tainter, J. A. (2006). Social complexity and sustainability. *Ecological Complexity, 6*, 547–553.

Tannon, J., & Campbell, C. (2009). *Renewable energy: Yes we can*. www.renewableenergyworld.com

The Arlington Institute. (2003). *A strategy: Moving America away from oil*. Arlington: The Arlington Institute.

The Ecologist. (1992, July–August). Whose common future?. *22*(4).

The Federal Government. (2009). *Founding an international renewable energy* agency (IRENA). The Federal Government of Germany.

The Pentagon. (2008). *The joint operating environment 2008*. United States Joint Forces Command.

The Pentagon. (2010). *The joint operating environment 2010*. United States Joint Forces Command.

Thompson, E. P. (1995). *Costumbres en común*. Barcelona: Crítica.

Thorpe, B. (1999). *Citizen's guide to clean production*. Lowell: Clean Production Network, University of Massachusetts Lowell.

Tietenberg, T. (1998, February–March). Ethical influences on the evaluation of the Us tradable permits approach to air pollution control. *Ecological Economics*.

Tinbergen, J., & Hueting, R. (1991). GNP and markets prices: Wrong signals for sustainable economic success that mask environmental destruction. In R. Goodland et al. (Eds.), *Environmentally sustainable economic development: Building on Brundtland*. Paris: UNESCO.

Tishner, U., et al. (2010). Product-service systems. In UNEP-Delft (Ed.), *Design for sustainability. A step-by-step approach*. Paris: UNEP.

Tolle, E. (2005). *A new earth: Awakening to your life's purpose*. New York: Penguin.

Tomlison, I. (2012). *Feeding the future: How organic farming can help feed the world*. Soil Asssociation. www.soilassociation.org

Toshkov, V. (2008, January 18). Bulgaria and Russia sign pipeline deal. *International Herald Tribune*.

Totnes EDP. (2010). *Totnes 2030, an energy descent action plan*. Transition Network. www.transitionnetwork.org/

Townsend, E. (2013). Commentary: Why peak oil threatens the International Monetary System, *ASPO-USA*. www.resilience.org

Transition US. (2011). Guiding principles of transition. www.transitionus.org

Transition Network. (2012). Transition initiatives directory. Transition Network. http://www.transitionnetwork.org/

TT-EDAP. (2010). *Totnes 2030, an energy descent action plan*. www.transitionnetwork.org

Tucker, A. (2007). Fostering change to sustainable consumption and production: An evidence based view. *Journal of Cleaner Production, 16*, 1218–1225.

Tukker, A., et al. (2006). *The Oslo Declaration on Sustainable Consumption, Journal of Industrial Ecology, 11(1–2)*.

TUV/SUD. (2009). *Hydrogen and fuel cells*. www.netinform.de/H2

Tverberg, G. (2009a). *Obama's energy policy announcement*. www.theoildrum.com/node/5021

Tverberg, G. (2009b). *Where we are headed: Peak oil and financial crisis*. www.theoildrum/node/5320

Tverberg, G. (2009c). *Science 1101–Petroleum and peak oil*. www.theoildrum.com/node/5969

Tverberg, G. (2010a). *Relocalization – Thinking about imports from less development world*. www.theoildrum.com/code/6648

Tverberg, G. (2010b). *World oil production – Looking for clues as to what may be ahead*. www.theoildrum.com/node/7258

Tverberg, G. (2011). *Review of oil limits, recession, and bumping against the growth ceiling*. www.theoildrum.com/node/8269

Tverberg, G. (2013). *Our energy predicament in charts*. www.theoildrum.com/node/9901

UCS (Union of Concerned Scientists). (1992). *World scientists' warning to humanity*. Cambridge: UCS Publications.
UCS (Union of Concerned Scientists). (2008). *Federal science and the public good. Securing the integrity of science in policy making*. Cambridge: UCS Publications.
UCS (Union of Concerned Scientists). (2009). *Failure to yield. Evaluating the performance of genetically engineered crops*. Cambridge: UCS Publications.
UCS (Union of Concerned Scientists). (2012). *Heads they win, tails we lose. How corporations corrupt science at the public's expense*. Cambridge: UCS Publications.
Udo de Haes, H. A. (2007). Industrial ecology and life cycle assessment. In R. U. Ayres & L. W. Ayres (Eds.), *Handbook of industrial ecology*. Cheltenham: Edward Elgar.
UIC (International Union of Railways). (2009). UIC world rail statistics for 2008. UIC Statistic Department. www.uic.org/
UIC-CER (International Union of Railways-Community of European Railway). (2008). *Rail transport and environment. Facts and figures*. UIC-CER.
UK Sustainable Development Commission. (2003). *People, planet, prosperity. Policies for sustainable consumption*. www.sd-commission.gov.uk
UNCED (United Nations Conference on Environment and Development). (1992). *Agenda 21*. www.unep.org
UNEP (United Nations Enviromental Program). (2001). *Enhancing synergies and mutual supportiveness of multilateral environmental agreements and the World Trade Organization*. Paris: UNEP.
UNEP (United Nations Enviromental Program). (2002a). *Industry as a partner for sustainable development. 10 years after Rio: The UNEP assessment*. Paris: UNEP.
UNEP (United Nations Enviromental Program). (2002b). *Sustainable consumption and cleaner production*. Paris: UNEP.
UNEP (United Nations Enviromental Program). (2002c). *Sustainable consumption. A global status report*. Paris: UNEP.
UNEP (United Nations Enviromental Program). (2003). *Environmentally sound technologies for sustainable development (revised draft)*. Paris: UNEP.
UNEP. (2004). *The use of economic instruments in environmental policy: Opportunities and challenges*, Geneva, Switzerland.
UNEP (United Nations Enviromental Program). (2008). *Planning for change. Guidelines for national programmes and sustainable consumption and production*. Paris: UNEP.
UNEP (United Nations Enviromental Program). (2009). *Recycling – From E-waste to resources*. Paris: UNEP.
UNEP. (2009). *Recycling. From E-waste to resources*. Paris: UNEP.
UNEP (United Nations Enviromental Program). (2010a). *Metal stocks in society – Scientific synthesis*. Paris: UNEP.
UNEP (United Nations Enviromental Program). (2010b). *Recycling rates of metals*. Paris: UNEP.
UNEP/AIE. (2002). *Reforming energy subsidies*. Nueva York: United Nations Publication.
UNEP/CI (Consumer International). *Tracking progress: Implementing sustainable consumption policies*. UNEP. www.unep.org
UNEP/IEA. (2002). *Reforming energy subsidies*. UNEP/IEA. www.unep.org
UNEP, & IIDS. (2002). *Manual de Medio Ambiente*. Paris: UNEP.
United Nations. (2000). *Millennium declaration*. New York: UN.
UNU/GEIC/IAS (United Nations University, Global Environmental Information Centre, Institute of Advanced Studies). (1998). *Global climate governance*. Tokyo: UNU/GEIC/IAS.
UNU/GEIC/IAS. (1999). *Global climate governance*. Tokyo: UNU.
USJFCOM (United States Joint Forces Command). (2010). *JOE 2010 (Joint Operating Environment)*. USJFCOM. www.forscom.army.mil/
Utchitelle, L. (2008). *Soaring fuel prices take a withering toll on truckers*. www.nytimes.com
UNWSSD. (2012). *Implementation plan*. UN.
Vaidya, S. (2008). The ripple effect of high oil prices. www.resilience.org

Vail, J. (2008a). *Geopolitical disruptions 1: Theory of disruptions to oil & resource supply.* www.theoildrum.com/node/4373

Vail, J. (2008b). *Oil demand destruction & brittle systems.* www.energybulletin.net/node/4411

Van Berkel, R. (2009). Comparability of industrial symbioses. *Journal of Industrial Ecology, 13*(4), 483–486.

Van Berkel, R. (2010). Quantifying sustainability. Benefits of industrial symbioses. *Journal of Industrial Ecology, 14*(3), 371–373.

Van der Velpen, B. (2009). *A win-win-win business model for producers and end-users of chemicals and the environment: "Chemical Leasing".* Chemical Leasing Research Project. www.royalhaskoningdhv.com

Van der Voet, H. A. (2007). Substance flow analysis methodology. In R. U. Ayres & L. W. Ayres (Eds.), *Handbook of industrial ecology*. Cheltenham: Edward Elgar.

Van Rossem, C., et al. (2006). *Extended producer responsibility.* Lund: International Institute for Industrial Environmental Economics, University of Lund.

Vancouver City Planning Commision. (2006). *Peak oil impact on cities, survival & culture: Dealing with community change under stress. Seeking out new leaders.* Vancouver: Metro Vancouver Planning Coalition.

Vernon, C. (2008). *What would $120 oil mean for the economy?* http://europe.theoildrum.com

Vernon, C. (2009). *Transport and adaptive capacity: An integrated approach to UK policy evaluation.* http://europe.theoildrum.com/node/5365

Vernon, C. (2010). *What difference would Nord Stream mean to European energy Supply.* http://europe.theoildrum.com/node/6157

Visnjic, I., & Van Looy, B. (2009). *Manufacturing firms diversifying into services: A conceptual and empirical assessment.* POMS 20th annual conference.

Wakeford, T. (2003, May). Who's in control? *The Ecologist.*

Walker, B. (2005). A resilience approach to integrated assessment. *The International Assessment Journal, 5*(1), 77–97.

Walker, B., et al. (2006). A handful of heuristics and some propositions for understanding resilience in social-ecological systems. *Ecology and Society, 11*(1), 13.

Walker, B., et al. (2008). Resilience thinking. Resilience Alliance. http://www.resalliance.org/

Wang, U. (2010). The rise of CIGS – Finally? *Renewable Energy World.* www.renewableenergyworld.com

WB (World Bank). (1993). *World development report 2012.* Washington, DC: World Bank.

WBCSD (Business Council for Sustainable Development). (1992). *Changing course.* Geneva: WBCSD.

WBCSD (Business Council for Sustainable Development). (2000). *Eco-efficiency.* Geneva: WBCSD.

WBCSD (Business Council for Sustainable Development). (2001). *Sustainability through the market. Seven keys to success.* Geneva: WBCSD.

WBCSD (Business Council for Sustainable Development). (2012). *A vision for sustainable consumption.* Geneva: WBCSD.

WBCSD. (2002). *The business case of sustainable development.* Geneva: WBCSD.

WBGU (Erman Advisory Council on Global Change). (2003). *World in transition. Towards sustainable energy systems.* Berlin: WBGU.

WCED (World Commission on Environment and Development). (1987). *Our common future: Report of the world commission on the environment and development.* Brundtland Report. UN General Assembly.

WCRE (World Council for the Renewable Energy). (2000). *Action plan for the global proliferation of renewable energy.* WCRE. www.wcre.de/

Webster, B. (2007, May 29). Holidaymakers forced to take slow boat. *The Times.*

Webster. (2010, March 10). Green fuels cause more harm than fossil fuels, according to report, *The Times.*

Weeramantry, C. G. (2007). *Islam, Buddhism, Hinduism and the environment.* www.energybulletin.net/print/32153

Westley, F., et al. (2002). Why systems of people and natures are not just social and environmental systems. In L. H. Gunderson & C. S. Holling (Eds.), *Panarchy. Understanding transformations in human and natural systems*. London: Island Press.
Weird, F. (2007). Kremlin extends grip on oil. *The Christian Science Monitor*. www.csmonitor.com
Weizsäcker, E. U. (1994). *Earth politics*. London: Zed Books.
Weizsäcker, E. U., et al. (2004). *Factor 4 (Informe del Club de Roma)*. Barcelona: Galaxia Gutermberg.
Western Australia Government. (2003). *The Western Australian State Sustainability Strategy, State Government of Western Australia.*
Wheatley, M., & Creneam, G. (2004). *Solving, not attacking, complex systems. A five-state approach based on an ancient practice*. www.margaretwheatley.com
White, G. (2009). *Do Saudis have the clout to destroy NYMEX?* www.telegraph.co.uk/finance/markets
Whitelegg, J. (2009). *On the wrong track: Why high-speed trains are not such a green alternative*. www.guardian.co.uk
WI-LBST (Wuppertal Institut & Ludwig-Bolkow-Systemtechnik). (2011). *Impact of shale gas and shale oil extraction on the environment and on human health*. Brussels: European Parliament IP/A/ENVI/ST/20011-07.
Willer, H., & Klicher, L. (2009). *The world of organic agriculture. Statistics and emerging trends 2009*. Bonn: IFOM.
Woodyard, C. (2010). *Will masses embrace electric cars despite high prices?* www.usatoday.com
World Coal Association (2010). *Coal Statistics*. www.worldcoal.org.
WTO. (2001). *Doha ministerial declaration*. www.wto.org
WTO (World Trade Organization). (2007). *Risks lie ahead following stronger trade in 2006. WTO reports*. Geneva: WTO.
Wurster, R. (2010). *Como llevar energías renovables al máximo éxito por medio de hidrógeno*, VIII Congreso Internacional sobre Innovación y Desarrollo Tecnológico CIINDET 2010, Cuernavaca, México.
Wurster, R. et al. (2009). *Energy infrastructure 21. Role of hydrogen in addressing the challenges in the new Global Energy System*. EHA (European Hydrogen Association). www.h2euro.org/
Wuppertal Institute, et al. (2008). *Eco-innovation: Putting the EU on the path to resource*. Wuppertal: Wuppertal Institute.
Wurster, R., Wuppertal Institute, et al. (2003). *Hidrógeno. Visión General y economía de H2*. Ottobrunn: L-B-Systemtechnick.
Yachnin & Associates. (2010). *Sustainability trough product service systems & servicizing*. Ottawa: Yachnin & Associates.
Yong, G., & Coté, R. P. (2001). *EMS as an opportunity for engaging China's economic development zones: The case of Dalian*. www.uneptie.org/pc/ind-estates
Yuanyuan, L. (2011). China increases target for wind power capacity to 1,000 GW by 2050. Renewable Energy World; Zenz, K., Johnson, A. (2010). The limits of energy storage technology. www.renewableenrgyworld.com.
Yuanyuan, L. (2012). China increases target for wind power capacity to 1,000 Gw by 2050. www.renewableenergyworld.com
Zenz, K., & Johnson, A. (2009). The limits of energy storage technology. *Bulletin of Atomic Scientists*, Web Edition.
Ziesemer, J. (2007). *Energy use in organic food systems*. Roma: FAO.
Zittel, W. (2012). *Futures for the common good. Energy transition. Paths in a period of increasing scarcity*. Ottobrunn: Ludwig-Bölkow-Systemtecnik.
Zittel, W., & Schindler, J. (2003). *Future oil supply*. www.lbst.de
Zittel, W., & Schindler, J. (2004). *The countdown for the peak production has began, but what are the views of the most important International Energy Agencies?* www.lbst.de
Zittel, W., & Schindler, J. (2007). *Crude oil, the supply outlook*. Energy Watch Group.